# FINITE MATHEMATICS

*SIXTH EDITION*

*Contributions by:*
*Department of Mathematics,* NDSU

**Kendall Hunt**
publishing company

publishing company

www.kendallhunt.com
*Send all inquiries to:*
4050 Westmark Drive
Dubuque, IA 52004-1840

Copyright © 2003, 2005, 2007, 2009, 2012, 2014 by Kendall Hunt Publishing Company

ISBN 978-1-4652-4010-1

Printed in the United States of America
10  9  8  7  6  5  4  3  2  1

# Contents

# Prelude: Problem Solving

As one could reasonably expect, you will be called upon to solve many problems over the course of your time in Finite Math. To that end, we propose a five-step process to employ when solving problems.

1. *Familiarize yourself with the problem situation.*

   Many times you will need to read a problem carefully, and sometimes more than once. It is often helpful to list the information given and to actually write down what, specifically, has been asked. If the problem involves unknown quantities, choose a variable (or variables) to represent the unknown quantity or quantities and clearly declare what a variable is supposed to be representing. Oftentimes a problem is easier to solve if you draw a diagram or construct a table that shows what is happening in the problem situation. It may even be beneficial to do some experimenting with numerical values just to garner some feeling for what is happening in the problem situation.

2. *Translate the problem into mathematical language.*

   We call this *modeling the problem*. Sometimes this is as simple as writing an equation, but often more is required. Your model should accurately reflect the information available.

3. *Carry out some process with your model.*

   Some process or sequence of manipulations will likely be necessary. Do this work carefully and in an organized manner. Don't skip steps. It is much easier to spot mistakes if there is a clear sequence of steps to follow, and this often prevents a need to start a problem anew.

4. *Check your solution.*

   Is your solution reasonable? Does your solution actually answer the question that was posed?

5. *Clearly state your answer.*

   Many times, the answer to a question will require some accompanying statement that relates your work back to the original question. Get in the habit of stating your answers in complete sentences. This also helps you determine if you have actually answered the question(s) posed in a problem.

# PART
# I

# Introductory Mathematical Modeling

# Mathematical Modeling with Lines and Functions

It has been said that "mathematics is the handmaiden of science". While that statement certainly seems reasonable, many people that regularly use mathematics are not scientists. Architects, retail managers, financial advisors, political leaders, and a host of others all use mathematics as part of decision-making processes. Mathematics helps in solving many problems encountered in business, industry, environment, and social management. Certainly some of the mathematics used is very complicated and difficult. However, much is straightforward and widely applicable to a variety of seemingly unrelated problems. We begin by investigating basic kinds of *mathematical models,* with which we endeavor to describe the relationships between related quantities. Mathematical models typically approximate real situations, but a good model can be used to help make decisions and evaluate consequences of decisions.-

## 1.1  LINEARITY

In order to truly understand how one quantity relates to another, one must be familiar with the concept of change. There are two measures of change that we will discuss in this section, *total change* and *average rate of change*.

### Total Change and Average Rate of Change

**Total change** is simply the complete change that has occurred in some quantity over a fixed reference period (time, for instance). Algebraically, it is the difference between the initial value and ending value of the quantity over that reference period. Geometrically, it is the vertical difference between the first and last points on a graph representing the values of the quantity over that reference period. For example, if a plant was two inches high at the beginning of the year and eight inches high at the end of the year, then we can say that the total change in height is eight minus two, or six inches over the course of the year.

Given two quantities, say $x$ and $y$, that are related in some way, we often use the symbol $\Delta y$ (read "delta y") to indicate the total change in the values of $y$ given a corresponding total change in the values of $x$, denoted by $\Delta x$.

Also, given two related quantities, it is often desirable to compute the *rate* at which one is changing with respect to the other. In particular, we are interested in the **average rate of change** of one quantity with respect to the other. If $x$ and $y$ are two related quantities, and $\Delta y$ is the total change in values of $y$ corresponding to a total change $\Delta x$ in values of $x$, then the corresponding average rate of change of $y$ with respect to $x$ is the quantity $\frac{\Delta y}{\Delta x}$. Notice that if units are attached to $x$ and $y$, then this average rate of change has units $\frac{\text{units of } y}{\text{units of } x}$, or (units of $y$) per (units of $x$).

For example, if the total change in height of the plant was six inches over the past year, then we could say that the average rate of growth over the past year was 6 inches/1 year or 6 inches per year. We could also say that the average rate of growth each month was 6 inches/12 months or $\frac{1}{2}$ inch per month, and so on.

If $x$ and $y$ are two related quantities, then as $x$ changes from $x = a$ to $x = b$, we summarize the ideas of total change and average rate of change as follows:

---

**Total Change and Average Rate of Change**

$$\text{total change in } y = \Delta y = (\text{value of } y \text{ when } x = b) - (\text{value of } y \text{ when } x = a)$$

$$\text{average rate of change of } y \text{ with respect to } x = \frac{\text{total change in } y}{\text{total change in } x} = \frac{\Delta y}{\Delta x}$$

---

To illustrate this topic, we will first explore population growth projections used by city planners.

## EXAMPLE 1 ◆ Population Growth

Suppose that a town of 15,000 people increases in population by 500 people per year.

(a) Create a table that shows the population of the town over fifteen years in three-year increments.

(b) Plot the population over time as found in (a).

(c) Find a formula for population in terms of time.

**Solution**

We let $P$ represent population and $t$ represent number of years, with $t = 0$ corresponding to the point in time when $P = 15,000$.

(a)

| Years ($t$) | Population ($P$) |
|:---:|:---:|
| 0 | 15,000 |
| 3 | 16,500 |
| 6 | 18,000 |
| 9 | 19,500 |
| 12 | 21,000 |
| 15 | 22,500 |

(b) Plotting $P$ in terms of $t$, we have the graph shown in Figure 1.1.

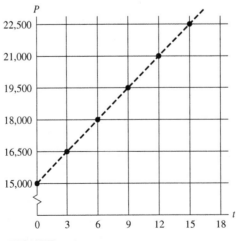

(Note: The dashed line is not part of the graph. It is just to illustrate that the points appear to lie on a line.)

*FIGURE 1.1*

(c) $P = 15,000 + 500t$

In the previous example, the average rate of change of 500 people per year remains constant, irrespective of which pairs of distinct years are used in the calculation (calculate the average rate of change over different periods of time, not all the same length). When this is the case, the relationship between the two quantities is called **linear**. Graphically, as you might expect, this phenomenon is illustrated by a line. So, in the last example the population $P$ is linearly related to time $t$. We have included a line through the plotted points as an illustration (see Figure 1.1). Any equation, graph, or table of values that describes two quantities that are related linearly is said to be a **linear model**.

## EXAMPLE 2 ♦ Creating Linear Models

A manufacturer of pianos has *fixed costs* (for building, machinery, recycling, etc.) of $1200 each day. These are expenses that must be paid regardless of the number of pianos produced. The *variable costs* (for material, labor, shipping, etc.) for making and selling a single piano amount to $1500. Let $P$ represent the number of pianos produced in a day.

(a) Calculate the factory's total cost, $C$, to make 1, 5, 8, and 12 pianos.

(b) Find a formula for $C$ in terms of $P$.

(c) What is the average rate of change of cost with respect to the number of pianos produced? How does this relate to the variable cost of producing a single piano?

**Solution**

As given, daily fixed costs are $1200. To calculate the total production costs for a day when pianos are produced, we simply multiply the number of pianos by the variable cost per piano and add the fixed costs.

(a)

| Pianos Produced ($P$) | Total Cost ($C$) |
|:---:|:---:|
| 1 | $1500(1) + 1200 = \$2700$ |
| 5 | $1500(5) + 1200 = \$8700$ |
| 8 | $1500(8) + 1200 = \$13{,}200$ |
| 12 | $1500(12) + 1200 = \$19{,}200$ |

(b) $C = 1500P + 1200$

(c) We will show that $C$ is linearly related to $P$ by demonstrating that the average rate of change of $C$ with respect to $P$ is always constant. Choose two arbitrary production levels, say $P = P_1$ and $P = P_2$ pianos. We calculate the average rate of change of $C$ with respect to $P$ as $P$ changes from to $P_1$ to $P_2$. Using our results from (b), we have

$$\frac{\Delta C}{\Delta P} = \frac{(1500P_2 + 1200) - (1500P_1 + 1200)}{P_2 - P_1}$$

$$\frac{\Delta C}{\Delta P} = \frac{1500P_2 + 1200 - 1500P_1 + 1200}{P_2 - P_1}$$

$$\frac{\Delta C}{\Delta P} = \frac{1500P_2 - 1500P_1}{P_2 - P_1}$$

$$\frac{\Delta C}{\Delta P} = \frac{1500(P_2 - P_1)}{(P_2 - P_1)}$$

$$\frac{\Delta C}{\Delta P} = 1500$$

So, the average rate of change of $C$ with respect to $P$ is always \$1500 per piano. Notice that this rate of change is exactly the variable cost associated with producing one piano.

---

The last two examples show some important characteristics of equations that model linear relationships between two quantities.

In Example 1,

current population = (average rate of growth per year) $\times$ (number of years) + initial population

In Example 2,

total costs = (average rate of increase per piano) $\times$ (number of pianos) + fixed costs

In general, if $x$ and $y$ are two linearly related quantities such that when $x = 0$ we have $y = b$ (call this the *initial value of y*) and the average rate of change of $y$ with respect to $x$ is constant, say $m$, then for $\Delta x \neq 0$, we have

$$m = \frac{\Delta y}{\Delta x}$$

So, given any value of $x$ with corresponding value of $y$, and letting

$$\Delta x = x - 0 \quad \text{and} \quad \Delta y = y - b$$

we have

$$m = \frac{\Delta y}{\Delta x} = \frac{y - b}{x - 0} = \frac{y - b}{x}$$

Solving for $y$ gives

$$y = mx + b$$

In other words,

value of $y$ = (average rate of change of $y$ with respect to $x$) $\times$ (value of $x$) + (initial value of $y$)

This is a way of describing linearly related quantities by an equation (formula), provided that we know the average rate of change and an initial value. In the next section we expand upon this idea.

# EXERCISES

1. Which of the following tables represents linearly related quantities? Check by determining if equal increments in $x$-values correspond to equal increments in $y$-values.

(a)
| $x$ | $y$ |
|---|---|
| 0 | 10 |
| 3 | 15 |
| 6 | 22 |
| 9 | 28 |

(b)
| $x$ | $y$ |
|---|---|
| 0 | 50 |
| 200 | 100 |
| 600 | 150 |
| 1200 | 200 |

(c)
| $x$ | $y$ |
|---|---|
| 0 | 10 |
| 10 | 20 |
| 20 | 25 |
| 30 | 35 |

(d)
| $x$ | $y$ |
|---|---|
| −3 | 10 |
| −1 | 2 |
| 0 | −2 |
| 3 | −14 |

2. The table below shows the cost of selling various amounts of iced tea per day.

| $x$ (cups/day) | 0 | 5 | 10 | 50 | 100 | 200 |
|---|---|---|---|---|---|---|
| $C$ (dollars) | 75.00 | 77.50 | 80.00 | 100.00 | 125.00 | 175.00 |

(a) Explain why the relationship between $C$ and $x$ appears to be linear.

(b) Using the values in the table, make a plot of $C$ against $x$.

(c) Find the average rate of change in cost with respect to the number of cups per day. Label your response with appropriate units. Explain what this means in the context of the given situation.

(d) Why should it cost \$75 to serve zero cups of iced tea?

3. The population, $P$, in millions, of a country in year $t$ is given by the formula

$$P = 14 + 0.2t.$$

(a) Construct a table of values for $P$ for $t = 0, 5, 10, 15, 20, 25$.

**(b)** Make a plot of $P$ against $t$ using the results from part (a).

**(c)** What is the country's initial population?

**(d)** Find the average rate of change of the population in millions of people/yr.

**4.** An office supply company produces staplers. The company has fixed monthly costs of $2000 and it costs $3.75 to produce each stapler. Write a formula that expresses the company's total cost, $C$, of producing $x$ staplers in a month.

**5.** In 1999, the population of a town was 4650 and growing by 23 people per year. Find a formula for $P$, the town's population, in terms of $t$, the number of years since 1999.

**6.** A new boat was purchased for $12,500. The boat's value depreciates linearly to $8600 in three years. Write a formula that expresses the boat's value, $V$, in terms of its age, $t$, in years.

**7.** Tuition cost, $T$, in dollars, for part-time students at Northeast Bible College is given by $T = 3000 + 200c$, where $c$ represents the number of credits taken.

**(a)** Find the tuition cost for six credits.

**(b)** How many credits were taken if the tuition was $6200?

**(c)** Make a table showing costs for taking from four to sixteen credits. For each value of $c$, give the tuition cost, $T$.

**(d)** Which of these values of $c$ has the smallest cost per credit?

**(e)** What does the 3000 represent in the formula for $T$?

**(f)** What does the 200 represent in the formula for $T$?

**8.** Survivor Sanctuary, a hidden island in the Pacific Ocean, experienced approximately linear population growth from 1980 to 2000. On the other hand, Afghanistan, a country in Asia, was torn by warfare in the 1980's and did not experience linear or near-linear growth.

| Year | 1980 | 1985 | 1990 | 1995 | 2000 |
|---|---|---|---|---|---|
| Pop. of Place A (millions) | 16 | 14 | 16 | 15.8 | 15.9 |
| Pop. of Place B (millions) | 12.2 | 13.5 | 14.7 | 15.8 | 17 |

**(a)** Which of these two places is Place $A$ and which is Place $B$? Explain.

**(b)** What is the approximate average rate of change in population with respect to time for the population that is growing approximately linearly?

**(c)** Using average rate of change, estimate the population of Survivor Sanctuary in 1988.

**9.** Outside the U.S., temperature readings are usually given in degrees Celsius; inside the U.S., they are often given in degrees Fahrenheit. The exact conversion from Celsius, $C$, to Fahrenheit, $F$, is given by the formula

$$F = \frac{9}{5}C + 32.$$

An approximate conversion is obtained by doubling the Celsius temperature and adding 30 to get the equivalent Fahrenheit temperature.

**(a)** Write a formula using $C$ and $F$ to express the approximate conversion.

**(b)** How far off is the approximation if the Celsius temperature is $-10°, 0°, 10°, 25°$?

**(c)** For what temperature (in degrees Celsius) does the approximation agree with the actual formula?

**10.** A woodworker goes into business selling birdhouses. His start-up costs, including tools, plans, and advertising, total $2000. Labor and materials for each birdhouse cost $35.

**(a)** Calculate the woodworker's total cost, $C$, to make 1, 3, 10, 15, and 20 birdhouses. Sketch a graph of $C$ versus $n$, the number of birdhouses that he makes.

**(b)** Find a formula for $C$ in terms of $n$.

**(c)** What is the average rate of change of $C$ with respect to $n$? What does this rate indicate about the woodworker's expenses?

**11.** A company has found that there is a linear relationship between the amount of money that it spends on marketing a product on the internet and the number of units that it sells. If the company spends no money on marketing, it still sells 400 units. For each $5000 spent on marketing, an additional 30 units are sold.

**(a)** If $x$ is the amount of money that the company spends on marketing, find a formula for $y$, the number of units sold, in terms of $x$.

**(b)** How many units does the firm sell if it spends $25,000 on marketing? $40,000?

**(c)** How much marketing money must be spent to sell 1000 units?

**(d)** What is the average rate of change of $y$ with respect to $x$? Give an interpretation that relates units sold and marketing costs.

**12.** At sea level, water boils at $212°F$. At an altitude of 1100 feet, water boils at $210°F$. The relationship between the boiling point and altitude is linear.

**(a)** Find the average rate of change in boiling point with respect to altitude.

**(b)** Find a formula that gives the boiling point of water in terms of altitude.

**(c)** Find the boiling point of water in each of the following cities: Minneapolis, MN (550 ft.); St. Joseph, MO (1300 ft.); Gillette, WY (3120 ft.)

**13.** A plane costs $100,000 new. After 10 years, its value is $25,000. Assuming linear depreciation, find the value of the plane when it is 8 years old and when it is 12 years old.

**14.** The profit, $P$, (in thousands of dollars) on $x$ thousand units of a specialty item is $P = 0.8x - 16.5$. The cost, $C$, (in thousands of dollars) of manufacturing $x$ thousand units is given by $C = 0.9x + 16.5$.

(a) Find a formula that gives the revenue $R$ from selling $x$ items, where $P = R - C$.

(b) How many items must be sold for the company to *break even* (zero profit)?

15. A doctor has a "rule of thumb" for determining the target weights for her patients, based on their heights. For women, the target weight is 100 lbs plus 5 lbs for each inch over 5 ft. in height. For men, it is 106 lbs plus 6 lbs for each inch over 5 ft. in height.

(a) Fill in the following tables:

### Women's Target Weight

| inches over 5 ft. | Weight (lbs.) |
|---|---|
| 0 | |
| 1 | |
| 2 | |
| 4 | |
| 6 | |
| 12 | |
| 15 | |

### Men's Target Weight

| inches over 5 ft. | Weight (lbs.) |
|---|---|
| 0 | |
| 1 | |
| 2 | |
| 4 | |
| 6 | |
| 12 | |
| 15 | |

(b) Find a formula that gives a woman's target weight, $y$, in terms of the number of inches, $x$, by which her height exceeds 5 ft.

(c) What is the target weight for a woman who is 69 inches tall?

(d) Do parts (b) and (c) for men.

16. The first edition of the textbook *Finite Math For Fun* was 372 pages long. By the time the ninth edition came along, the text had grown to 596 pages in length. Assume that the length $L$ of the book is linearly related to the edition number $n$.

(a) Write an equation that gives $L$ in terms of $n$.

(b) If the length trend continues, what edition of the book will be the first to have 1000 pages?

17. According to production records, on a certain production line the number of defective radio components produced is linearly related to the total number of these components produced. Suppose that one day 10 components out of a total of 300 produced are defective, and on another day 15 components out of a total of 425 produced are defective. How many defective components are expected on a day when a total of 500 components are produced?

18. A certain car can be rented according to one of two different rates:

- $40 per day and $0.30 per mile driven
- $30 per day and $0.50 per mile driven

(a) For each rate, write an equation that gives the cost, $C$ (in dollars), of driving $x$ miles in a day.

(b) Which rate is the least expensive if only 30 miles are to be driven on a single day?

19. Suppose that the cost of an airline ticket is linearly related to the distance traveled. The cost of a 200-mile flight is $76 and the cost of a 350-mile flight is $100. Let $C$ be the cost (in dollars) of a flight of $x$ miles.

(a) Write an equation that gives $C$ in terms of $x$.

(b) Determine the cost of a 275-mile flight.

(c) How many miles are flow if the cost is $524?

20. The height, $h$, of a falling sheet of paper, in feet above the ground, $t$ seconds after it is dropped is given by $h = -1.8t + 9$.

(a) How fast is the piece of paper falling?

(b) How far above ground is the paper after 4 seconds?

(c) How long will it take for the paper to reach the ground?

21. A soft-drink manufacturer can produce 1000 cases of soda in a week at a total cost of $6000, and 1500 cases of soda in a week at a total cost of $8500. Determine the manufacturers weekly fixed costs and weekly variable costs.

22. Suppose that a new tractor is worth $120,000 right now and is expected to have a useful life of 20 years, at which time the tractor will have a scrap value of $6000. Suppose that the value of the tractor decreases at the same rate each year. Calculate the rate at which the value decreases and also determine how long into its useful life the tractor is if its value has dropped below $80,000.

# Interlude:  Practice Problems 1.1

A small business that manufactures picnic tables has determined that total monthly production cost, $C$, and the number of tables produced, $x$, are linearly related quantities. The business has also established that monthly revenue, $R$, and $x$ are also linearly related. The company sells every table it produces each month. The graph shown in Figure 1.2 has been generated to show these relationships.

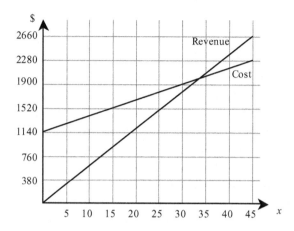

FIGURE 1.2

1.  Calculate the average rate of change of $C$ with respect to $x$. Calculate the average rate of change of $R$ with respect to $x$. Label both with appropriate units.

2.  Find formulas for both $C$ and $R$ in terms of $x$ (fill in the blanks).

    $C =$ _____

    $R =$ _____

9

3.  Estimate the coordinates of the point on the graph where the two lines meet. What is the significance of that point for the small business? Explain in regards to number of picnic tables produced and sold.

4.  Let $P$ denote the monthly profit that the business realizes from selling $x$ picnic tables. Find a formula for $P$ in terms of $x$ (fill in the blank). Sketch a graph of $P$ against $x$ on the provided grid. Put scale on the vertical axis.

    $P = $ _____

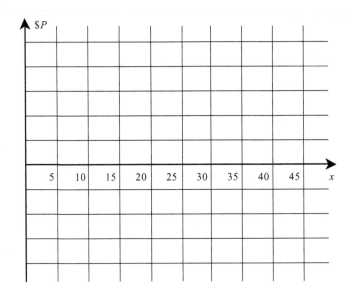

# 1.2  LINES: A QUICK REVIEW

In the previous section, we saw that if two quantities are linearly related, then the relationship can be illustrated graphically with a line. We also saw that average rate of change is one way to measure the rate at which one quantity is changing with respect to the other. In this section, we generalize the idea of average rate of change with the notion of *slope of a line*.

## Slope of a Line

When moving between point $S$ and point $T$ on a line lying in the $xy$-plane, as shown in Figure 1.1, we move vertically (parallel to the $y$-axis) and horizontally (parallel to the $x$-axis).

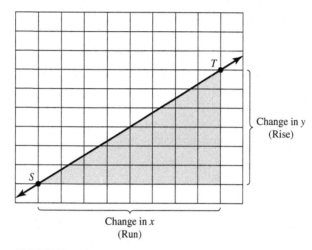

*FIGURE 1.1*

Evaluating the ratio

$$\frac{\text{change in } y}{\text{change in } x}$$

gives a measure of the steepness of the line. The larger the magnitude of this ratio, the steeper the line. It also indicates if the line is climbing from left to right (positive ratio) or falling from left to right (negative ratio). This ratio is called the **slope** of the line. A common interpretation of slope is "rise over run," as illustrated in Figure 1.1.

Consider two points in the $xy$-plane with coordinates $(x_1, y_1)$ and $(x_2, y_2)$. We can then say that the change in $x$, or "run," is the difference between $x_1$ and $x_2$, and the change in $y$, or "rise," is the difference between $y_1$ and $y_2$. There is a correspondence in change, in that as $x_1$ changes to $x_2$, $y_1$ changes to $y_2$, and vice versa. Consequently, we define the slope, $m$, of the line through these two points, assuming $x_1 \neq x_2$, as follows:

---

### Slope of a Line

The slope, $m$, of a nonvertical line through the points $(x_1, y_1)$ and $(x_2, y_2)$ is

$$m = \frac{\text{rise}}{\text{run}} = \frac{\text{change in } y}{\text{change in } x} = \frac{\Delta y}{\Delta x} = \frac{y_2 - y_1}{x_2 - x_1}$$

## EXAMPLE 1 ◆ Calculating Slopes

(a) Find the slope of each of the lines in Figure 1.2. Assume that the lines lie in the *xy*-plane and that the gridlines each mark one unit.

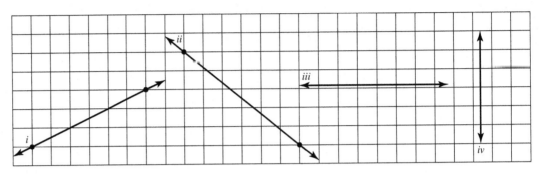

FIGURE 1.2

(b) Graph the line containing the point *p* with slope *m*.

   i.   $p(1, -3), m = -\dfrac{2}{5}$

   ii.  $p(-1, 4), m = \dfrac{3}{2}$

(c) Find the slope of the line through the given points in the *xy*-plane.

   i.   $(-2, 1)$ and $(5, 7)$

   ii.  $\left(\dfrac{1}{2}, 3\right)$ and $(2, -1)$

**Solution**

(a)  i.   The line is rising from left to right, so the slope is positive. Also, $\Delta y = 3$ while $\Delta x = 6$.

Therefore, the slope is $\dfrac{\Delta y}{\Delta x} = \dfrac{3}{6} = \dfrac{1}{2}$.

   ii.  The line is falling from left to right, so the slope is negative. Also, $\Delta y = -5$ while $\Delta x = 6$.

Therefore, the slope is $\dfrac{\Delta y}{\Delta x} = \dfrac{-5}{6} = -\dfrac{5}{6}$.

   iii. The line is horizontal. Consequently, $\Delta x$ can be any nonzero number, depending on the points at which we choose to look, but $\Delta y = 0$ between any two distinct points.

Therefore, the slope is $\dfrac{\Delta y}{\Delta x} = \dfrac{0}{\Delta x} = 0$.

   iv.  The line is vertical. Also notice that $\Delta x = 0$ regardless of the points at which we choose to look.

Since $\dfrac{\Delta y}{\Delta x}$ is undefined if $\Delta x = 0$, the slope of this line is undefined.

(b) See Figure 1.3.

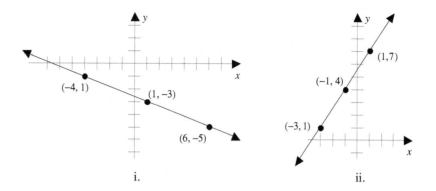

*FIGURE 1.3*

(c)  Slope, $m$, is given by $m = \dfrac{y_2 - y_1}{x_2 - x_1}$.

i.   The slope of the line through $(-2, 1)$ and $(5, 7)$ is

$$m = \frac{7 - 1}{5 - (-2)} = \frac{6}{7}$$

ii.  The slope of the line through $\left(\dfrac{1}{2}, 3\right)$ and $(2, -1)$ is

$$m = \frac{-1 - 3}{2 - \dfrac{1}{2}} = \frac{-4}{\dfrac{3}{2}} = -4\left(\frac{2}{3}\right) = -\frac{8}{3}$$

It should be noted here that, given any nonvertical line, any two distinct points on the line can be used to calculate the slope of the line. It can be shown, using similar triangles, that we will always get the same value for slope, regardless of which two points are used. More relevant to our previous discussion, though, is that the slope of a line in the $xy$-plane is just the average rate of change of $y$ with respect to $x$, which we have already shown is constant when the relationship between $x$ and $y$ is linear.

Consider the graphs shown in Figure 1.4. The first graph contains *parallel* lines $L_1$ and $L_2$. The second graph contains *perpendicular* lines $L_3$ and $L_4$.

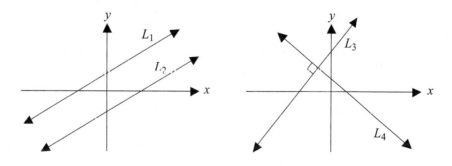

*FIGURE 1.4*

Two lines are parallel if they have the same slope. It can be shown that the slopes of two perpendicular lines are negative reciprocals of each other; that is, if a line $L_1$ that is neither horizontal nor vertical has slope $m$ and $L_2$ is perpendicular to $L_1$, then

$$\text{slope of } L_2 = -\frac{1}{m}.$$

## Equations for Lines

We now turn our attention to an algebraic treatment of lines, which means we will consider equations that are related to lines. **The graph of an equation** in the variables $x$ and $y$ consists of all points in the $xy$-plane whose coordinates $(x, y)$ satisfy the equation. So, when we speak of **an equation for a line,** we mean an equation whose graph is the given line. We have already encountered one such type of equation in the previous section when we investigated equations of the form $y = mx + b$.

Suppose that $L$ is a nonvertical line in the $xy$-plane that passes through the point $(0, b)$ and has slope $m$. If we let $(x, y)$ denote the coordinates of any other point on the line, then we know that

$$m = \frac{\Delta y}{\Delta x} = \frac{y - b}{x - 0}$$

Solving for $y$ gives

$$y = mx + b$$

This gives us one way of writing an equation whose graph is the line through the point $(0, b)$ with slope $m$. We call this equation the **slope-intercept form equation** for that line.

---

### Slope-Intercept Form (Equation for a Line)

Given a nonvertical line in the $xy$-plane with slope $m$ that passes through the point $(0, b)$, the slope-intercept form equation for the line is

$$y = mx + b$$

The number $b$ is called the **$y$-intercept** of the line.

---

Notice that this equation is exactly the same as the equation we developed at the end of the last section, given two linearly related quantities $x$ and $y$ with constant average rate of change of $y$ with respect to $x$ being $m$.

---

### EXAMPLE 2 ◆ Using Slope-Intercept Form

(a) Write the equation $3y - 2x = 6$ in slope-intercept form and sketch a graph of the equation.

(b) Write an equation for the line with slope $\frac{1}{2}$ that passes through the point $(2, 4)$.

(c) The math department buys a new calculator for $130. Three years later, its value is $70. Assuming *linear depreciation* (calculator is losing value in a linear fashion), what is the value of the calculator 6 months after it is purchased? How about after one year? What does this mean graphically?

**Solution**

(a) Writing the given equation in slope-intercept form, we have

$$3y - 2x = 6$$

$$3y = 2x + 6$$

$$y = \frac{2x + 6}{3}$$

$$y = \frac{2}{3}x + 2$$

Figure 1.5 shows the graph of the equation $3y - 2x = 6$.

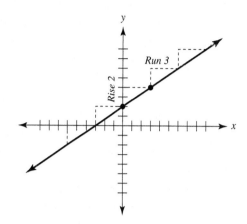

**FIGURE 1.5**

(b) We will write an equation for the line in slope-intercept form. Since $m = \frac{1}{2}$ we have

$$y = \frac{1}{2}x + b$$

Now, we will find $b$, the $y$-intercept. Notice that we can find $b$ by substituting in the $x$ and $y$ values of 2 and 4, respectively. Doing so, we get

$$4 = \frac{1}{2}(2) + b$$

$$4 = 1 + b$$

$$3 = b$$

Therefore, an equation for the line with slope $\frac{1}{2}$ that passes through the point $(2, 4)$ is

$$y = \frac{1}{2}x + 3$$

(c) Let $x$ represent the number of years since the calculator was purchased, and let $y$ represent the corresponding value of the calculator. Since the values of $x$ and $y$ are related in a linear fashion, we can represent the relationship between these two quantities by an equation of the form $y = mx + b$. Since we know that $y = 130$ when $x = 0$, we have $y = mx + 130$ for some number $m$. We note that when $x = 3, y = 70$. So,

$$m = \frac{\Delta y}{\Delta x} = \frac{70 - 130}{3 - 0} = -20$$

Consequently, $y = -20x + 130$, and

- after six months ($\frac{1}{2}$ year), the value is $-20(1/2) + 130 = \$120$.
- after one year, the value of the calculator is $-20(1) + 130 = \$110$.

The number $m$ calculated above was just the average rate of change of $y$ with respect to $x$. Graphically, this means that a graph of the value of the calculator ($y$) against years since purchase ($x$) is a line with slope $-20$ and $y$-intercept 130. Also, the points $(\frac{1}{2}, 120)$ and $(1, 110)$ lie on that line.

While slope-intercept form is a very common way to write an equation for a line, it is not the only way. Consider the line in Figure 1.6, where $(x, y)$ represents an arbitrary point on the line in the coordinate plane, and $(x_1, y_1)$ is some fixed point on the line.

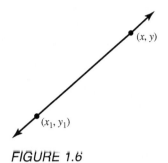

The slope of this line is $m = \dfrac{\Delta y}{\Delta x} = \dfrac{y - y_1}{x - x_1}$.

Multiplying both sides of the above equation by $x - x_1$ gives

$$y - y_1 = m(x - x_1).$$

This equation is called the **point-slope form equation** for the given line.

*FIGURE 1.6*

---

**Point-Slope Form (Equation for a Line)**

Given a nonvertical line in the $xy$-plane with slope $m$ that passes through the point $(x_1, y_1)$, the point-slope form equation for the line is

$$y - y_1 = m(x - x_1)$$

---

We most often use the point-slope form of an equation for a line in the following situations:

- When we are given the slope and a point on the line. Often we then rewrite the equation in slope-intercept form.
- When we are given two points on the line. Since only two points are needed to determine the slope of a line, we are able to find the slope. Our task is then to use the slope that we found along with one of the points (it doesn't matter which) to find the equation for the line. Again, we will most likely find it useful to then write the equation in slope-intercept form.

**EXAMPLE 3 ◆ Using Point-Slope Form**

(a) Find an equation for the line with slope 3 that passes through the point $(-5, 1)$. Write the equation in slope-intercept form.

(b) The table below gives the velocity, $v$, of a softball $t$ seconds after being thrown straight up in the air. Find a formula for $v$ in terms of $t$. Interpret the value of $v$ predicted when $t = 0$ and the average rate of change.

**Velocity of a softball $t$ seconds after being thrown in the air**

| $t$ (sec) | 1 | 2 | 3 | 4 |
|---|---|---|---|---|
| $v$ (ft/sec) | 58 | 26 | $-6$ | $-38$ |

(c) Find equations for each line shown in Figure 1.7 and interpret the slopes in terms of temperature.

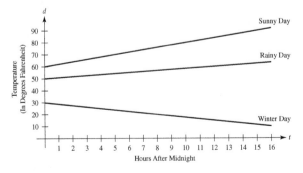

*FIGURE 1.7*

(d) Sales of a cosmetics company increased linearly from \$550,000 in 2002 to \$2.5 million in 2012. Find an equation that expresses the sales, $y$, in year $x$, where $x = 0$ corresponds to 2002, and use the equation to estimate the sales in 2009.

**Solution**

(a) Since the slope and a point on the line are given, we use the point-slope form of an equation for a line, $y - y_1 = m(x - x_1)$. Using 3 for the slope and the point $(-5, 1)$ for $(x_1, y_1)$ gives

$$y - 1 = 3(x - (-5))$$
$$y - 1 = 3(x + 5)$$
$$y - 1 = 3x + 15$$
$$y = 3x + 16$$

(b) Notice that in the data table, for each second of time that passes by, the velocity decreases by 32 feet per second. Since the rate of decrease in velocity is constant, we can model the behavior by a linear equation using a slope of $-32$. (This constant rate of decrease is the slope.) We are also given four data points in the table for $(t, v)$: $(1, 58)$, $(2, 26)$, $(3, -6)$ and $(4, -38)$. Since we only need one point to use the point-slope form of equation for a line, any point can be chosen. We arbitrarily choose $(1, 58)$, obtaining

$$v - 58 = -32(t - 1)$$
$$v - 58 = -32t + 32$$
$$v = -32t + 90$$

So, the softball started out (at time $t = 0$) traveling 90 feet per second and its velocity is decreasing at a rate of 32 feet per second each second.

(c) On a typical sunny day, the temperature starts out at 60 degrees and has a constant rate of increase (slope) of 30 degrees/16 hours = 15/8 or 1.875 degrees per hour. Therefore, an equation of the line is

$$d = \frac{15}{8}t + 60$$

On a rainy day, the temperature starts out at 50 degrees and has a constant rate of increase (slope) of 10 degrees/12 hours = 5/6 (about 0.83) degrees per hour. Therefore, an equation of the line is

$$d = \frac{5}{6}t + 50$$

On a winter day, the temperature starts out at 30 degrees and has a constant rate of decrease (slope) of $-10$ degrees/8 hours = $-5/4$ or $-1.25$ degrees per hour. Therefore, an equation of the line is

$$d = -\frac{5}{4}t + 30$$

Notice that slope here is negative, indicating that the temperature is decreasing at the described rate throughout the day.

(d) If we set 2012 as year $x = 0$ and put the sales $y$ in thousands of dollars, then we have the points $(0, 550)$ and $(10, 2500)$ on the line that is the graph of $y$ against $x$. Given these two points, we are now able to find an equation of the line modeling the data using the point-slope form. First, we find the slope, $m$.

$$m = \frac{y_2 - y_1}{x_2 - x_1} = \frac{2500 - 550}{10 - 0} = \frac{1950}{10} = 195$$

The slope indicates that sales are increasing at a rate of $195,000 per year. We could use either of the points with the slope that we found to find the equation of the line and we would get the same result. However, since we have an initial value for $y$, we see that

$$y = 195x + 550$$

Since 2009 is 7 years after our beginning year of 1992, we find the value of $y$ when $x = 7$, yielding

$$y = 195(7) + 550$$
$$y - 1915$$

We opted to let $y$ represent thousands of dollars in sales, so the estimated sales in 2009 are $1,915,000.

---

As we have seen, the defining characteristic of linearly related quantities is a constant average rate of change. Graphically, this resulted in using lines to describe how the quantities were related. However, many situations we encounter show a greater degree of variability in that two quantities of interest may be related in a nonlinear fashion. In the next section, we investigate the notion of using lines and linear equations to *approximately* represent nonlinear relationships.

## More on Intercepts and Equations for Lines

This will be discussed in more detail later, but if the graph of an equation in the xy-plane intersects the $x$-axis at the point $(a,0)$, then we say that the number $a$ is an $x$-**intercept** for the graph. Clearly, the graph of a non-horizontal line in the $xy$-plane can have at most one $x$-intercept.

For instance, refer to Figure 1.5, which shows the graph of the line given by the equation $3y - 2x = 6$. From the graph, it appears that this line has $x$-intercept $-3$. We can verify this algebraically, as the point of intersection with the $x$-axis has $y$-coordinate $y = 0$. So, we have

$$3(0) - 2x = 6$$
$$-2x = 6$$
$$x = -3$$

From the slope-intercept form of equation for a line, we see that a horizontal line (slope $m = 0$) that passes through the point $(0,b)$ has an equation of the form $y = b$. This reinforces the fact that every point on the line has y-coordinate $b$, irrespective of x-coordinate.

Similarly, each point of a vertical line that passes through the point $(a,0)$ has x-coordinate $a$, irrespective of y-coordinate. Consequently, the graph of the equation $x = a$ is exactly that line. Notice that no slope-intercept (or point-slope) form of equation for a vertical line exists, as the slope of a vertical line is undefined.

We briefly summarize some basic equations for graphs of lines in the $xy$-plane.

Equations for Lines
- Vertical line through $(a,0)$: $x = a$
- Horizontal line through $(0,b)$: $y = b$
- Slope-intercept form: $y = mx + b$
- Point-slope form: $y - y_1 = m(x - x_1)$

# EXERCISES

*For problems 1–5 determine whether each statement is true or false. If it is false, correct the statement to make it true.*

1. The slope of a line is another way of describing the average rate of change.

2. The graph of $2y - 8 = 3x$ has a y-intercept of 4.

3. The lines $3x + 4y = 12$ and $4x + 3y = 12$ are perpendicular.

4. Slope is not defined for horizontal lines.

5. The line $3x - y = 4$ rises more steeply from left to right than the line $4x - y = -1$.

*Graphing a linear equation is often easier if the equation is in slope-intercept form. If possible, rewrite the equations in problems 6–10 in this form.*

6. $2x + 9y = 18$

7. $y - 0.5 = 3(x - 0.1)$

8. $7(x + y) = 6$

9. $4x - 2y + 1 = 0$

10. $y = 3$

*Find equations for the lines with the properties in problems 11–14.*

11. Slope 5 and y-intercept 6

12. Slope $-3$ and x-intercept 8

13. Passes through points $(-1, 6)$ and $(3, -2)$ $-1$

14. Has x-intercept 5 and y-intercept $-7$  $\frac{7}{5}$

15. Write an equation for the line that passes through the origin and is horizontal.  $y = 0$

16. Write an equation for the horizontal line that passes through the point $(-5, 2)$.  $y = 2$

17. Figure 1.8 shows five different line segments. Without using a calculator, match each line segment to one of the following equations.

$$y_1 = 15 + 2x \qquad y_2 = 15 + 4x$$
$$y_3 = 2x - 20 \qquad y_4 = 70 - x$$
$$y_5 = 70 - 2x$$

**FIGURE 1.8 (NOT DRAWN TO SCALE)**

18. Using the standard window size $-10 \le x \le 10$ by $-10 \le y \le 10$, graph $y = x$, $y = 5x$, $y = 50x$, $y = 200x$ and $y = 1000x$ on your graphing calculator. Explain what happens to the graphs of the lines as the slopes become larger.

19. Without using a calculator, match the following formulas to the lines shown in Figure 1.9.

$$y_1 = 3 + 3x \qquad y_2 = -5 + 3x$$
$$y_3 = 3 + 4x \qquad y_4 = 3 - 3x$$
$$y_5 = 3 - 4x$$

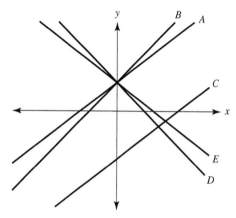

**FIGURE 1.9 (NOT DRAWN TO SCALE)**

20. Sketch, by hand, the graphs of $y = -2$ and $x = -2$. Describe each graph in general terms.

21. Line $l$ is given by $y = 5 - \frac{1}{3}x$ and point $p$ has coordinates $(4, 3)$.

    (a) Find the equation of the line containing $p$ that is parallel to $l$.

    (b) Find the equation of the line containing $p$ that is perpendicular to $l$.

    (c) Graph the equations in parts (a) and (b).

22. You need to rent a van and compare the charges of three different companies $A$, $B$, and $C$, in terms of $x$, the distance driven, in miles, in one day. Company $A$ charges 20 cents per mile plus $40 per day, company $B$ charges 10 cents per mile plus $60 per day, and company $C$ charges $110 per day with no mileage charges.

    (a) Find formulas for the cost of driving vans rented from companies $A$, $B$, and $C$, in terms of $x$.

    (b) Graph the cost for each company for $0 \le x \le 600$ and $0 \le y \le 150$. Put all three graphs on the same set of axes.

    (c) What do the slope and the y-intercept tell you in each situation?

    (d) Use the graph in part (b) to find the circumstances under which company $A$ would be the cheapest. Company $B$? Company $C$? Explain.

23. The cost of a Hotbox oven is $840 and it depreciates $40 each year. The cost of an Equator-bake oven is $1,400 and it depreciates $75 per year.

    (a) If a Hotbox and an Equator-bake are bought at the same time, when do the two ovens have equal value?

    (b) If the depreciation continues at the same rate, what happens to the values of the ovens in 20 years time?

24. John wants to buy a dozen muffins. The local bakery sells blueberry and poppy seed muffins for the same price.

    (a) Make a table of all the possible combinations of muffins if he buys a dozen, where $b$ is the number of blueberry muffins and $p$ is the number of poppy seed muffins.

    (b) Find a formula for $p$ in terms of $b$.

    (c) Graph your equation.

25. A theme park manager graphed weekly profits in terms of the number of visitors, and found that the relationship was linear. One week the profit was $990,000 when 132,400 visitors attended. Another week 174,525 visitors produced a profit of $1,327,000.

    (a) Find an equation for weekly profit, $p$, in terms of the number of visitors, $n$.

    (b) Interpret the slope and the $p$-intercept.

    (c) What is the number of visitors for which there is zero profit?

    (d) Find a formula for the number of visitors in terms of profit.

    (e) If the weekly profit was $125,000, how many visitors attended the theme park?

26. According to one economic model, the demand for gasoline in terms of price is a linear equation. If the price of gasoline is $p = \$1.30$ per gallon, then the quantity demanded in a fixed period is $q = 50$ gallons. If the price rises to $\$1.55$ per gallon, then the quantity demanded falls to 40 gallons in that period.

    (a) Find a formula for $q$ in terms of $p$.

    (b) Explain the economic significance of the slope used to generate your formula.

    (c) Explain the economic significance of the $q$-axis and $p$-axis intercepts.

$$999000 - 132400x = 1327000 - 174525x$$

$$42125x = 328000$$

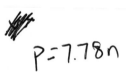

$$P = 7.78n$$

$$\begin{array}{r} 174525 \\ 132400 \\ \hline 42125 \end{array}$$

# Interlude: Practice Problems 1.2

The table below shows the account balance of a certain bank account over a period of several months, where the balance is calculated based on how many months have passed.

| Months ($m$) | 0 | 1 | 2 | 3 | 4 | 5 |
|---|---|---|---|---|---|---|
| Balance ($B$) | $700 | $1425 | $1795.50 | $1917.30 | 3694.25 | $4354.10 |

1.  Explain why $m$ and $B$ are not linearly related.

2.  On the scatterplot below (from the data above), sketch two lines, one through the first and last points on the plot and one through the last two points on the plot. Calculate the slope of each line.

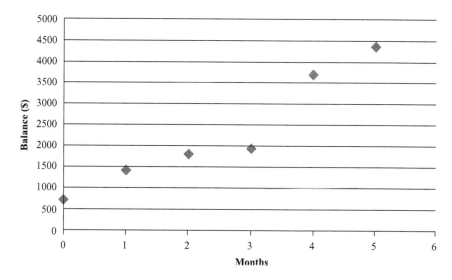

3.  Interpret the slopes you found in 2. in terms of the account balance and time. Label each appropriately.

4.  Find equations for the lines from 2. Use each equation to predict the account balance 8 months after the initial deposit was made. Which prediction do you think is more accurate? Explain your choice.

# 1.3  FUNCTIONS AND MODELING

An idea that is central to almost all of mathematics is that of a function. Functions are used to describe relationships between two quantities when one quantity depends on the other. We have already looked at situations where two quantities are linearly related, so we have, by default, worked with *linear functions*. Now, we turn our attention to models that describe relationships that are not necessarily linear.

## The Function Concept

We define the mathematical idea of a function in three parts. A **function** consists of:

1.  A collection of real number inputs called the **domain** of the function
2.  A collection of real number outputs called the **range** of the function
3.  A rule of correspondence between inputs and outputs such that any given input corresponds to precisely one output

It is important to note here that the **rule** of a function must be such that exactly one output is generated from a given input, but different inputs may, in fact, generate the same output.

## Describing Functions

There are four principal ways we describe or define particular functions—verbally, numerically (charts and tables), graphically, and algebraically (formulas). In each case we need to be able to identify the three components that comprise the function. Practically speaking, most functions we work with will be defined either numerically, graphically, or algebraically.

**EXAMPLE 1 ◆ Defining a Function Numerically**

Consider the following table.

| Input | $-3$ | $-2$ | 0 | 1 | 3 | 4 | 7 | 10 |
|---|---|---|---|---|---|---|---|---|
| Output | 18 | 14 | 10 | 4 | $-6$ | $-13$ | $-124$ | $3\pi + 4$ |

Here, we see that the domain (collection of inputs) of the function being described is the set $\{-3, -2, 0, 1, 3, 4, 7, 10\}$ and the range (corresponding outputs) is the collection $\{18, 14, 10, 4, -6, -13, -124, 3\pi + 4\}$. Notice that one of the outputs is an irrational number. The rule of correspondence is given by the table.

**EXAMPLE 2 ◆ Defining a Function Graphically**

Consider the graph shown in Figure 1.1.

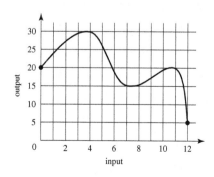

*FIGURE 1.1*

The domain of the function defined by the graph is the interval [0, 12] and the range is the interval [5, 30]. Points on the graph give the rule of correspondence. For instance, the point (7, 15) lying on the graph indicates that an input of 7 generates a corresponding output of 15. Notice that no input corresponds to more than one output, but outputs between 15 and 20 are each generated by three different inputs.

## EXAMPLE 3 ◆ Defining a Function Algebraically

Consider the familiar equation

$$A = \pi r^2$$

If we think of the variable $r$ as representing input and the variable $A$ as representing corresponding output, then this equation gives a very concise and unambiguous description of the relationship between $r$ and $A$; that is, the equation itself gives the rule of correspondence between input and output. We would say in this situation that $A$ **is a function of** $r$, indicating how we view both variables in terms of input and output. It is also common in this situation to call $r$ the **independent variable,** and to call $A$ the **dependent variable** (values of $A$ depend on the values of $r$).

The question of domain and range becomes somewhat more involved when we consider functions defined by equations. In the previous example, the formula $A = \pi r^2$ typically appears within a very specific context, namely computing the area of a circle of radius $r$. In that case, it is quite proper to think of the domain of the function that is being described as all positive numbers, in which case the range will also be all positive numbers. However, the formula, in and of itself, produces real number values for $A$ even if negative values of $r$ are used as inputs. With that in mind, we adopt a convention for dealing with the issue of domains of functions defined by equations.

**Domain Convention:** When dealing with functions defined by equations, we will assume that the domain is the set of all real number values that can be taken on by the independent variable so that the equation will produce real number values for the dependent variable, unless the specific situation being described by the function dictates otherwise.

## EXAMPLE 4 ◆ Domains

Consider the equation

$$y = \sqrt{5x - 8}$$

Treating $y$ as a function of $x$, we see that the equation will only yield real number values for $y$ if the expression under the radical is nonnegative. In other words, any values of the independent variable $x$ must satisfy the inequality

$$5x - 8 \geq 0$$

so

$$x \geq \frac{8}{5}$$

Thus the domain of this function, written in interval notation, is $[\frac{8}{5}, \infty)$.

## Functional Notation

We often use a special type of notation, called **functional notation,** when working with functions. Given some function, we "name" the function, typically with a letter, say $f$. If we let the variable $x$ represent input, then the symbol $f(x)$ (read "$f$ of $x$") denotes the resulting output under the rule of correspondence for the function we have chosen to call $f$.

For instance, consider the equation

$$f(x) = \frac{x + 1}{2x}$$

This equation defines a function, which we have denoted by the letter $f$. This function takes as an input any nonzero real number $x$ and yields as output the quantity $\frac{x+1}{2x}$.

Functional notation is convenient in that it gives us a compact way of describing what is happening when we are **evaluating** a given function at a particular input. Given the function $g$ defined by

$$g(x) = 2x^2 - 3x + 8$$

we indicate the output corresponding to an input of $-3$ by $g(-3)$, and calculate

$$g(-3) = 2(-3)^2 - 3(-3) + 8 = 18 + 9 + 8 = 35$$

according to the rule for $g$ given in the above equation.

### EXAMPLE 5 ♦ Evaluating a Function

Let $H$ be the function defined by

$$H(t) = 1 - t^2$$

Then if $a$ and $b$ are any real numbers, we have

$$H(2a) = 1 - (2a)^2 = 1 - 4a^2$$

and

$$H(a + b) = 1 - (a + b)^2 = 1 - (a^2 + 2ab + b^2) = 1 - a^2 - 2ab - b^2$$

We leave it as an exercise to verify that $H(a + b) \neq H(a) + H(b)$.

## Graphs of Functions

Given a function $f$, defined algebraically, it is often convenient to investigate the behavior of $f$ by looking at a graphical representation of the interplay between inputs and corresponding outputs. The **graph of a function** $f$ is the collection of all points in the Cartesian coordinate plane of the form $(x, f(x))$, where $x$ is any number in the domain of $f$.

The graphs of most functions defined by formulas can be obtained very easily using a graphing calculator. However, it is important to know about the properties of the function you are working with so that you do not misinterpret the results a graphing calculator may produce. In particular, it is not uncommon for a graphing calculator to produce an image that doesn't show all of the critical behavior of the graph of a function because of inappropriate viewing window settings or because of a lack of resolution sufficient to show significant features.

**EXAMPLE 6  ◆  The Graph of a Function**

Consider the graph shown in Figure 1.2. This is a sketch of the graph of the function $f$ defined by

$$f(x) = x^2$$

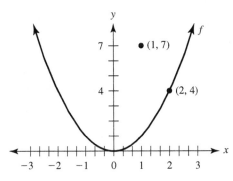

FIGURE 1.2

Notice that the point $(2, 4)$ is on the graph of $f$ since $f(2) = 4$ but the point $(1, 7)$ is not on the graph because $f(1) \neq 7$. The graph also reflects what a quick analysis of $f$ would reveal about domain and range. The domain of $f$ is the collection of all real numbers, while the range of $f$ is the collection of all nonnegative real numbers.

## Features of Graphs of Functions

Given the graph of a function, there are several features of the graph to which we pay particularly close attention. Here, we will look briefly at intercepts and local maximum and minimum values.

Suppose that $f$ is a function. Given the graph of $y = f(x)$, the $x$-coordinates of points where the graph of $f$ intersects the $x$-axis are called the $x$-**intercepts** of the graph. The $y$-coordinate of a point where the graph intersects the $y$-axis is called the $y$-**intercept** of the graph.

It should be noted here that the existence of $x$-intercepts or a $y$-intercept is dependent on the function involved. The graph of a given function may have several $x$-intercepts (or none), and will have at most one $y$-intercept.

**EXAMPLE 7  ◆  The Graph of a Function**

The graph shown in Figure 1.3 defines a function $f$ with domain $[-2, 6]$ and range $[-3, 4]$.

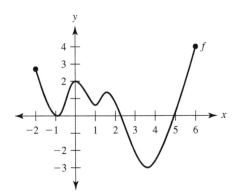

FIGURE 1.3

Estimating, it appears that the graph of $f$ has a $y$-intercept of 2 and $x$-intercepts of approximately $-1$, 2.3, and 4.9.

---

If $f$ is defined algebraically, then the $x$-intercepts of the graph of $f$ are precisely those values of $x$ for which $f(x) = 0$, while the $y$-intercept is $f(0)$, provided zero is in the domain of $f$. When the rule of $f$ is complicated, finding $x$-intercepts algebraically can be very challenging. In practice, a good approximation of these numbers is usually just as valuable as an exact value. Most graphing calculators have built-in functions designed for this purpose.

As was evident in the graph of the function $f$ from the previous example, the graph of a function may have several "peaks" and "valleys." These features are also important, and the behavior of the graph between peaks and valleys is something to which we frequently pay attention. Peaks are "local" high points on the graph, and so the $y$-coordinate of such a point is consequently an output that is bigger than all "nearby" outputs. So, if the graph of a function $f$ has a peak at the point $(a, f(a))$, then we say that $f$ has a **local maximum value at** $a$. Analogously, if the graph of a function $f$ has a valley at the point $(b, f(b))$, then we say that say that $f$ has a **local minimum value at** $b$.

## EXAMPLE 8 ◆ Local Maximum and Minimum Values

The graph shown in Figure 1.4 is a sketch of the graph of

$$f(x) = x^3 + 3x^2 - 24x + 3$$

obtained from a graphing calculator. The coordinates listed are the coordinates of the respective peak and valley.

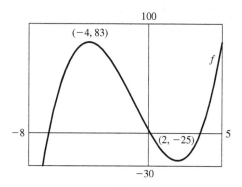

FIGURE 1.4

So, $f$ has a local maximum value at $x = -4$, and $f(-4) = 83$ is called a **local maximum value** for $f$. Similarly, $f$ has a local minimum value at $x = 2$, and $f(2) = -25$ is called a **local minimum value** for $f$.

---

Finding the exact locations of local maximum and minimum values for a given function can be very difficult. This is a topic that receives quite a bit of attention in the study of calculus. For now, we often rely on routines that graphing calculators have available to approximate the locations of local maximum or local minimum values.

## Modeling with Functions

The idea of a function is very general. However, since a function is essentially a description of the relationship between two quantities, it is very natural to use functions to describe or model situations that occur in the world or theoretical settings. Analyzing the behavior of a function that models the relationship between two quantities is often a key ingredient in making decisions about how to treat the quantities involved.

**EXAMPLE 9 ◆ Modeling a Relationship Between Two Quantities**

The field cricket is often called the "thermometer cricket" because it demonstrates a remarkably reliable relationship between chirp rate and ambient air temperature. The temperature, in degrees Fahrenheit, can be estimated by counting the number of times a field cricket chirps in 15 seconds and adding 38 to the total. So, if $T$ denotes the estimated temperature and $R$ is the chirp rate, in chirps per minute ($0 \leq R \leq 160$), then the equation

$$T = \frac{1}{4}R + 38$$

describes $T$ as a function of $R$. So, if a field cricket is chirping at a rate of 100 chirps per minute, then this model gives an estimated temperature of

$$T = \frac{1}{4}(100) + 38 = 63, \text{ or } 63°F$$

Notice also that the equation given can be rewritten in such a way that chirp rate is explicitly written in terms of temperature (we also have $R$ as a function of $T$).

$$T = \frac{1}{4}R + 38$$

$$4T = R + 152$$

$$R = 4T - 152$$

**EXAMPLE 10 ◆ Using a Function Model to Maximize a Quantity**

An open box is to be made from a square sheet of tin 14 centimeters on a side by cutting small squares of equal size from each of the corners and turning up the edges (see Figure 1.5). Suppose that it is deemed desirable to create a box of maximum possible volume. A function can be used to describe the volume of the box.

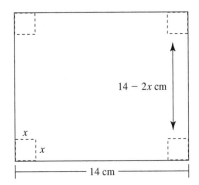
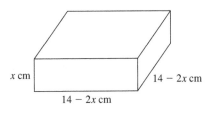

*FIGURE 1.5*

If we let $V(x)$ denote the volume (in cubic cm) of the box that results from cutting out squares of side length $x$ cm from each corner, then

$$V(x) = x(14 - 2x)^2 \quad \text{for} \quad 0 < x < 7$$

Figure 1.6 shows a sketch of the graph of the function $V$.

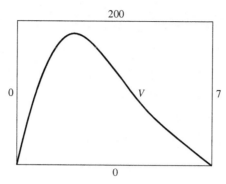

*FIGURE 1.6*

Using a graphing calculator routine, the peak in the above graph can be shown to be at the point where $x = \frac{7}{3}$. Consequently, the box of largest possible volume must measure $\frac{7}{3}$ cm $\times \frac{28}{3}$ cm $\times \frac{28}{3}$ cm.

---

Any company that manufactures some product will have **costs of production** and **revenues** from producing and selling the product. If we let $C$ represent production costs and let $R$ represent revenues, there are three possible scenarios describing how $C$ and $R$ can be compared.

- If $R > C$, then the company experiences a **profit**.
- If $R < C$, then the company experiences a **loss**.
- If $R = C$, then the company **breaks even**.

We typically think of production costs and revenues as functions of the number of units of the product produced and sold. Let $x$ denote the number of units of the product produced and sold (we assume that every unit produced is sold), and let $C(x)$ and $R(x)$ denote the corresponding **cost** and **revenue functions.**

Costs of production can be placed into two categories. **Fixed costs** are costs that are incurred irrespective of the number of units produced (rent, fixed salaries, insurance, etc.). **Variable costs** depend on the number of units produced (production material, packaging, shipping, etc.). Using the notation from above,

$$C(x) = \text{fixed costs} + \text{variable costs}$$

Revenue depends on two quantities, the number of units sold and the selling price per item. If $x$ units are sold at a price of $\$p$ per unit, then the corresponding revenue is given by

$$R(x) = (\text{number sold}) \times (\text{price per unit})$$

$$R(x) = xp$$

Based on cost and revenue functions, we can also construct a **profit function.** Let $P(x)$ denote the profit from producing and selling $x$ units of product. Profit is the difference between revenue and cost, so

$$P(x) = R(x) - C(x)$$

Analyzing the relationship between $C(x)$, $R(x)$, and $P(x)$ is called **profit-loss analysis.**

**EXAMPLE 11 ◆ Cost, Revenue, and Profit Functions**

Minntronics produces and sells a powerful model of scientific calculator. Monthly fixed costs amount to $17,604 and the cost to produce each calculator is $9.30. Each calculator produced is sold for $17.45.

(a) Find the cost, revenue, and profit functions.

(b) Determine the minimum number of calculators that must be produced and sold each month in order for the company to be profitable.

**Solution**

Let $x$ represent the number of calculators produced and sold in a month.

(a) The cost function is given by

$$C(x) = \text{fixed costs} + \text{variable costs}$$
$$C(x) = 17{,}604 + 9.3x \text{ (dollars)}$$

The revenue function is given by

$$R(x) = (\text{number sold}) \times (\text{price per calculator})$$
$$R(x) = x \times 17.45$$
$$R(x) = 17.45x \text{ (dollars)}$$

The profit function is given by

$$P(x) = R(x) - C(x)$$
$$P(x) = 17.45x - (17{,}604 + 9.3x)$$
$$P(x) = 8.15x - 17{,}604 \text{ (dollars)}$$

(b) Notice that cost, revenue, and profit are all modeled by linear functions in this case. We have two options for ascertaining the level of production and sales needed to attain profitability. We can

- Determine the values of $x$ so that $R(x) > C(x)$, or
- Determine the values of $x$ so that $P(x) > 0$

We will deal algebraically with the inequality $R(x) > C(x)$ and illustrate what is happening with graphs of the linear functions (see Figure 1.7).

$$R(x) > C(x)$$
$$17.45x > 17{,}604 + 9.3x$$
$$8.15x > 17{,}604$$
$$x > \frac{17{,}604}{8.15}$$
$$x > 2160$$

So, Minntech must produce and sell at least 2161 calculators each month to be profitable. Notice that a production and sales level of 2160 calculators is a **break-even point**—revenue is equal to cost and so profit is zero.

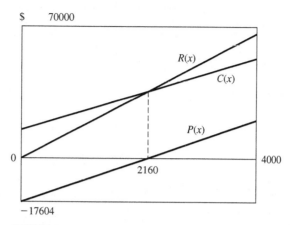

*FIGURE 1.7*

In the last example, the price at which calculators were sold was fixed at $17.45 per calculator regardless of the number produced and sold. Quite often there is a dynamic relationship between selling price of a product and the number of units of the product that consumers are willing to purchase at that price. Financial departments frequently model this relationship by a **price-demand function,** where typically the price per unit is given as a function of the number of units sold.

## EXAMPLE 12 ♦ Price-Demand and Revenue

The research department in a company that manufactures and sells metal shelving kits has established, based on historical data, that the price at which $x$ hundred kits can be sold is given by the price-demand function

$$p(x) = 50 - 1.25x \text{ (dollars per kit)}$$

for $0 \le x \le 40$.

(a) State the revenue function.

(b) Calculate the selling price and corresponding revenue for sales of 1000, 1500, 2000, 2500, and 3000 shelving kits.

**Solution**

Note here that $x$ is in hundreds of kits. The condition $0 \le x \le 40$ means that the models we will be using are only valid for levels of sales between 0 and 4000 kits.

(a) Revenue is given by

$$R(x) = xp(x)$$

$$R(x) = x(50 - 1.25x) \text{ (hundreds of dollars)}$$

The revenue $R(x)$ is in hundreds of dollars because $x$ is in hundreds of kits, while $p(x)$ is in dollars per kit.

(b) Consider the following table. Remember that $x$ is in hundreds of shelving kits, so a level of sales of 1000 kits corresponds to a value of $x = 10$, for instance.

| Hundreds of kits ($x$) | Selling price per kit ($) | Revenue (hundreds of $) |
|---|---|---|
| 10 | $50 - 1.25(10) = 37.50$ | $10 \times 37.50 = 375$ |
| 15 | $50 - 1.25(15) = 31.25$ | $15 \times 31.25 = 468.75$ |
| 20 | $50 - 1.25(20) = 25$ | $20 \times 25 = 500$ |
| 25 | $50 - 1.25(25) = 18.75$ | $25 \times 18.75 = 468.75$ |
| 30 | $50 - 1.25(30) = 12.50$ | $30 \times 12.50 = 375$ |
| 35 | $50 - 1.25(35) = 6.25$ | $35 \times 6.25 = 218.75$ |

What is indicated in the table, for example, is that 1500 kits can be sold at a price of $31.25 per kit, generating revenue of $46,875. Notice that selling more kits doesn't necessarily guarantee more revenue. For example, more revenue is earned selling 2000 kits than is earned selling 2500 kits.

## EXAMPLE 13 ◆ Revenue

Given the revenue function from the previous example, determine the following.

(a) The number of kits that should be sold in order to generate revenue of at least $40,000.

(b) The number of kits that should be manufactured and sold in order to generate maximum possible revenue.

### Solution

We have established that revenue is given by $R(x) = x(50 - 1.25x)$ (hundreds of dollars). Both questions can be answered using algebraic methods, but we opt instead to take a graphical approach. Since one thing we seek is the values of $x$ for which $R(x) \geq 400$ (at least $40,000 revenue), we will graph the following equations on a graphing calculator and determine where the graph of the revenue function is showing outputs of 400 or higher. We graph

$$y_1 = x(50 - 1.25x)$$

$$y_2 = 400$$

The table from the previous example gives helpful information on setting an appropriate viewing window (see Figure 1.8). Notice that $0 \leq x \leq 40$ and $R(x) \geq 0$.

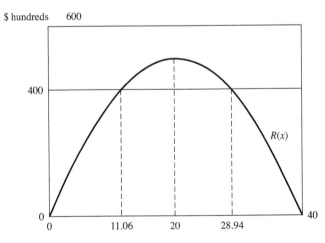

FIGURE 1.8

Using an intersection routine on the calculator, we find that the graph of the revenue function intersects the horizontal line corresponding to $40,000 revenue when $x \approx 11.06$ and $x \approx 28.94$. Since $x$ is in hundreds of kits, these have been rounded to two decimal places. As can be seen from the graph, selling anywhere from 1106 kits to 2894 kits will guarantee revenue of at least $40,000. Also, using a maximum finder on the calculator, the peak of the graph of the revenue function occurs when $x = 20$, so revenue is maximized when 2000 shelving kits are sold. From the last example we see that this level of sales generates a revenue of $50,000.

We conclude with an example involving analysis of cost, revenue, and profit functions.

## EXAMPLE 14 ♦ Cost, Revenue, and Profit

Suppose that the company in the last two examples has established that the cost of producing $x$ hundred metal shelving kits is given by the cost function

$$C(x) = 160 + 10x \text{ (hundreds of dollars)}$$

Sketch graphs of the cost, revenue, and profit functions on the same set of axes and find all break-even production levels.

### Solution

We have established that revenue is given by $R(x) = x(50 - 1.25x)$, so profit, $P(x)$, is given by

$$P(x) = R(x) - C(x)$$
$$P(x) = x(50 - 1.25x) - (160 + 10x)$$

We elect to leave the rule for $P(x)$ in the form above since we do not need to do any manipulations with $P(x)$. Again, all of our work will be done graphically. We will graph the following equations on a graphing calculator (see Figure 1.9).

$$y_1 = 160 + 10x$$
$$y_2 = x(50 - 1.25x)$$
$$y_3 = x(50 - 1.25x) - (160 + 10x)$$

It will be helpful for setting an appropriate viewing window to use our previous results and to notice that

$$P(0) = R(0) - C(0)$$
$$P(0) = 0 - 160$$
$$P(0) = -160$$

and

$$P(40) = R(40) - C(40)$$
$$P(40) = 40(50 - 1.25(40)) - (160 + 10(40))$$
$$P(40) = -560$$

(See Figure 1.9)

Using an intersection routine, we find that the cost and revenue graphs intersect when $x \approx 4.69$ and when $x \approx 27.31$. We have rounded to two decimal places since $x$ is in hundreds of metal shelving kits. So, the company has break-even production levels at about 469 kits and 2731 kits. In between those production levels, the company experiences a profit. Producing and selling less than 469 kits or more than 2731 kits results in a loss. We would arrive at the same results using a routine of the calculator to estimate the $x$-intercepts of the graph of the profit function. Finally, notice that although maximum revenue occurs when 2000 kits are produced and sold, maximum profit occurs at a different level of sales, namely, 1600 kits (check this).

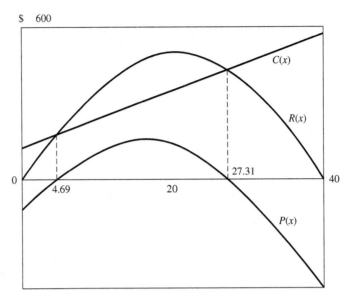

$ 600

$C(x)$

$R(x)$

27.31

40

0

4.69

20

$P(x)$

FIGURE 1.9

# EXERCISES

1. Consider the equation $3p + 6q^2 = 2$. Does this equation define $p$ as a function of $q$? Does it define $q$ as a function of $p$?

2. A dance studio charges $80 per student for a series of four 2-hour dance lessons. The studio costs are $30 per hour for the instructor, $15 per lesson for room rental, and $3 per student for miscellaneous expenses. If $x$ is the number of students enrolled in the class, express the studio's profit, $P$, as a function of $x$, and calculate the profit if 10 students enroll in the class.

3. Let $f(x) = \dfrac{3x - 2}{\sqrt{x - 5}}$. Determine the domain of $f$.

4. The value of a delivery truck is a function of its age, in years, as given by

$$v(x) = -2300x + 16500$$

where $x$ is the age of the truck in years and $v(x)$ is the corresponding value.

   (a) What did the truck cost when new?

   (b) If the truck has a useful life of seven years, what is the scrap value of the truck?

5. Based on data from testing, an automobile tire manufacturer has modeled tread life of one particular line of tires as a function of tire pressure. If $p$ is the tire pressure (in psi) and $T(p)$ is the corresponding tread life (in thousands of miles), then the model is

$$T(p) = -0.5128p^2 + 33.3p - 481$$

Here, $25 \le p \le 40$. Given this model, what air pressure gives the longest tread life?

6. The market research department of a small company that manufactures beer steins for Oktoberfest has determined the following price-demand equation, where $p$ is the price (in dollars) at which $x$ hundred steins can be sold and $0 \le x \le 11$.

$$p(x) = 28 - 2.2x$$

In addition, the research department has determined the following cost function:

$$C(x) = 35 + 8x \text{ (in hundreds of dollars).}$$

   (a) Write a formula for $P(x)$, the profit generated by selling $x$ hundred steins (no need to simplify).

   (b) Determine how many steins must be manufactured and sold in order for the company to break even.

   (c) For what levels of production is it true that $P(x) > 0$?

7. A rectangular box is to have a square base and a volume of 20 cubic feet. The material for the base costs 30 cents per square foot, the material for the top costs 20 cents per square foot, and the material for the sides costs 10 cents per square foot. If $x$ is the length of one side of the base, in feet, write a function $C(x)$ that gives the total cost of the material for the box, in dollars, in terms of $x$. Indicate the domain of this function.

8. A function $f$ is defined by the following rule: "For a given input $x$, the corresponding output $f(x)$ is generated by first multiplying the input by 4, subtracting 1, and then finding the reciprocal of the result." Find a formula for $f(x)$.

9. Let $g(x) = 0.4x^3 - 2x^2 + 1.5x + 1$. Graph this function in the standard viewing window on your graphing calculator.

Use the graph of $g$ to answer the following questions. Give all estimates correct to three decimal places.

(a) Find all values of $x$ in the interval $[-10, 10]$ such that $g(x) = 4$.

(b) Find the largest output for $g(x)$ on the interval $[-10, 10]$.

(c) Solve the inequality $g(x) \leq 1$. Give the solution in interval notation.

(d) Estimate the local maximum and minimum values of $g(x)$ in the interval $[-10, 10]$.

10. Cliff operates a small business that manufactures miniature wax figurines. He has fixed weekly costs of $500 and it costs him $8 to manufacture a single figurine. Cliff sells the figurines for $25 apiece. Let $x$ be the number of figurines Cliff manufactures and sells each week, and assume that $0 \leq x \leq 200$.

(a) Write a formula for $C(x)$ the cost incurred by manufacturing $x$ figurines in a week.

(b) Write a formula for $R(x)$ the revenue earned by selling $x$ figurines in a week.

(c) Write a formula for $P(x)$ the profit earned by selling $x$ figurines in a week.

(d) What is the minimum number of figurines Cliff must sell in a week to guarantee a profit of at least $2400?

11. A company that manufactures electric popcorn poppers has the following cost and revenue functions.

$$C(x) = 20 + 2.5x \quad \text{and} \quad R(x) = x(16 - 0.8x)$$

Here $x$ is in thousands of poppers and $C(x)$ and $R(x)$ are in thousands of dollars. Assume that $0 \leq x \leq 20$.

(a) Write a formula for the profit function, $P(x)$.

(b) Determine the level of production at which maximum profit occurs, to the nearest popcorn popper.

(c) What is the maximum possible profit, to the nearest dollar?

12. Figure 1.10 shows the graphs of four linear functions. Three of the graphs represent cost, revenue, and profit functions, respectively, from a particular situation, and one of the graphs is unrelated.

(a) Which of the graphs represents the cost function?

(b) Which of the graphs represents the revenue function?

(c) Which of the graphs represents the profit function?

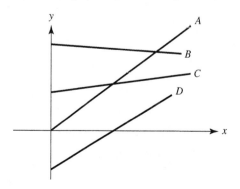

**FIGURE 1.10**

13. The research department for Spanky's Electronics established the following cost and revenue functions for their coffee makers.

$$C(x) = 80 + 15x \quad \text{and} \quad R(x) = x(60 - 1.75x)$$
$$\text{for} \quad 0 \leq x \leq 30$$

Here, $x$ is in thousands of units and both $C(x)$ and $R(x)$ are in thousands of dollars.

(a) Give a formula for $P(x)$, the profit from selling $x$ thousand coffee makers.

(b) What is the minimum number of coffee makers that must be sold in order to guarantee a profit of $150,000? (Give your solution to the nearest coffee maker.)

14. The graphs in Figure 1.11 show the cost of a hospital stay under three different health insurance plans ($A$, $B$, and $C$). The cost under each plan is a function of the number of days spent in the hospital.

(a) Which plan is least expensive for a stay of three days?

(b) Suppose a person had selected plan $B$ and was billed $800 for a hospital stay. For how long was the person hospitalized?

(c) Under what circumstances would plan $B$ be the least expensive choice?

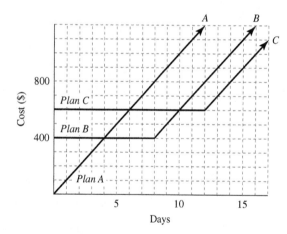

**FIGURE 1.11**

15. According to data from the National Traffic Safety Institute, the stopping distance $y$ (in feet) of a car traveling $x$ mph can be reasonably approximated by

$$y = 0.056057x^2 + 1.06657x$$

(a) Find the stopping distance for a car traveling 25 mph.

(b) How fast can you drive if you need to be certain of stopping within 150 feet?

16. One traffic safety study suggests that the accident rate as a function of age of the driver is approximated by

$$r(t) = 60 - 2.28t + 0.0232t^2$$

for $16 \leq t \leq 35$, where t is the age of the driver in years. Given this, find the age at which the accident rate is at a minimum and the age at which the accident rate is at a maximum.

17. Suppose that each weekly edition of a college newspaper incurs fixed production costs of $70 and variable costs of $0.40 per copy. The newspaper sells for $0.50 per copy. Suppose that $x$ represents the number of copies of an edition produced and sold in a given week (assume that all copies produced are sold), and let $C(x)$, $R(x)$, and $P(x)$ denote the corresponding cost, revenue, and profit functions (in dollars), respectively.

    (a) Give formulas for $C(x)$, $R(x)$, and $P(x)$.

    (b) What is the profit (or loss) from the sale of 500 copies of the newspaper?

    (c) How many copies need to be produced and sold in order to break even?

18. The graphs in Figure 1.12 are graphs of a cost function, $C(x)$, and a revenue function, $R(x)$, for $0 \leq x \leq 70$. (You must decide which graph goes with each function.) Use the graphs to answer the questions that follow. It is important that you estimate carefully.

    (a) What does the $y$-intercept of the graph of the cost function indicate?

    (b) Estimate the break-even production level(s).

    (c) Calculate the profit earned at a production level of 50 units.

    (d) Give a formula for $C(x)$.

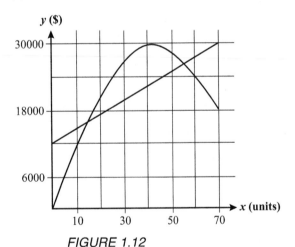

FIGURE 1.12

19. Entomologists have discovered that a linear relationship exists between the number of chirps of crickets of a certain species and the air temperature. When the temperature is $70°F$, the crickets chirp at the rate of 120 times/minute, and when the temperature is $80°F$, they chirp at a rate of 160 times/minute. Let $N$ denote the number of chirps per minute and let $T$ denote the temperature. Find a formula for $N$ in terms of $T$ and use the formula to predict the rate at which the crickets chirp when the temperature is $98°F$.

20. Slippery Rock Real Estate is building a new housing development. Market analysis indicates that if 40 houses are built in the development, each can be sold for $200,000. However, if 60 houses are built, each will sell for $160,000. Suppose that the relationship between price and demand is linear.

    (a) Write a demand equation giving selling price in terms of number of houses built.

    (b) Write a revenue function giving revenue from house sales in terms of number of houses built.

    (c) Determine the number of houses that should be built in order for the company to maximize revenue from house sales in the development.

21. At a price of $6 each, a manufacturer can sell 3000 rubber trashcans per month. If the price is $8, only 2750 trashcans can be sold each month. Suppose that demand and price are linearly related. Determine the price at which the rubber trashcans should be set in order to maximize monthly revenue.

# Graphing Calculator Exercises

**Notes:**

1. It is assumed at this point that you can access your calculator's graphing utility.
2. Your owner's manual is useful for finding the locations of the common commands and routines that you are expected to use. Another resource for Texas Instruments graphing calculators can be found at

   *http://education.ti.com/calculators/pd/US/Online-Learning/Tutorials*

## PART I: SETTING YOUR VIEWING WINDOW APPROPRIATELY (NO "ZOOMING" ALLOWED)

1. The graph of the equation $y = 0.6x^5 + 1.5x^4 - 13x^3 - 21x^2 + 72x + 1$ has two "peaks" and two "valleys." Find appropriate viewing window settings to clearly view these features on your calculator. Sketch the picture you see and list your **Xmin, Xmax, Ymin,** and **Ymax** values.

   **Xmin:**

   **Xmax:**

   **Ymin:**

   **Ymax:**

2. The graph of $y = 0.0025x^4 - 0.5x^2 + 9$ has one "peak" and two "valleys." Find appropriate viewing window settings to clearly view these features on your calculator. Sketch the picture you see and list your **Xmin, Xmax, Ymin,** and **Ymax** values.

   **Xmin:**

   **Xmax:**

   **Ymin:**

   **Ymax:**

3.  Graph the equation $y = -0.9x^4 + 1.8x^2 - 0.5$. Find appropriate viewing window settings to obtain a picture on your calculator like the one shown. List your **Xmin, Xmax, Ymin,** and **Ymax** values.

    **Xmin:**

    **Xmax:**

    **Ymin:**

    **Ymax:**

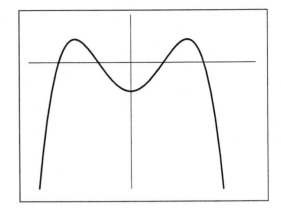

4.  Graph the equation $y = x(24 - 3x)$. Find appropriate viewing window settings to obtain a picture on your calculator like the one shown. List your **Xmin, Xmax, Ymin,** and **Ymax** values.

    **Xmin:**

    **Xmax:**

    **Ymin:**

    **Ymax:**

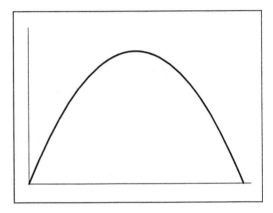

## PART II:   USING SOME BUILT-IN ROUTINES (DO NOT "ZOOM AND TRACE")

There are several built-in routines available on the Texas Instruments graphing calculators that you will be expected to use on occasion. In particular, you should be acquainted with your machine's root/zero finder, intersect feature, and maximum/minimum finder.

1.  Graph the equation $y = 0.2x^3 + x - 1$ in the standard viewing window and make a sketch. Use the root/zero finding feature of your calculator to find the $x$-intercept. Give the value of the $x$-intercept, accurate to four decimal places.

    **Remark:** The $x$-coordinate you found is the approximate solution to the equation $0.2x^3 + x - 1 = 0$. (Check this)

    **Estimate:**

2. Graph the following two equations in the standard viewing window and make a sketch.

$$y = -2x^2 + 3x + 1$$

$$y = 3x - 2$$

Use the intersect routine on your machine to find the points of intersection of the two graphs. Give the coordinates accurate to four decimal places.

**Remark:** Notice that the $x$-coordinates of the points of intersection are the approximate solutions to the equation
$-2x^2 + 3x + 1 = 3x - 2$. (Check this)

**Estimates:**

3. Graph the equation $y = 0.8x^3 - 3.9x^2 + 5x - 3$ in the standard viewing window and make a sketch. You should notice that the graph has one "peak" and one "valley." both located in the fourth quadrant. First, use the maximum finder on your machine to find the coordinates of the peak. Repeat this to find the coordinates of the valley, except this time use the minimum finder. Record the coordinates of both points, accurate to four decimal places.

**Peak:**

**Valley:**

# Interlude: Practice Problems 1.3, Part 1

Blurry Vision, Inc. makes DVD players. The annual operating cost for producing and selling x thousand units of a particular model of DVD player is given by

$$C(x) = 200 + 15x \quad \text{thousand dollars.}$$

Also, the annual revenue from selling $x$ thousand of these players is given by

$$R(x) = x(52 - 0.3x) \quad \text{thousand dollars.}$$

Here, $0 \leq x \leq 130$.

1. Find all levels of sales that will generate a revenue of at least $1,500,000, and give your solution in terms of numbers of DVD players. Support your result graphically.

2. Determine the price(s) at which players should be set in order to generate a revenue of exactly $2,000,000. Support your result graphically.

3. Determine the number of DVD players that must be sold to generate the maximum possible revenue, and also the corresponding maximum possible revenue. Give your results to the nearest DVD player and the nearest dollar, respectively. Support your result graphically.

4.  Write $P(x)$, the profit function.

5.  Find all levels of sales that will generate a profit of exactly $300,000, and give your solution(s) as a number of DVD players. Support your result graphically.

6.  Find all levels of sales where the company breaks even, and give your solution(s) as a number of DVD players. Support your result graphically.

# Interlude: Practice Problems 1.3, Part 2

1. Suppose that there is a linear relationship between demand for crude oil and the price of a barrel of crude oil. The daily demand for crude oil is 92.6 million barrels when the price is $98.62 per barrel and this demand drops to 89.3 million barrels when the price rises to $102.62. Let $q$ denote the daily demand for crude oil, in millions of barrels, and let $p$ denote the price of a barrel of crude oil, in dollars.

    (a) Write an equation (the demand equation) that gives $q$ in terms of $p$.

    (b) Suppose that the daily supply of crude oil also varies with price according to the *supply equation* $q = 0.2p + 66.9$. What price (to the nearest cent) for a barrel of oil would result in a daily supply of 83.4 million barrels?

    (c) Graph both the supply and demand equations on the same set of axes and find the point of intersection of those graphs (this point is called the *equilibrium point*). What is the practical significance of that point in terms of price per barrel and number of barrels of crude oil?

2.  When a popular style of running shoe is priced at $80, Runner's Emporium sells an average of 96 pairs per week. Based on comparative research, management believes that for each decrease of $2.50 in price, four additional pairs of shoes could be sold weekly.

    (a)  Let $n$ denote the number of $2.50 decreases in price and let $q$ denote the number of pairs of shoes sold. Write an equation that gives $n$ in terms of $q$.

    (b)  Write a function $R(p)$ that gives the expected weekly revenue from sales of this particular style of shoe as a function of $p$, the price per pair (in dollars).

    (c)  Determine the selling price per pair that will result in maximum possible revenue, as well as the maximum possible revenue.

    (d)  Determine the lowest price (to the nearest cent) that will result in weekly revenue of at least $7000 from sales of the shoes.

# Chapter 1  Progress Check I

1. At a market price of $260, the demand for a type of printer is 8000 units, but at a price of $200, the demand increases to 10,000 units. The manufacturer of this printer will not sell any units if the market price is $100 (or lower) but for each $50 increase in price above $100, will put an additional 1000 units on the market. Let $p$ denote the market price in dollars of one of these printers and let $q$ denote the number of printers.

   (a) Write the supply equation, giving $p$ in terms of $q$.

   (b) Assume that the relationship between price and demand is linear. Write the demand equation, giving $p$ in terms of $q$.

   (c) Find the equilibrium point. Identify both the price and quantity at the equilibrium point.

   (d) Suppose that it is proposed that maximum possible revenue for sales of this type of printer occurs when the printers are priced at the price indicated by the equilibrium point. Is this true? If so, explain why. If not, what price actually maximizes revenue?

2. Putrescence Punishing Products makes a very effective water filter at a plant in rural Clay County. The plant has fixed monthly costs of $20,000 and production costs of $20 per filter. Each filter produced is sold for $38. The plant can produce at most 4000 filters each month.

   (a) Give a formula for $C(x)$, the cost of producing $x$ filters per month.

   (b) Give a formula for $R(x)$, the revenue earned by selling $x$ filters per month.

   (c) Give a formula for $P(x)$, the profit earned by selling $x$ filters per month.

   (d) What is the minimum number of filters the plant must produce and sell in a month to guarantee a profit?

   (e) What is the minimum number of filters the plant must produce and sell in a month to generate a revenue of at least $90,000?

   (f) What is the minimum number of filters the plant must produce and sell in a month to generate a profit of at least $30,000?

# Chapter 1 Progress Check II

Suppose that the price at which $x$ **thousand** flash drives can be sold is given by the demand function

$$p(x) = 12 - 0.4x \quad \text{dollars per drive,}$$

where $0 \le x \le 22$. In addition, the cost of producing $x$ **thousand** drives is given by the cost function

$$C(x) = 26 + 0.85x \quad \text{thousand dollars.}$$

1. Determine the selling price that results in a demand of 18,000 drives.

2. Determine the production level that results in production costs of $40,000, to the nearest flash drive.

3. Let $R(x)$ denote the corresponding revenue function. Write a formula for $R(x)$ and sketch graphs of both $R(x)$ and $C(x)$ on the same set of axes. Show only what is relevant, label the axes appropriately, and provide scale on the axes.

4. Find the level of production (to the nearest flash drive) that maximizes revenue. Illustrate your result graphically.

5.  What is the selling price (per flash drive) that generates maximum revenue?

6.  What is the maximum possible revenue?

7.  Find all break-even production levels, to the nearest flash drive. Illustrate your result graphically.

8.  Determine the level of sales (numbers of flash drives) that results in revenue of at least $85,000. Illustrate your results graphically.

# Chapter 1 Progress Check III

1. A website that purports to have copies of every Math 104 exam ever written along with exam keys charges a log-on fee. Operators of the site found that when the fee was $2, the site received 280 log-ons per month. When the fee was lowered to $1.50, the site received 560 log-ons per month. Suppose that the relationship between log-ons per month and log-on fee is linear.

   (a) Let $x$ be the log-on fee, in dollars. Give a formula for $R(x)$, the monthly revenue the site receives from log-on fees.

   (b) Suppose that for maintaining the site, the operators pay a flat rate of $30 per month. Give a formula for $P(x)$, the monthly profit for the site based on log-on fees.

   (c) Determine the log-on fee that results in maximum monthly profit for the site, and the number of log-ons received at that fee.

2.  A strawberry farmer will receive $4 per bushel of strawberries if the strawberries are harvested this week. Each week that passes, the value drops by $0.10 per bushel. The farmer estimates that there are approximately 120 bushels of strawberries in the fields, and that the crop is increasing at a rate of 4 bushels per week.

    (a)  Write an equation that gives the expected revenue, $R$, from the strawberry harvest in term of $n$, the number of weeks that the farmer waits to harvest.

    (b)  When should the farmer harvest the strawberries to maximize the revenue from the harvest? Support your result graphically.

    (c)  How many bushels of strawberries will yield the maximum possible revenue? What is the maximum revenue?

# ANSWERS TO SELECTED EXERCISES

## Section 1.1

1. The fourth table only.

3. (a) See table below
   (b) Make sure you labeled and showed a scale on both axes.
   (c) 14 million
   (d) 0.2 million people/year or 200,000 people/year

   | $t$ | $P$ |
   |-----|-----|
   | 0   | 14  |
   | 5   | 15  |
   | 10  | 16  |
   | 15  | 17  |
   | 20  | 18  |
   | 25  | 19  |

5. $P = 23t + 4650$

7. (a) $T = \$4200$
   (b) 16 credits
   (c) In your table for $c = 4$ to $c = 16$, $T = 3800$, 4000, 4200, 4400, 4600, 4800, 5000, 5200, 5400, 5600, 5800, 6000, and 6200, consecutively.
   (d) Sixteen credits
   (e) Initial cost of $3,000 dollars. It does not depend on the number of credits taken.
   (f) Variable costs of $200 per credit. It is the average rate of change.

9. (a) $F = 2C + 30$
   (b) See table:

   | $C$ | Actual | Approximate |
   |-----|--------|-------------|
   | $-10$ | 14   | 10          |
   | 0   | 32     | 30          |
   | 10  | 50     | 50          |
   | 25  | 77     | 80          |

   (c) 10°

11. (a) $y = \dfrac{3}{500}x + 400$
    (b) $y = 550$; $y = 640$
    (c) $100,000
    (d) This rate of change means an increase of 3 units are sold for each additional $500 spent on marketing

13. 8 yrs.: $40,000, 12 yrs.: $10,000

15. (a) values in women's table: 100, 105, 110, 120, 130, 160, 175
    values in men's table: 106, 112, 118, 130, 142, 178, 196
    (b) $y = 5x + 100$
    (c) 145 lbs.
    (d) $y = 6x + 106$; 160 lbs.

17. 18 defective components

19. (a) $C = \dfrac{4}{25}x + 44$
    (b) $88
    (c) 3000 miles

21. fixed costs: $1000; variable costs: $5 per case

## Section 1.2

1. True

3. False. The lines are not perpendicular.

5. False. The line given by $4x - y = -1$ has a bigger positive slope than the line given by $3x - y = -1$ and so rises more steeply from left to right.

7. $y = 3x + 0.2$

9. $y = 2x + \dfrac{1}{2}$

11. $y = 5x + 6$

13. $y = -2x + 4$

15. $y = 0$

17. $A \leftrightarrow y_2, B \leftrightarrow y_1; C \leftrightarrow y_3; D \leftrightarrow y_4; E \leftrightarrow y_5$

19. $A \leftrightarrow y_1, B \leftrightarrow y_3; C \leftrightarrow y_2; D \leftrightarrow y_5; E \leftrightarrow y_4$

21. (a) $y = -\dfrac{1}{3}x + \dfrac{13}{3}$
    (b) $y = 3x - 9$

23. (a) after 16 years
    (b) A Hotbox over is worth $40, while an Equator-bake oven is worth $0.

25. (a) $p = 8n - 69{,}200$
    (b) Slope: $8 profit per visitor
    $p$-intercept: if there are no visitors, loss of $69,200
    (c) $8650
    (d) $n = \dfrac{p + 69{,}200}{8} = \dfrac{1}{8}p + 8650$
    (e) 24,275 visitors

## Section 1.3

1. Yes. No.

3. $(5, \infty)$

5. approximately 32.47 psi

7. $C(x) = 0.5x^2 + \dfrac{8}{x}$

9. (a) 4.538
   (b) $[-10, 0] \cup [0.919, 1.081]$
   (c) local max: 1.307, local min: $-1.714$

11. (a) $P(x) = x(16 - 0.8x) - (20 + 2.5x)$
    (b) 8438 poppers
    (c) $36,953

13. (a) $P(x) = x(60 - 1.75x) - (80 + 15x)$
    (b) 7037 coffee makers

15. (a) 61.700 feet
    (b) less than 43.083 mph

17. (a) $C(x) = 70 + 0.4x; R(x) = 0.5x; P(x) = 0.1x - 70$
    (b) loss of $20
    (c) 700 copies

19. $N = 4T - 160$; when $T = 98, N = 232$ chirps per minute

21. $15 per trashcan

# Consumer Finance: Saving and Borrowing

It is a fact that our modern economy depends on borrowed money. If you have a student loan, you have first-hand experience with borrowing. When you use a credit card, money is being loaned. Borrowed money allows students to pay for education, families to purchase homes, businesses to purchase equipment. Many forms of saving are really just instances of borrowing—a savings account is just a loan, as the bank or savings institution is borrowing money from the holder of the account. On almost every type of loan, a fee is charged for the use of that money. In this chapter we investigate two aspects of consumer finance: saving and borrowing.

## 2.1 SIMPLE INTEREST AND COMPOUND INTEREST

If you deposit money into a bank account or if you borrow money from a lending agent, a fee is usually issued or charged for the use of the money. The money that is deposited or borrowed is called principal. The fee is called **interest**. Interest is usually computed as a percentage of the principal over a given period of time. Under one method of paying interest, called **simple interest**, interest is paid only on the original principal amount, regardless of the amount of time that passes. Another way of paying interest is to pay interest on both principal and on interest that has already been paid on the principal. Interest of this sort is called **compound interest**, which we will investigate later. For now, we will consider interest paid on lump sum deposits, where the only modifications to an account balance come from interest payments rather than regular deposits, payments, or withdrawals. In particular, we will begin by looking at simple interest.

### Simple Interest

**Simple interest** is often described as "a percentage of the principal charged on an annual basis". More specifically, simple interest is based on a fixed principal amount, a fixed annual interest rate, and a fixed period of time, and is given by the following formula:

> Simple Interest
>
> $$I = Prt,$$
>
> where $P =$ principal
>
> $r =$ annual simple interest rate (written as a decimal)
>
> $t =$ time in years
>
> $I =$ simple interest earned

For example, the simple interest on a loan of \$100 at an annual rate of 8% for 9 months would be

$$I = Prt$$

$$I = 100 \times 0.08 \times \frac{9}{12} = \$6$$

So, at the end of 9 months, the borrower would repay the $100 principal plus the $6 interest, for a total of $106.

When you deposit money into a bank account (or some other sort of investment) that pays interest, you are the lender, so there is no difference in calculation of interest from that perspective. Whether you are a borrower or a lender, it is common to refer to the principal as the **present value** of the loan or account.

If you deposit $P into an account that pays simple interest at an annual rate, $r$, for a period of $t$ years, then it is natural to be interested in how much total money you will have available to you at the end of the time period. We call this total the **future value** of your deposit. If we denote this quantity by $A$, then clearly $A$ is the sum of the principal and the interest paid on the principal, as given by the following formula:

---

Future Value (Simple Interest)

$$A = P + Prt \quad \text{or} \quad A = P(1 + rt),$$

where $P$ = principal (present value)

      $r$ = annual simple interest rate (written as a decimal)

      $t$ = time in years

      $A$ = future value (total amount)

---

### EXAMPLE 1 ♦ Future Value of a Deposit

Find the future value of a deposit of $1500 that earns simple interest at an annual rate of 7% over the course of 18 months and calculate the amount of interest earned in that time.

**Solution**

We use $A = P(1 + rt)$, remembering to convert 18 months into years.

$$A = 1500\left(1 + (0.07)\left(\frac{18}{12}\right)\right)$$

$$A = 1500(1.105)$$

$$A = \$1657.50$$

So, the future value of the deposit is $1657.50, meaning that $1657.50 - 1500 = \$157.50$ in interest was earned.

---

**Remark:** When doing calculations like those in the previous example, it is often the case that our numbers are not "nice," in that we need to deal with numbers that don't have a finite decimal expansion. For instance, if the time period in question had been 14 months rather than 18 months, then in the previous example we would have had

$$A = 1500\left(1 + (0.07)\left(\frac{14}{12}\right)\right),$$

where $(1 + (0.07)(\frac{14}{12})) = 1.081666666\ldots$

Care must be taken in this situation not to round or truncate numbers in such a way as to seriously endanger the accuracy of the calculation. It is usually preferable to allow the memory of your calculator to keep track of decimal places until you need to make a final assessment of your results.

---

Often it is desirable to determine how much money should be deposited into an interest-earning account in order to meet some specific future financial goal or obligation.

## EXAMPLE 2 ◆ Present Value

Suppose that you wish to make a single deposit into an account that earns simple interest at an annual rate of 5.5% with the intent of having $10,000 available to you in 2 years. How much should you deposit right now to achieve this goal?

**Solution**

We use $A = P(1 + rt)$, with the objective being to find $P$.

$$10,000 = P(1 + (0.055)(2))$$

$$\frac{10,000}{(1 + (0.055)(2))} = P$$

$$P \approx \$9009.01$$

Adjusted **up** to the nearest penny, this gives an initial deposit of $9009.01. Notice that $P$ was adjusted up to ensure that the goal of a future value of $10,000 was actually met (this will become more of an issue later).

## EXAMPLE 3 ◆ Calculating an Interest Rate

What annual interest rate, to the nearest one hundredth of a percent, would an account that pays simple interest need to offer in order for a deposit of $600 to have a future value of $650 in 180 days?

**Solution**

Notice that the deposit in question earns $50 in simple interest, so we use $I = Prt$, with the objective being to find $r$.

$$50 = (600)(r)\left(\frac{180}{365}\right)$$

$$\frac{50}{(600)\left(\dfrac{180}{365}\right)} = r$$

$$r \approx 0.16898$$

So, adjusting **up** to the nearest one hundredth of a percent, an annual interest rate of 16.90% is needed.

## EXAMPLE 4 ◆ Treasury Notes

One of the ways that the U.S. Treasury Department finances public debt is through the sale of T-notes (Treasury notes). Interestingly enough, these are sold at public auction. A T-note has a *face value* of at least $100 and a *maturity term* of one of 2, 3, 5, 7, or 10 years. The maturity term is the length of time from issuance of the note until its face value is paid (typically, notes are actually purchased for a price close to the face value). A given note will have a stated interest rate and the holder of the note receives semi-annual interest payments based on that rate and the face value of the note. For instance, a 7-year note with a face value of $1000 with an issue date of April 30, 2012 could be purchased for $993.54 and bore an annual interest rate of 1.25%. The holder of such a note would receive interest payments of $6.25 every six months until April 30, 2019 and would then be paid the face value of the note as well.

## Compound Interest

In essence, **compound interest** is just repeated payment of simple interest with a changing principal. The principal changes from period to period by the amount of interest earned during that period of time. We will consider compound interest where the interest is paid at the end of fixed and equal periods of time, called **compounding periods**.

Suppose that you deposit $1000 into a savings account that advertises an annual interest rate of 12%, with interest compounded quarterly (four times per year). Admittedly, this is not a realistic interest rate, but it helps make calculations easier. Assuming that you don't make any withdrawals from the account, how much money will be in the account after two years? In other words, what is the future value of your principal of $1000 at the end of the given period of time? The answer to that question can be found by a sequence of simple interest calculations. In this case, interest is paid to the account every quarter (about three months) and the interest that is paid is added to the account balance. See the following table. Calculations are given to the nearest cent when necessary.

| Compounding Periods (Quarters) | Beginning Account Balance (Principal) | Interest Earned | Ending Account Balance |
|:---:|:---:|:---:|:---:|
| 1 | 1000 | $1000 \times 0.12 \times \frac{1}{4} = 30$ | 1030 |
| 2 | 1030 | $1030 \times 0.12 \times \frac{1}{4} = 30.90$ | 1060.90 |
| 3 | 1060.90 | $1060.90 \times 0.12 \times \frac{1}{4} \approx 31.83$ | 1092.73 |
| 4 | 1092.73 | $1092.73 \times 0.12 \times \frac{1}{4} \approx 32.78$ | 1125.51 |
| 5 | 1125.51 | $1125.51 \times 0.12 \times \frac{1}{4} \approx 33.77$ | 1159.28 |
| 6 | 1159.28 | $1159.28 \times 0.12 \times \frac{1}{4} \approx 34.78$ | 1194.06 |
| 7 | 1194.06 | $1194.06 \times 0.12 \times \frac{1}{4} \approx 35.82$ | 1229.88 |
| 8 | 1229.88 | $1229.88 \times 0.12 \times \frac{1}{4} \approx 36.90$ | 1266.78 |

Notice that on each successive compounding period the interest paid to the account increases, as we would expect given that interest payments are added to the given account balance. The total amount of interest earned in this case is $266.78. By contrast, over the same period of time, if the $1000 deposit earned only simple interest, the interest earned would be $240.

Another important thing to notice from the calculations above is how interest is calculated each compounding period. Each compounding period, the interest that is paid to the account balance is one-fourth of the annual interest rate, namely 3% of the account balance. In this case we would say that the account is earning a **periodic rate** (or **rate per compounding period**) of 3% per quarter.

Let's look at the same situation again, except rather than concentrating on interest earned each compounding period, we will only consider the account balance at the end of each compounding period. We will employ repeated applications of the formula, $A = P(1 + rt)$, where we have $1 + rt = 1 + (0.12)(\frac{1}{4})$ $= 1.03$. Notice that each compounding period we calculate the ending account balance simply by multiplying the beginning account balance by 1.03.

| Compounding Periods (Quarters) | Beginning Account Balance (Principal) | Ending Account Balance |
|:---:|:---:|:---:|
| 1 | 1000 | $1000(1.03)$ |
| 2 | $1000(1.03)$ | $[1000(1.03)](1.03) = 1000(1.03)^2$ |
| 3 | $1000(1.03)^2$ | $[1000(1.03)^2](1.03) = 1000(1.03)^3$ |
| 4 | $1000(1.03)^3$ | $[1000(1.03)^3](1.03) = 1000(1.03)^4$ |
| 5 | $1000(1.03)^4$ | $[1000(1.03)^4](1.03) = 1000(1.03)^5$ |
| 6 | $1000(1.03)^5$ | $[1000(1.03)^5](1.03) = 1000(1.03)^6$ |
| 7 | $1000(1.03)^6$ | $[1000(1.03)^6](1.03) = 1000(1.03)^7$ |
| 8 | $1000(1.03)^7$ | $[1000(1.03)^7](1.03) = 1000(1.03)^8$ |

Rounding to the nearest penny, we see an account balance of $1000(1.03)^8 \approx \$1266.77$, which very nearly matches the results from earlier (why is there a discrepancy?). The pattern that is evident from this table suggests an important result. In general, given a lump sum deposit, $P$, and annual interest rate, $r$, with interest compounded $m$ times per year for a total of $t$ years, we have a periodic interest rate, $i = r/m$, and total number of compounding periods, $n = mt$, and we calculate the future value, $A$, of the deposit by the following formula:

---

**Future Value (Compound Interest)**

$$A = P(1 + i)^n,$$

where $P =$ principal (present value)

$\quad\quad i =$ periodic interest rate (written as a decimal)

$\quad\quad n =$ total number of compounding periods

$\quad\quad A =$ future value (total amount)

---

## EXAMPLE 5 ◆ Future Value of a Deposit

Find the future value of a deposit of $1500 that earns interest at an annual rate of 7% compounded monthly over the course of 18 months, and calculate the amount of interest earned in that time.

**Solution**

We use $A = P(1 + i)^n$, keeping in mind that a compounding period is one month, so the corresponding monthly interest rate is one-twelfth of the annual interest rate.

$$A = 1500\left(1 + \frac{0.07}{12}\right)^{18}$$

$$A \approx \$1665.56$$

Rounded to the nearest cent, the future value of the deposit is $1665.56, meaning that $1665.56 - 1500 = \$165.56$ in interest was earned. Compare this result to Example 1.

---

## EXAMPLE 6 ◆ Present Value

Suppose that you wish to make a single deposit into an account earning an annual interest rate of 5.5%, compounded weekly, with the intent of having $10,000 available to you in 2 years. How much should you deposit right now to achieve this goal?

**Solution**

We use $A = P(1 + i)^n$, with the objective of finding $P$. Since interest is compounded weekly and there are 52 weeks in a year, we have

$$10{,}000 = P\left(1 + \frac{0.055}{52}\right)^{104}$$

$$\frac{10{,}000}{\left(1 + \dfrac{0.055}{52}\right)^{104}} = P$$

$$P \approx \$8958.87$$

Notice again that $P$ was adjusted **up** to the nearest penny to ensure that the goal of a future value of $10,000 was actually met. Compare this result to Example 2.

**EXAMPLE 7 ◆ Calculating an Interest Rate**

What annual interest rate, to the nearest one hundredth of a percent, would an account that pays monthly compounded interest need to offer in order for a deposit of $600 to have a future value of $650 in 6 months?

**Solution**

We use $A = P(1 + i)^n$, keeping in mind that $i$ is one twelfth of the annual interest rate we seek since interest is compounded monthly. Let $r$ denote the annual interest rate, so that $r = 12i$. We will take two different approaches to finding $r$: an algebraic approach and a graphical approach.

*Algebraic Approach:*

$$650 = 600(1 + i)^6$$

$$\frac{650}{600} = (1 + i)^6$$

$$\sqrt[6]{\frac{650}{600}} = 1 + i$$

$$\sqrt[6]{\frac{650}{600}} - 1 = i$$

Since $r = 12i$, we have

$$r = 12\left(\sqrt[6]{\frac{650}{600}} - 1\right)$$

$$r \approx .016116$$

Adjusting up to the nearest one hundredth of a percentage, an annual interest rate of 16.12% is needed. Compare this result to Example 3.

*Graphical Approach:* In order to minimize the number of calculations, we will graph the equations

$$y_1 = 600\left(1 + \frac{x}{12}\right)^6$$

$$y_2 = 650$$

on a graphing calculator and approximate the point of intersection of the graphs (see Figure 2.1). Note here that $x$ represents the annual interest rate we seek.

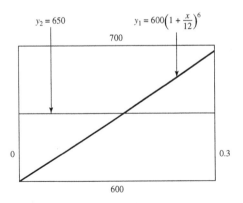

*FIGURE 2.1*

Using an intersection routine on the calculator, correct to 5 decimal places, we find $x \approx 0.16116$, yielding an annual interest rate of 16.12%, adjusted up to the nearest one hundredth of a percent. Compare these results to Example 3.

Example 7 illustrates why, when it became necessary to approximate a result, we did not necessarily just round to the desired degree of accuracy, but we approximated to a result that matched the objective stated in the problem. You should check to see what would happen to a $600 deposit that earns 16.1% interest compounded monthly for six months.

## EXAMPLE 8 ♦ Growth Time

How long would it take for a deposit of $5000 to grow to $7000, if the deposit earns 5% interest compounded daily?

**Solution**

We use $A = P(1 + i)^n$, where $n$ is what we seek. We take a graphical approach to solving this problem. In particular, we need to solve for $n$ in the equation

$$7000 = 5000\left(1 + \frac{0.05}{365}\right)^n$$

There are algebraic methods as well, but they are beyond the scope of this text. We will graph the equations

$$y_1 = 5000\left(1 + \frac{0.05}{365}\right)^x$$

$$y_2 = 7000$$

on a graphing calculator and approximate the point of intersection of the graphs (see Figure 2.2). Note here that $x$ represents a number of days.

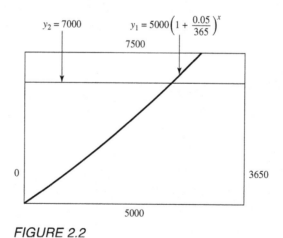

*FIGURE 2.2*

Using an intersection routine on the calculator, correct to one decimal place, we find $x \approx 2456.4$, yielding a growth time of 2457 days, adjusted up to the nearest day. We adjust our result **up** to the nearest day since interest is paid at the end of each day.

## Interest Rate Terminology

Thus far, we have looked predominantly at interest in situations where an annual interest rate was given. For instance, in the discussion that began the discourse on compound interest, we investigated a situation given an annual interest rate of 12%, with interest compounded quarterly. (This resulted in a quarterly interest rate of 3%.) Notice, though, that after four quarters (one year), the deposit of $1000 has grown in value by more than 12%: $1000(1.03)^4 \approx \$1125.51$, so, in fact, the principal has grown by slightly more than 12.55% over the course of one year.

The stated annual rate of 12% in this case is what is known as the **annual nominal rate** specific to the situation. The actual percentage growth of the principal over the course of one year is called the **annual percentage yield (APY)** or **effective annual rate (EAR).** So, in this case, we would have an APY of about 12.55%. APY is just the rate of simple interest that would realize the same amount of interest on the given principal over a one-year period of time. Notice that the nominal rate does not, by itself, indicate or take into account any sort of interest compounding. The APY, on the other hand, does actually demonstrate the effect of interest compounding over a fixed period of time. Sometimes the APY is called the "true interest rate" as it gives a more readily available assessment of the actual percentage growth of the principal over the course of one year.

Suppose, then, that a principal, $P$, is to be invested at an annual nominal rate, $r$, with interest compounded $m$ times per year. We know that after one year the principal has grown to a value $A$, where

$$A = P\left(1 + \frac{r}{m}\right)^m.$$

Since the APY is the simple interest rate that yields the exact same value for $A$ over the course of one year, we also have

$$A = P(1 + \text{APY})$$

Consequently,

$$
\underset{\substack{\text{amount at simple} \\ \text{interest after 1 year}}}{P(1 + \text{APY})} = \underset{\substack{\text{amount at compounded} \\ \text{interest after 1 year}}}{P\left(1 + \frac{r}{m}\right)^m}
$$

so

$$1 + \text{APY} = \left(1 + \frac{r}{m}\right)^m$$

$$\text{APY} = \left(1 + \frac{r}{m}\right)^m - 1$$

We summarize this as follows:

---

**Annual Percentage Yield (Effective Annual Rate)**

$$\text{APY} = \left(1 + \frac{r}{m}\right)^m - 1,$$

where $r$ = annual nominal rate (written as a decimal)

$\quad m$ = number of compounding periods per year

$\quad\text{APY}$ = annual percentage yield or effective annual rate

---

**EXAMPLE 9 ◆ Calculating APY**

Find the APY given an annual nominal rate of 8% compounded monthly.

**Solution**

We use $\text{APY} = \left(1 + \frac{r}{m}\right)^m - 1$.

$$\text{APY} = \left(1 + \frac{0.08}{12}\right)^{12} - 1$$

$$\text{APY} \approx 0.0829995$$

So, an annual interest rate of 8% compounded monthly gives an APY of about 8.3%, rounded to the nearest one-tenth of a percent.

---

Sometimes investments are advertised strictly by APY. Given this information, we can find the corresponding annual nominal rate of interest.

## EXAMPLE 10 ♦ Finding a Nominal Rate

A bank is advertising an APY of 3.5% on savings accounts. If interest on the savings account is compounded daily, what is the annual nominal interest rate?

**Solution**

We use $\text{APY} = \left(1 + \frac{r}{m}\right)^m - 1$, with the goal of solving for $r$. We can approach this either algebraically or graphically.

*Algebraic Approach:*

$$0.035 = \left(1 + \frac{r}{365}\right)^{365} - 1$$

$$1.035 = \left(1 + \frac{r}{365}\right)^{365}$$

$$\sqrt[365]{1.035} = 1 + \frac{r}{365}$$

$$\sqrt[365]{1.035} - 1 = \frac{r}{365}$$

$$365\left(\sqrt[365]{1.035} - 1\right) = r$$

$$r \approx 0.0344$$

So, rounded to the nearest one hundredth of a percent, we have a corresponding annual rate of about 3.44%.

*Graphical Approach:* We will graph the equations

$$y_1 = \left(1 + \frac{x}{365}\right)^{365} - 1$$

$$y_2 = 0.035$$

on a graphing calculator and approximate the point of intersection of the graphs (see Figure 2.3). Note here that $x$ represents the annual interest rate we seek.

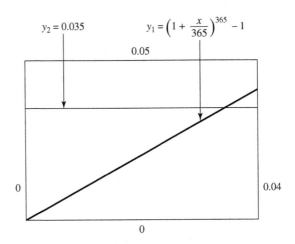

FIGURE 2.3

Using an intersection routine on the calculator, correct to 5 decimal places, we find $x \approx 0.03440$, yielding an annual interest rate, to the nearest one-hundredth of a percent, of about 3.44%.

## EXERCISES

*For the following exercises, assume a 365-day year unless directed otherwise. Also, when calculating interest rates, give your result as a percentage, correct to one hundredth of a percent.*

1. Calculate the amount of simple interest paid to a principal of $700 given an annual interest rate of

   (a) 6% for 16 months
   (b) 4.5% for 2 years
   (c) 8.99% for 60 weeks
   (d) 8.1% for 36 months
   (e) 18.9% for 3 years and 4 months
   (f) 0.25% for 1000 days

2. Calculate the future value of a deposit of $1500 that earns simple interest at an annual rate of

   (a) 5.25% for 6 years
   (b) 3.8% for 3 years and 3 months
   (c) 7% for 300 days
   (d) 2.9% for 72 weeks
   (e) 1.99% for 18 months
   (f) 4.56% for one month

3. Jane deposited some money into an account 3 months ago. She just withdrew her money, which had earned simple interest while in the bank. Her withdrawal was $226.60, of which $6.60 was interest. How much was Jane's deposit, and what was the annual interest rate earned by her deposit?

4. An investment that earned simple interest at an annual rate of 6.5% for 54 months is now worth $878.90. How much was initially invested?

5. Bob deposits $2500 into an account that pays simple interest at an annual rate of 8%. How long will it take for his deposit to double in value?

6. Julie purchased a corporate bond for $10,000. The bond will be paid in full (*come to maturity*) in five years. The bond earns an annual interest rate of 8.4%, and interest payments are made semiannually. Calculate the total amount of interest Julie will be paid over the life of the bond.

7. Determine how much a principal of $8000 will be worth and how much interest it will have earned after three years given an annual interest rate of

   (a) 6% compounded annually
   (b) 5.4% compounded monthly
   (c) 3.85% compounded daily
   (d) 6.85% compounded weekly
   (e) 4.25% compounded quarterly
   (f) 7.99% compounded semimonthly (twice a month)

8. Ron wishes to make a single deposit into a bank account that earns a nominal rate of 6.3% compounded quarterly. If he wants the deposit to be worth $5800 in two years, then how much should he deposit right now?

9. What annual interest rate, compounded monthly, would be needed for a principal of $14,000 to grow in value to $18,000 in 8.5 years?

10. Monica is going to deposit $5000 into a savings account that earns 4.45% interest compounded daily. When the account has $8000 in it she will withdraw all of the money

and use it to purchase a motorcycle. How long will Monica have to wait to purchase her bike?

**11.** Calculate the APY given an annual nominal rate of 5% if interest is compounded

  **(a)** annually
  **(b)** semiannually
  **(c)** quarterly
  **(d)** monthly
  **(e)** daily
  **(f)** semimonthly (twice a month)

**12.** A friend has asked your advice on the purchase of a certificate of deposit (CD). Myrtle's Money Mart offers a CD that pays an interest rate of 7.2% compounded quarterly while Sudsville First National offers a CD that pays an annual interest rate of 7.1% compounded monthly. Which CD do you recommend? Why?

**13.** Which periodic rate has a higher APY: a weekly rate of 0.2865%, or a monthly rate of 1.23%?

**14.** Given an APY of 7.48%, find the corresponding annual interest rate if interest is compounded semiannually.

*Problems 15. and 16. deal with* **zero-coupon bonds**. *A zero-coupon bond is a bond where the holder does not receive interest payments. Rather, the bond is sold at a discount and at maturity the face value of the bond is paid. The difference between the face value and the purchase price is the interest earned by the holder.*

**15.** If you purchase a 20-year zero-coupon bond with a face value of $30,000 for $10,516, what is the annually compounded rate of return?

**16.** A 15-year zero-coupon bond has a face value of $50,000. What should the bond be purchased for in order to realize an annually compounded rate of return of 10%?

**17.** What annual interest rate would an investment of $8000 earning simple interest need to have in order to double in value in the same length of time it takes for $8000 to double at a monthly rate of 0.75%?

**18.** Jackie has $10,000 in an investment account that guarantees an interest rate of 4.2% compounded quarterly. She plans to withdraw the money for a down payment on a new home when the account balance reaches $18,000. When will she be able to make that withdrawal?

**19.** How long would it take for an investment of $1500 earning 6% compounded daily to be worth more than an investment of $2000 that earns 5% compounded daily?

**20.** Nine years ago, a single deposit was made into an account that is now worth $14,125. In that time, a total of $3545 in interest has been earned. What annual interest rate has been paid on the account if interest has been compounded monthly?

**21.** A savings account pays an annual interest rate of 4.25% compounded quarterly. At the end of the first compounding period, how much interest is earned by each dollar that was initially deposited?

**22.** How long will it take for a single deposit of $6000 to at least triple in value if it earns an annual interest rate of 4.9% with interest compounded weekly (assume a 52-week year)?

**23.** (See Example 4) Suppose that a 10-year Treasury note with a face value of $20,000 sells for $18,400.

  **(a)** If the semi-annual interest payments are $275, then what annual interest rate does the T-note carry?

  **(b)** Calculate the future value of the T-note when it comes to maturity.

  **(c)** Suppose that you could invest $18,400 into an account that earns interest compounded semiannually. Calculate the annual interest rate necessary to have that investment accumulate to the value from (b) over the course of 10 years as well as the corresponding APY. Give both percentages accurate to two decimal places.

**24.** Suppose that a tax preparation firm offers to give you an advance on the tax refund you are expected to receive in six weeks. You expect a refund of $1500 and the firm offers to give you the $1500 less a $50 processing fee right now, and you in turn will sign over your refund check to the firm when it is received. Thinking of the advance as a loan being charged simple interest, what annual interest rate is the firm charging you?

**25.** For each of the following compounding frequencies, calculate the annual interest rate that results in the same APY as an annual interest rate of 5.8% compounded semiannually.

  **(a)** Quarterly
  **(b)** Monthly
  **(c)** Weekly
  **(d)** Daily

**26.** Which will grow to $5000 faster: $3000 invested at an annual rate of 5.8% compounded monthly, or $3000 invested at an annual rate of 5.6% compounded daily?

# Interlude: Practice Problems 2.1, Part 1

1. What annual interest rate, compounded semiannually, would result in the same APY as an annual rate of 8% compounded weekly? Give your result as a percentage, correct to two decimal places.

2. Consider a single deposit of $4000 that earns a simple interest rate of 8% and a single deposit of $2000 that earns an annual interest rate of 8% compounded monthly. How long will it take for the value of the $2000 deposit to overtake the value of the $4000 deposit?

3. How long will it take a single deposit of $11,000 grow in value to at least $15,000 if it earns interest at 4% compounded weekly?

4. A certificate of deposit (CD) offers an annual interest rate of 2.75% compounded monthly and a term of 20 months. If you wish to purchase one of these CDs with the expectation that it be redeemed for $5000 at the end of the term, what should the purchase price be?

# Interlude: Practice Problems 2.1, Part 2

1.  Consider a single deposit of $5000 that earns an annual interest rate of 4.6% compounded monthly for four years and then earns an annual interest rate of 5.2% compounded quarterly for six years. After that 10 years, what is the deposit worth?

2.  A T-note with a term of one year and a face value of $10,000 carrying an interest rate of 2.25% is purchased for $9980. Given this, calculate the annual **yield** of this T-note (the actual simple interest rate earned on the purchase price). Give the percentage accurate to two decimal places.

3.  Consider an initial principal of $50,000 and an annual interest rate of 10%. Fill in the following table, giving results to the nearest cent when appropriate.

| compounding frequency | compounding periods/year | value after one year |
|---|---|---|
| Quarterly | | |
| Monthly | | |
| Weekly | | |
| Daily | | |
| Every hour | | |
| Every minute | | |
| Every second | | |

4.  Consider an annual interest rate of 9%. Fill in the following table, giving the APYs as percentages accurate to six decimal places.

| compounding frequency | compounding periods/year | corresponding APY |
|---|---|---|
| Quarterly | | |
| Monthly | | |
| Weekly | | |
| Daily | | |
| Every hour | | |
| Every minute | | |
| Every second | | |

5.  What do your results above suggest about the effect of increased compounding frequency given a fixed annual interest rate?

## 2.2  FUTURE VALUE OF AN ANNUITY

In the previous section we established how to determine what happens over time to a single deposit that is earning interest. However, another important question involves the ultimate fate of a series of fixed deposits made on a regular basis. We call such a sequence of periodic deposits an **annuity.** More generally, any sequence of equal periodic deposits or payments is referred to as an annuity. In particular, we will investigate what are called **ordinary annuities.** These are annuities where the periodic deposits are made at the end of the prescribed period (month, quarter, year, etc.). In addition, these will earn interest that is compounded with the same frequency as deposits are made.

### Future Value of an Ordinary Annuity

As an example, suppose that an individual wishes to make monthly deposits of $100 into an account that earns 6% interest compounded monthly. How much money will the account contain after 6 months of deposits? Assume that the first deposit is made one month from now.

The question that has been posed is more complicated than a "typical" future value question involving compound interest simply because we need to keep track of six separate deposits that each earn compound interest for different lengths of time. We summarize this in the following table. The total accumulated amount of money in the account has been rounded to the nearest cent. Notice that the monthly interest rate is 0.5%.

| Deposit Number | Deposit Amount | Future Value |
|:---:|:---:|:---:|
| 1 | 100 | $100(1.005)^5$ |
| 2 | 100 | $100(1.005)^4$ |
| 3 | 100 | $100(1.005)^3$ |
| 4 | 100 | $100(1.005)^2$ |
| 5 | 100 | $100(1.005)^1$ |
| 6 | 100 | 100 |
| **Total Amount** | | $607.55 |

Notice that the last deposit has not earned any interest since it was made at the end of the six-month period of time. The total amount shown is the sum of the future values of each deposit. We will call this total amount the **future value of the annuity.** If we denote this quantity by *FV*, then we have

$$FV = 100 + 100(1.005)^1 + 100(1.005)^2 + 100(1.005)^3 + 100(1.005)^4 + 100(1.005)^5$$

Rather than computing each term in this sum, notice that

$$1.005FV = 100(1.005)^1 + 100(1.005)^2 + 100(1.005)^3 + 100(1.005)^4 + 100(1.005)^5 + 100(1.005)^6$$

so

$$\underbrace{1.005FV - FV}_{0.005FV} = (100(1.005) + 100(1.005)^2 + \cdots + 100(1.005)^6)$$
$$-(100 + 100(1.005)^1 + \cdots + 100(1.005)^5) = 100(1.005)^6 - 100$$

Since $0.005FV = 100(1.005)^6 - 100$, we see that

$$FV = \frac{100(1.005)^6 - 100}{0.005} = 100\left[\frac{(1.005)^6 - 1}{0.005}\right]$$

Notice that the major components of the right-hand side of the equation above correspond to the pertinent facts about the annuity in question: $100 periodic deposits, 6 deposits, and periodic interest rate 0.005. Finishing the calculation and rounding to the nearest cent gives $FV \approx 607.55$.

The actual numbers used in the previous scenario are not important. The last equation, however, is important. We can generalize the above computation to get a formula for calculating the future value of an ordinary annuity:

Future Value (Ordinary Annuity)

$$FV = d\left[\frac{(1 + i)^n - 1}{i}\right],$$

where $d$ = amount of periodic deposits (payments)

$i$ = periodic interest rate (written as a decimal)

$n$ = total number of deposits (payments)

$FV$ = future value (total accumulation of money)

Notice here that the periodic deposits coincide with the frequency with which interest is compounded.

## EXAMPLE 1 ◆ Future Value and Interest Earned

Suppose that quarterly deposits of $600 are made into an account that earns an interest rate of 5% compounded quarterly. Determine how much money will be in the account after three years. How much of that total is interest?

**Solution**

We use $FV = d[\frac{(1 + i)^n - 1}{i}]$. Note that three years is 12 quarters, so we have

$$FV = 600\left[\frac{\left(1 + \dfrac{0.05}{4}\right)^{12} - 1}{\left(\dfrac{0.05}{4}\right)}\right]$$

$$FV \approx \$7716.22$$

Rounded to the nearest cent, we expect the account to have $7716.22 after three years. Since 12 deposits of $600 have been made, $7200 of that total amount is from deposits, meaning that $7716.22 - 7200 = \$516.22$ in interest has been earned.

In order to further illustrate how the account balance in the previous example is changing over time, consider the following table. Calculations have been rounded to the nearest cent when necessary.

| Quarter | Account Balance Before Deposit ($B$) | Periodic Interest Earned on Balance ($I$) | Deposit ($d$) | Ending Account Balance ($B + I + d$) |
|---------|--------------------------------------|--------------------------------------------|---------------|--------------------------------------|
| 1  | 0       | $0 \times \frac{0.05}{4} = 0$          | 600 | 600     |
| 2  | 600     | $600 \times \frac{0.05}{4} = 7.50$     | 600 | 1207.50 |
| 3  | 1207.50 | $127.50 \times \frac{0.05}{4} \approx 15.09$  | 600 | 1822.59 |
| 4  | 1822.59 | $1822.59 \times \frac{0.05}{4} \approx 22.78$ | 600 | 2445.37 |
| 5  | 2445.37 | $2445.37 \times \frac{0.05}{4} \approx 30.57$ | 600 | 3075.94 |
| 6  | 3075.94 | $3075.94 \times \frac{0.05}{4} \approx 38.45$ | 600 | 3714.39 |
| 7  | 3714.39 | $3714.39 \times \frac{0.05}{4} \approx 46.43$ | 600 | 4360.82 |
| 8  | 4360.82 | $4360.82 \times \frac{0.05}{4} \approx 54.51$ | 600 | 5015.33 |
| 9  | 5015.33 | $5015.33 \times \frac{0.05}{4} \approx 62.69$ | 600 | 5678.02 |
| 10 | 5678.02 | $5678.02 \times \frac{0.05}{4} \approx 70.98$ | 600 | 6349    |
| 11 | 6349    | $6349 \times \frac{0.05}{4} \approx 79.36$    | 600 | 7028.36 |
| 12 | 7028.36 | $7028.36 \times \frac{0.05}{4} \approx 87.85$ | 600 | 7716.21 |

The final account balance shown in the table differs from the results we obtained in the example because of the need to round several times in the table calculations.

## Sinking Funds

Often, it is desirable to have a savings plan accumulate to some fixed amount by a certain point in time, perhaps in order to meet some future financial goal or obligation. Such a savings plan is called a **sinking fund.** We have investigated such a fund where a single deposit is made into an account and allowed to earn interest for some set period of time. We now turn our attention to a savings plan involving an ordinary annuity.

**EXAMPLE 2 ♦ Calculating Deposits for a Sinking Fund**

Henry wants to set aside a fixed amount of money into a savings account each month for the next 30 years, at which point he will withdraw the money and take a trip to celebrate his retirement. He wants to have $20,000 available for the trip. If the savings account earns an annual interest rate of 4% compounded monthly, then how much should Henry deposit each month?

**Solution**

The account is an ordinary annuity with a known future value, so we use $FV = d[\frac{(1 + i)^n - 1}{i}]$, where $d$ is what we seek. Note that 30 years results in 360 deposits.

$$20000 = d\left[\frac{\left(1 + \frac{0.04}{12}\right)^{360} - 1}{\left(\frac{0.04}{12}\right)}\right]$$

The bracketed expression above looks intimidating, but it is really just a fraction, so multiplying both sides of the equation by the reciprocal of that fraction allows us to solve for $d$.

$$20000\left[\frac{\left(\frac{0.04}{12}\right)}{\left(1 + \frac{0.04}{12}\right)^{360} - 1}\right] = d$$

$$d \approx \$28.82$$

Adjusting up to the nearest cent, Henry should deposit $28.82 into the savings account each month.

Notice that in the previous example, the total amount of money that is expected to be in the savings account after 360 deposits is given by

$$FV = \$28.82\left[\frac{\left(1 + \frac{0.04}{12}\right)^{360} - 1}{\left(\frac{0.04}{12}\right)}\right]$$

$$FV \approx \$20,002.50$$

Henry's contributions total $28.82 \times 360 = \$10,375.20$, so just under half of the money in the account is interest that has been earned over the course of 30 years.

## EXAMPLE 3 ◆ Number of Deposits Needed

Suppose that you wish to start putting away a fixed amount of money each week to save up for the purchase of a new vehicle. If you want to have $15,000 available to you when you go car shopping and you can make weekly deposits of $40 into a savings account that earns 5.6% interest compounded weekly, how long will it be before you can make your purchase?

**Solution**

We use $FV = d[\frac{(1 + i)^n - 1}{i}]$, where $n$ is what we seek.

$$15000 = 40\left[\frac{\left(1 + \dfrac{0.056}{52}\right)^n - 1}{\left(\dfrac{0.056}{52}\right)}\right]$$

Solving for $n$ using algebraic techniques is beyond the scope of this text, so we opt instead for a graphical approach. We will graph the equations

$$y_1 = 40\left(\left(1 + \frac{0.056}{52}\right)^x - 1\right)\bigg/\left(\frac{0.056}{52}\right)$$

$$y_2 = 15000$$

on a graphing calculator and approximate the point of intersection of the graphs (see Figure 2.4). Note here that $x$ represents a number of weeks.

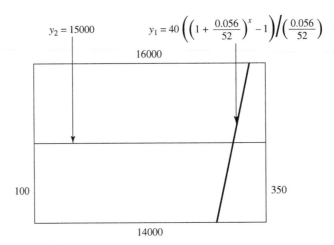

FIGURE 2.4

Using an intersection routine on the calculator we find $x \approx 315.15$, yielding a total number of deposits needed of 316. This number of deposits will actually put the account balance over $15,000, but 315 deposits will leave the account short of the desired total.

**EXAMPLE 4 ◆ Approximating an Interest Rate**

Mitch knows that in three years he will need to buy a new computer to replace the one he is currently using. He can set aside $60 each month and hopes to have $2400 available after three years to make the computer purchase. If he is going to make these $60 deposits into an interest-earning account, what annual interest rate compounded monthly would the account need to pay in order for Mitch to reach his goal?

**Solution**

We use $FV = d[\frac{(1 + i)^n - 1}{i}]$, where $i$ is what we seek, keeping in mind that $12i$ is the annual interest rate required.

$$2400 = 60\left[\frac{(1 + i)^{36} - 1}{i}\right]$$

Approximating $i$ using algebraic techniques is very challenging, so we once again opt for a graphical approach. Since $i = \frac{r}{12}$ and we are looking for $r$, we will graph the equations

$$y_1 = 60\left(\left(1 + \frac{x}{12}\right)^{36} - 1\right)\Big/\left(\frac{x}{12}\right)$$

$$y_2 = 2400$$

on a graphing calculator and approximate the point of intersection of the graphs (see Figure 2.5). Note here that $x$ represents the annual interest rate.

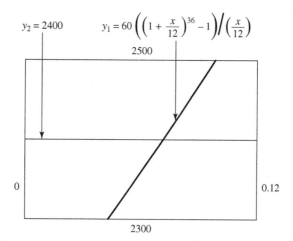

FIGURE 2.5

Using an intersection routine on the calculator we find $x \approx 0.07116$, yielding an annual interest rate of 7.12%, adjusted up to the nearest one hundredth of a percent.

# EXERCISES

*For the following exercises, assume a 365-day year unless directed otherwise. Also, when calculating interest rates, give your result as a percentage, correct to one-hundredth of a percent. All questions involving annuities are dealing with ordinary annuities, and interest is compounded with the same frequency as deposits or payments are made. Also, unless specified otherwise, stated interest rates are annual rates.*

1. Calculate the future value of an annuity with deposits of $3000 made

   (a) annually at 6% for 5 years
   (b) quarterly at 4.5% for 8.5 years
   (c) monthly at 2.99% for 28.25 years
   (d) weekly at 8.25% for 10 years

2. Calculate how much interest is earned by an annuity with deposits of $500 made

   (a) semiannually at 4% for 8 years
   (b) monthly at 5.28% for 4.75 years
   (c) daily at 0.75% for 1000 days
   (d) weekly at 7.9% for 4 years

3. For each of the following you are given the future value of an annuity, an annual interest rate, and an amount of time. Find the corresponding periodic deposits.

   (a) $12,000; 8% compounded monthly; 5 years
   (b) $350,000; 5.99% compounded quarterly; 25 years
   (c) $58,500; 3.45% compounded annually; 36 years
   (d) $1,234,456.78; 4% compounded weekly; 15 years

4. Cooper wants to have $3000 saved up in three years. He will make equal deposits each quarter into an account that earns 3.8% compounded quarterly. How much should he deposit each quarter?

5. A savings plan offers 7.2% interest compounded monthly. Find the value of the account if you make a $50 deposit into the account each month for 6 years.

6. Sally wishes to have $50,000 in savings when her daughter starts college in 16 years. If the local bank offers an interest rate of 7% compounded quarterly, what size quarterly deposits should she make to accomplish her goal?

7. John started making annual deposits into his IRA account when he turned 36. Each year he deposits $2000 into the account. Find the value of the IRA after John turns 66, assuming that the account earns 6.4% compounded annually.

8. Suppose that you can afford to set aside $200 each month towards the purchase of a boat valued at $5600. If you can earn 5.4% compounded monthly on your deposits, how long will it take to save up enough money to purchase the boat?

9. Bill opens a savings account and deposits $50 each week into an account that earns 6% compounded weekly for a total of 18 years. At that time he stops making deposits, but leaves the money in the account. Find the value of the account after a total of 40 years has passed since the account was opened.

10. Chris wants to make annual deposits of $2000 into a retirement account with a goal of having $125,000 in the account after 25 years. What annual interest rate compounded annually does the account need to earn in order for this goal to be met?

11. Suppose that an investment account guarantees an interest rate of 5% compounded monthly. Find the amount of money that would need to be invested in a single lump sum in such an account to match the accumulated value of the account if monthly deposits of $50 are made over the course of 10 years.

12. Vince received a cash gift of $2500 and deposited that money into an account that pays 4% interest compounded quarterly. Three months later, Ted started making quarterly deposits of $75 into an account that earns 5.2% interest compounded quarterly. How long will it take for the balance of Ted's account to exceed the balance of Vince's account?

13. A city has issued bonds to finance a new convention center. The bonds have a total face value of $750,000 and are payable in eight years. A sinking fund has been opened to meet this obligation. If the interest rate on the fund is 6% compounded semiannually, how much needs to be deposited into the fund every six months to meet the future obligation?

14. In an effort to investigate the effect of starting age in building a retirement account, suppose that it is desired to have a retirement fund of $500,000 at age 70. If annual deposits earn 7% interest compounded annually, find the amount of the necessary annual deposits for the following cases.

   (a) deposits are made for 25 years
   (b) deposits are made for 30 years
   (c) deposits are made for 40 years
   (d) deposits are made for 50 years

15. Emily invested $2000 each year for ten years into an IRA that earned 6% interest compounded annually. At the end of ten years she quit making deposits and let the money in the account continue to earn interest for another 25 years. At the time Emily stopped making deposits, her soul mate Nick opened an IRA that also earned 6% compounded annually. He made annual deposits of $2000 for 25 years.

   (a) Calculate how much money is in each IRA account after Nick has made his 25th deposit. Determine the amount of interest that has been earned in each case.

   (b) What does this exercise suggest to you as the best time to start saving for retirement or for some future financial goal or obligation?

# Interlude: Practice Problems 2.2, Part 1

1. Consider an ordinary annuity with semiannual deposits of $1800 and an annual interest rate of 2.8% compounded semiannually.

   (a) Find the future value of the annuity after 12 years.

   (b) Determine the amount of interest that has been earned after 12 years.

2. Maxx has elected to have money taken directly out of his monthly paycheck and deposited in a supplemental retirement account that guarantees a fixed annual interest rate of 3.25% with interest compounded monthly. He wants to have at least $50,000 in this account when he hits his target retirement age in 20 years. How much will need to be taken out of his paycheck each month to accomplish this goal?

3.  Katie is setting aside money for a remodeling project she has planned for the future. She will make monthly deposits of $200 into an account that earns interest at a rate of 5.2% compounded monthly. She will begin the project once she has at least $8000 saved. How many deposits does she need to make before the project can begin?

4.  I am going to start putting money away every week to save up for a trip to Greenville, PA one year from now. How much do I need to deposit each week into an account that earns an annual interest rate of 4.1% compounded weekly so that I have at least $4000 saved up for my trip?

# Interlude: Practice Problems 2.2, Part 2

A couple wishes to start saving for their child's college education. The child is eight years old now and will leave for college in exactly 10 years. The parents estimate that the total cost of the college education will be $150,000.

1. If the couple can make monthly deposits into an account that earns 3.8% interest compounded monthly, then how much should their deposits be in order to ensure that the account balance will be at least $150,000 in 10 years?

2. Suppose that the couple will make monthly deposits for the next six years, and then cease making deposits. What should be deposited in the account each month so that the account balance will be at least $150,000 in 10 years?

3.  Suppose that the couple decides to make equal-sized but somewhat larger deposits on a biannual basis (5 in total). How much needs to be deposited biannually so that the account balance will be at least $150,000 in 10 years?

4.  In each of the scenarios above, how much interest is earned on the account during the 10 years?

5.  If the couple can only afford to make monthly deposits of $800 for the next 10 years, then what annual interest rate, compounded monthly, would the account need to offer so that the account balance would be at least $150,000 in 10 years? Give your result as a percentage, accurate to two decimal places.

## 2.3 PRESENT VALUE OF AN ANNUITY

Thus far, we have investigated scenarios where either a single deposit or a sequence of equal deposits is made into some sort of savings investment with a goal of ascertaining the accumulated value of the account after some period of time. Now, we look at a variation on the idea of an ordinary annuity where the goal is not to accumulate money, but rather to distribute money.

### Present Value of an Annuity

Consider the following problem.

> A benefactor would like to set up a fund that will assist you in paying for college expenses. Starting in three months, you will receive quarterly payments of $500 from the account. The fund will be established with a single deposit, and after six payments have been made, the account will be empty. The account earns an annual interest rate of 5% compounded quarterly. How much does this benefactor need to deposit now to establish this fund?

It is apparent here that the initial deposit should be less than $3000, although that is ultimately how much will come to you from the account. It is helpful here to think of the initial deposit as consisting of six parts, each part being the amount of money needed to make a corresponding payment of $500. Let $P_i$ be the amount of money needed to make the $i$th payment of $500. Notice that each of these parts will earn compound interest for a different amount of time but must ultimately have a future value of $500. So, each of these can be considered the present value (principal) of a $500 payment and we can calculate the amounts using the compound interest formula $A = P(1 + i)^n$. Since we wish to calculate present value, we will rewrite that formula in the form

$$P = \frac{A}{(1 + i)^n} \quad \text{or} \quad P = A(1 + i)^{-n}.$$

At the time of the initial deposit, the part $P_1$ has only one quarter to earn interest and will be worth $500 after that quarter has passed. So, we have

$$P_1 = 500\left(1 + \frac{0.05}{4}\right)^{-1} = 500(1.0125)^{-1}$$

The part $P_2$ will earn interest for two quarters and ultimately be worth $500, so

$$P_2 = 500(1.0125)^{-2}$$

Similarly, we find

$$P_3 = 500(1.0125)^{-3}$$
$$P_4 = 500(1.0125)^{-4}$$
$$P_5 = 500(1.0125)^{-5}$$
$$P_6 = 500(1.0125)^{-6}$$

The sum $P_1 + P_2 + P_3 + P_4 + P_5 + P_6$ then gives the total amount of money that needs to be deposited initially to establish the fund. Since the fund is just a sequence of equal periodic payments, with payments made at the end of the payment period, it is an example of an ordinary annuity, so we call this initial deposit amount the **present value** of the annuity. We summarize the calculations in the following table, where amounts have been rounded to the nearest cent when necessary.

| Payment | Amount Needed Initially |
|:-------:|:-----------------------|
| 1 | $P_1 = 500(1.0125)^{-1} \approx 493.83$ |
| 2 | $P_2 = 500(1.0125)^{-2} \approx 487.73$ |
| 3 | $P_3 = 500(1.0125)^{-3} \approx 481.71$ |
| 4 | $P_4 = 500(1.0125)^{-4} \approx 475.76$ |
| 5 | $P_5 = 500(1.0125)^{-5} \approx 469.89$ |
| 6 | $P_6 = 500(1.0125)^{-6} \approx 464.09$ |
| Total | $P_1 + P_2 + \cdots + P_6 \approx 2873.01$ |

The present value of the annuity, then, is \$2873.01. Denote this quantity by $PV$, so that

$$PV = P_1 + P_2 + P_3 + P_4 + P_5 + P_6$$

$$PV = 500(1.0125)^{-1} + 500(1.0125)^{-2} + \cdots + 500(1.0125)^{-6}$$

Taking a similar approach to the one we used in the previous section when calculating the future value of an ordinary annuity, we see that

$$1.0125PV = 500 + 500(1.0125)^{-1} + \cdots + 500(1.0125)^{-5}$$

and so

$$\underbrace{1.0125PV - PV}_{0.0125PV} = (500 + 500(1.0125)^{-1} + 500(1.0125)^{-2} + \cdots + 500(1.0125)^{-5})$$
$$- (500(1.0125)^{-1} + 500(1.0125)^{-2} + \cdots + 500(1.0125)^{-6}) = 500 - 500(1.0125)^{-6}$$

Since $0.0125PV = 500 - 500(1.0125)^{-6}$, we see that

$$PV = \frac{500 - 500(1.0125)^{-6}}{0.0125} = 500\left[\frac{1 - (1.0125)^{-6}}{0.0125}\right]$$

Notice again that the major components of the right-hand side of the equation above correspond to the pertinent facts about the annuity in question: \$500 periodic payments, 6 payments, and periodic interest rate 0.0125. Finishing the calculation and rounding to the nearest cent gives $PV \approx \$2873.00$, which differs from our earlier calculation by one cent due to rounding.

The actual numbers used in the previous scenario are not important. The last equation, however, is important. We can generalize the above computation to get a formula for calculating the present value of an ordinary annuity:

---

**Present Value (Ordinary Annuity)**

$$PV = PMT\left[\frac{1 - (1 + i)^{-n}}{i}\right],$$

where $PMT =$ amount of periodic payments

$i =$ periodic interest rate (written as a decimal)

$n =$ total number of payments

$PV =$ present value (starting amount of money needed)

---

To reiterate, the present value of an ordinary annuity is the amount of money that would need to be deposited into an account initially in order for the prescribed equal periodic payments to be made out of the account for the indicated period of time. When the final payment is made, the account balance is to be \$0.

As further illustration, we will summarize the activity of the account set up by your benefactor. Rounding will be to the nearest penny when necessary. We will use the starting amount of money given in the most recent present value calculation.

| Payment Period | Beginning Account Balance ($B$) | Periodic Interest Earned on Balance ($I$) | Payment ($PMT$) | Ending Account Balance ($B + I - PMT$) |
|---|---|---|---|---|
| 1 | 2873 | $2873 \times \frac{0.05}{4} \approx 35.91$ | 500 | 2408.91 |
| 2 | 2408.91 | $2408.91 \times \frac{0.05}{4} \approx 30.11$ | 500 | 1939.02 |
| 3 | 1939.02 | $1939.02 \times \frac{0.05}{4} \approx 24.24$ | 500 | 1463.26 |
| 4 | 1463.26 | $1463.26 \times \frac{0.05}{4} \approx 18.29$ | 500 | 981.55 |
| 5 | 981.55 | $981.55 \times \frac{0.05}{4} \approx 12.27$ | 500 | 493.82 |
| 6 | 493.82 | $493.82 \times \frac{0.05}{4} \approx 6.17$ | 499.99 | 0 |

Notice that the last payment was only $499.99. This is because the account balance at the time of the final payment was $493.82 + 6.17 = \$499.99$.

## EXAMPLE 1 ◆ Present Value and Interest Earned

Determine the present value of an annuity with semiannual payments of $2500 for a total of four years if the money earns interest at an annual rate of 4% compounded semiannually. Also, calculate how much of the total amount of money paid out is due to interest.

**Solution**

We use $PV = PMT\left[\frac{1 - (1 + i)^{-n}}{i}\right]$. Note that four years yields a total of eight payments, so we have

$$PV = 2500\left[\frac{1 - \left(1 + \dfrac{0.04}{2}\right)^{-8}}{\left(\dfrac{0.04}{2}\right)}\right]$$

$$PV \approx \$18313.71$$

Adjusting up to the nearest cent, we find a present value of $18,313.71. This is the amount of money that must be deposited into the annuity account initially. Since a total of $8 \times 2500 = \$20{,}000$ is ultimately paid out of the annuity but only $18,313.71 was put in, $1686.29 of the $20,000 came from interest.

## EXAMPLE 2 ◆ Calculating Annuity Payments

Johanna received a high school graduation gift of $6000 from her grandparents. She will deposit the money into an account that earns 3% interest compounded quarterly and use the money to help pay for college expenses. Starting in three months, she will make equal quarterly withdrawals from the account for three years, after which the account will be empty. How much will she be able to withdraw each quarter?

**Solution**

We use $PV = PMT\left[\frac{1 - (1 + i)^{-n}}{i}\right]$, where $PMT$ is what we seek. The annuity has a present value of $6000, so

$$6000 = PMT\left[\frac{1 - \left(1 + \dfrac{0.03}{4}\right)^{-12}}{\left(\dfrac{0.03}{4}\right)}\right]$$

The bracketed expression above looks intimidating, but it is really just a fraction, so multiplying both sides of the equation by the reciprocal of that fraction allows us to solve for *PMT*.

$$6000 \left[ \frac{\left(\dfrac{0.03}{4}\right)}{1 - \left(1 + \dfrac{0.03}{4}\right)^{-8}} \right] = PMT$$

$$PMT \approx \$524.71$$

Rounding to the nearest cent, Johanna will be able to make twelve quarterly withdrawals of $524.71 from the account.

## EXAMPLE 3 ◆ Balance of an Annuity Account

Suppose that an annuity is set up to provide monthly payments for one year. The annuity is funded with an initial deposit of $8000 and interest is 4% compounded monthly, so that the monthly payments are $681.20. Given this, how much money is left in the account after the seventh payment has been made?

**Solution**

At first glance, it may seem that this question would require the construction of a balance sheet and calculating the annuity account balance for seven payments. However, notice that whatever the account balance is after the seventh payment, that amount of money needs to be adequate to fund five additional monthly payments of $681.20. In other words, the account balance after the seventh payment is exactly the single deposit amount necessary to fund a 5-payment annuity with the given monthly payments and interest rate, so the account balance is just the present value of an annuity consisting of five payments. So, we use $PV = PMT\left[\frac{1 - (1 + i)^{-n}}{i}\right]$. Here,

$$PV = 681.20 \left[ \frac{1 - \left(1 + \dfrac{0.04}{12}\right)^{-5}}{\left(\dfrac{0.04}{12}\right)} \right]$$

$$PV \approx \$3372.20$$

Rounding to the nearest cent, the annuity account balance is $3372.20. To further illustrate, consider the following table. Rounding is to the nearest cent.

| Payment Period | Beginning Account Balance (B) | Periodic Interest Earned on Balance (I) | Payment (PMT) | Ending Account Balance (B + I − PMT) |
|---|---|---|---|---|
| 1 | 8000 | $8000 \times \frac{0.04}{12} \approx 26.67$ | 681.20 | 7345.47 |
| 2 | 7345.47 | $7345.47 \times \frac{0.04}{12} \approx 24.48$ | 681.20 | 6688.75 |
| 3 | 6688.75 | $6688.75 \times \frac{0.04}{12} \approx 22.30$ | 681.20 | 6029.85 |
| 4 | 6029.85 | $6029.85 \times \frac{0.04}{12} \approx 20.10$ | 681.20 | 5368.75 |
| 5 | 5368.75 | $5368.75 \times \frac{0.04}{12} \approx 17.90$ | 681.20 | 4705.45 |
| 6 | 4705.45 | $4705.45 \times \frac{0.04}{12} \approx 15.68$ | 681.20 | 4039.93 |
| 7 | 4039.93 | $4039.93 \times \frac{0.04}{12} \approx 13.47$ | 681.20 | 3372.20 |

The idea illustrated in Example 3 will prove to be very helpful in the next section.

## Another View of Present Value

Consider the following scenario (compare to Exercise #12 in the previous section).

> Monique and Erin are going to start saving up money to take a ski trip. Each plans to have the same amount of money for the trip, which will take place in nine months. Monique is going to make a single deposit into a savings account that earns 6% interest compounded monthly, while Erin is going to start making monthly $100 deposits into an account that earns the same interest rate, with the first deposit made one month after Monique makes her deposit. How much should Monique deposit right now so that both of the women have the same amount of money in their respective accounts when it is time for the trip?

To answer the question posed, notice that the amount of money Monique has after nine months is given by

$$A = P\left(1 + \frac{0.06}{12}\right)^9$$

where $P$ is her deposit. On the other hand, the amount of money in Erin's account after nine months is given by

$$FV = 100\left[\frac{\left(1 + \dfrac{0.06}{12}\right)^9 - 1}{\left(\dfrac{0.06}{12}\right)}\right]$$

Since both accounts have accumulated to the same total amount of money after nine months, $A = FV$, or

$$P\left(1 + \frac{0.06}{12}\right)^9 = 100\left[\frac{\left(1 + \dfrac{0.06}{12}\right)^9 - 1}{\left(\dfrac{0.06}{12}\right)}\right]$$

Solving for $P$ gives

$$P = 100\left[\frac{\left(1 + \dfrac{0.06}{12}\right)^9 - 1}{\left(\dfrac{0.06}{12}\right)}\right]\left(1 + \frac{0.06}{12}\right)^{-9}$$

$$P = 100\left[\frac{1 - \left(1 + \dfrac{0.06}{12}\right)^{-9}}{\left(\dfrac{0.06}{12}\right)}\right]$$

$$P \approx \$877.91$$

In other words, the amount of money Monique should deposit right now is exactly the present value of an ordinary annuity with nine monthly payments of $100 with interest at 6% compounded monthly. Adjusted up to the nearest cent, this means she should deposit $877.91 right now.

We now have two interpretations of present value of an ordinary annuity; in the next section we will encounter a seemingly dissimilar situation that will relate back to this same notion.

# EXERCISES

*For the following exercises, assume a 365-day year unless directed otherwise. Also, when calculating interest rates, give your result as a percentage, correct to one hundredth of a percent. All questions involving annuities are dealing with ordinary annuities, and interest is compounded with the same frequency as deposits or payments are made. Also, unless specified otherwise, stated interest rates are annual rates.*

1. Calculate the present value of an annuity with payments of $1000 made

   (a) annually at 3.8% for 6 years
   (b) quarterly at 4.75% for 12.5 years
   (c) monthly at 5.99% for 7 years
   (d) weekly at 3% for 5 years

2. Calculate how much of the total of the annuity payments is due to interest earned by the annuity if payments of $800 are made

   (a) semiannually at 6% for 3.5 years
   (b) monthly at 5.65% for 10 years
   (c) quarterly at 8.8% for 6.5 years
   (d) weekly at 2.99% for 3 years

3. For each of the following you are given the present value of an annuity, an annual interest rate, and an amount of time. Find the corresponding periodic payments.

   (a) $120,000; 5% compounded monthly; 8 years
   (b) $38,000; 6.2% compounded quarterly; 12 years
   (c) $11,500; 3.29% compounded annually; 6 years
   (d) $987,654.32; 10% compounded weekly; 30 years

4. An account is started with a single deposit of $10,000. Interest is 7.7% compounded quarterly. The account will pay out equal quarterly payments for 6 years, at which point the account balance will be $0.

   (a) Determine the amount of each quarterly payment.
   (b) Calculate the account balance after the eighth payment has been made.

5. Pat wants to set up an annuity that will provide monthly payments of $75 for a total payout of $4500. The annuity will be funded with an initial deposit of $4000. What annual interest rate, compounded monthly, does the annuity need to earn to make this possible?

6. Fred just won $32 million playing the lottery. He has opted to collect his winnings over the course of 20 years with equal annual payments of $1.6 million (before taxes, of course). The lottery commission will set up an annuity to handle paying Fred. Assuming an interest rate of 4.25% compounded annually, determine how much money the commission needs to have right now and how much of the $32 million Fred receives is from interest earned by the annuity.

7. A couple makes a single deposit of $1000 into an investment account that earns a fixed interest rate of 3.5% compounded monthly. The money in the account will be used to fund their newborn's college education in 18 years. At that point in time, monthly withdrawals from the account will be initiated. After a total of 60 equal monthly withdrawals, the account should be empty. Find the amount of each monthly withdrawal and total amount of money that will be generated from the withdrawals.

8. Beginning at age 30, Trish invests $2000 each year into an IRA account. Assume that the account earns 5% compounded annually. When she retires, she plans to make equal annual withdrawals that will deplete the account when she turns 80. Find the amount of her annual withdrawals for each of the following retirement ages.

   (a) 65 years old
   (b) 67 years old
   (c) 68 years old
   (d) 70 years old

9. How much should be invested each year for ten years at 8% compounded annually so that for the subsequent 20 years annual withdrawals of $4000 can be made?

10. What single deposit that can earn 6.4% compounded quarterly will match the total accumulated value of a sequence of 16 quarterly deposits of $330 earning the same interest rate?

11. Consider an ordinary annuity with the stated present value, annual interest rate, and number of payments, and calculate the amount of money remaining in the annuity after the 12th payment has been issued.

    (a) $40,000; 4% compounded quarterly; 20
    (b) $1200; 6.4% compounded semiannually; 16
    (c) $2,000,000; 9.9% compounded annually; 30
    (d) $350,000; 3.99% compounded monthly; 240

# Interlude: Practice Problems 2.3

1.  A fund is being set up with the intent of offering annual scholarships of $10,000 for the next 25 years. If the fund earns an annual interest rate of 6% compounded annually, then what single initial deposit into the fund is needed to provide for the scholarships?

2.  A couple can make monthly deposits of $400 into an account that earns an annual interest rate of 3% compounded monthly. In six years, when one of their children leaves for college, they will cease making payments into the account. Starting one month after that last deposit, equal monthly payments will be issued from the account for four years to help pay college expenses. After the final payment, the account should be empty.

    (a) Calculate the quarterly payments that will be issued.

    (b) How much interest does the money in the account ultimately earn?

3.  A company has set aside $1,000,000 to pay for a development project. The company will issue semiannual payments of $150,000 to the development team for expenses over the length of the project, which is four years. At the end of the four years, the project is complete and the funding money is to be gone. What annual interest rate, compounded semiannually, does the company need to earn on their funding money to fund the project? Give the percentage accurate to two decimal places.

4.  A trust fund was set up with a deposit of $100,000 in an account that earns an annual interest rate of 8.2% compounded quarterly. Quarterly payments of $4000 have been issued from the fund for six years. How much money remains in the fund?

## 2.4  BORROWING

It is quite likely that at some time in your life you will need to borrow money, perhaps to fund an education, purchase a home, or just pay your bills. It is natural to expect that any lender would wish to charge a fee for the money that you borrow; as before, this fee is called **interest.** While the types of loans and schemes for charging interest and organizing repayment of the loan are myriad, we will investigate some of the more common situations that you might encounter as a borrower.

### Add-On Loans

We first consider what is perhaps the most basic scenario for a loan. The type of loan to be illustrated is called an **add-on loan**, because the interest charged to the loan is simply added to the amount borrowed to give the total amount of money needed to repay the loan. The interest charge is simple interest based on the amount borrowed, an annual interest rate, and the **term** of the loan (amount of time before the loan and interest is repaid in full). This is completely analogous to the situation we encountered in Section 2.1 when looking at the future value of a deposit that earned simple interest. As such, we can use the same formulas describing future value, present value, and interest in the context of simple interest calculations.

### EXAMPLE 1  ◆  Single Payment Add-On Loan

Susie loans Jon $500 to pay for repairs to his car, with the expectation that he would repay the $500 plus 8% simple interest after six months. Given this, how much interest will Jon pay on his loan, and how much will he pay Susie in six months?

**Solution**

We can calculate the amount of interest Jon pays on the $500 loan by using the simple interest formula, $I = Prt$. Here, $P$ denotes the amount Jon wishes to borrow.

$$I = 500 \times 0.08 \times \frac{6}{12}$$

$$I = \$20$$

Jon winds up paying $20 in interest, so he will pay Susie $520 in six months.

### EXAMPLE 2  ◆  Multiple Payment Add-On Loan

Lonnie the Loan Shark is happy to provide emergency loans to college students. He offers add-on loans of up to $1000 with an annual interest rate of 30%. While this rate may seem intimidating, he does offer borrowers the opportunity to repay the loan (plus interest) with equal monthly payments for up to one year. Given this, Adam comes to Lonnie to borrow $1000. Adam agrees to pay off his debt in 10 equal monthly payments. How much will Adam ultimately pay for his loan, and how much are his monthly payments?

**Solution**

We begin by calculating the amount of interest Adam pays on the loan by using the simple interest formula.

$$I = 1000 \times 0.30 \times \frac{10}{12}$$

$$I = \$250$$

Adam pays $250 in interest on the loan, so he will ultimately pay Lonnie $1250. Since he has agreed to repay this in 10 equal monthly payments, he will make monthly payments of $125.

Although this is not necessarily a universal rule, it is generally the case that in any loan where a *single* interest charge is assessed, if the loan is repaid with equal payments, then each payment pays the same fraction of loan and interest. For instance, in the previous example, each payment pays one tenth of the total interest on the loan as well as one tenth of the actual amount borrowed. Unless directed otherwise, you should assume that this is the case for all of the loan situations discussed in this text where a single simple interest charge is involved (we will contrast this later with a type of loan that involves a series of interest charges).

## Discounted Loans

In Section 2.1, the notion of **zero-coupon bonds** was put forth (see Exercises #15, #16). From the perspective of the issuer of a bond, the bond is a loan, with the purchaser serving as the lender. Recall that zero-coupon bonds are sold at a **discount**; the selling price of the bond is less than the face value of the bond and no interest payments are made during the life of the bond. At maturity, the face value of the bond is paid to the bearer of the bond and the difference between purchase price and face value is interest. Considering this type of bond as a loan, we call the type of loan illustrated a **discounted loan**. This terminology may be rather misleading, as what is discounted is not the cost of the loan, but rather how much the borrower actually receives. Given an annual interest rate, $r$ (called the **discount rate**), and a loan repayment term of t years, suppose that the discounted loan is for $\$P$ (this is called the **value** of the loan). The interest on the loan, $I$, is simple interest based on the value of the loan, so $I = Prt$. However, what the borrower gets, $D$, is actually the *difference* between the loan amount and the interest, namely

$$D = P - I$$

$$D = P - Prt$$

$$D = P(1 - rt)$$

Notice how this compares to the zero-coupon bonds. We summarize as follows:

---

**Discounted Loan**

$$D = P - Prt \quad \text{or} \quad D = P(1 - rt),$$

where $P =$ total amount to be repaid (value of the loan)

$r =$ annual simple interest rate (written as a decimal)

$t =$ time in years

$D =$ discounted loan amount (amount actually borrowed)

---

### EXAMPLE 3 ◆ Amount of a Discounted Loan

Scott goes to a local lending agent in hopes of securing a loan for a snowmobile. The lending agent offers Scott a discounted loan of $2000 carrying an annual interest rate of 5%, repayable with equal monthly payments over the course of the next 16 months. How much will Scott actually have to use for his snowmobile purchase, and what will his monthly payments be?

**Solution**

Since the loan being offered is a discounted loan, the $2000 offer is the total that Scott must repay over the course of 16 months, yielding monthly payments of $125. Of that $2000, we can calculate the amount that Scott will actually be able to use by $D = P(1 - rt)$

$$D = 2000\left[1 - (0.05)\left(\frac{16}{12}\right)\right]$$

$$D \approx \$1866.67$$

Rounded to the nearest cent, Scott will actually have $1866.67 to use for his purchase.

At casual glance, it may appear that a discounted loan is really no different than an add-on loan except for the fact that in the case of the add-on loan, the actual amount of the loan that is available for use is typically known before the total amount to be repaid is specified. However, this is not the case, as the next example illustrates.

## EXAMPLE 4 ♦ Comparing an Add-On Loan to a Discounted Loan

Suppose that Scott opts to shop around for a loan before shopping for a snowmobile. He already has an offer for a 5% discounted loan of $2000 (Example 3), but would like to entertain other offers. He checks with a neighborhood credit union and finds that he qualifies for a 5% add-on loan, repayable in 16 equal monthly payments. Is this a better offer than the 5% discounted loan? Assume here that Scott does not want to pay more than $2000 in total payments over the course of 16 months.

**Solution**

As seen earlier, we can use the usual simple interest formulas to work with add-on loans. Here we use $A = P(1 + rt)$, where $P$ is what we seek. We will assume a total amount to be repaid of $2000.

$$2000 = P\left[1 + (0.05)\left(\frac{16}{12}\right)\right]$$

$$\frac{2000}{1 + (0.05)\left(\frac{16}{12}\right)} = P$$

$$P = \$1875$$

With the add-on loan, Scott will actually have $1875 to use for his purchase, as compared to $1866.67 available with the 5% discounted loan, so the add-on loan is a better offer.

## Installment Loans

Imagine that you have just purchased a car and needed to borrow $5000 to do so. A finance agent cheerfully tells you that you qualify for a 48-month installment loan carrying an annual interest rate of 6%, and for $117.43 each month, you can drive away with your new set of wheels. As far as repaying your loan goes, what have you just been told?

The repayment scheme and method for calculating interest in this situation are different from what we looked at with the previously discussed types of loans. Even though you are repaying the loan with equal periodic payments, a quick calculation shows that something much different is happening here (compare to Example 2).

First off, notice that under the terms given to you, you wind up paying a total of $117.43 \times 48 = \$5636.64$ to repay the loan. Consequently, you have been charged $636.64 in interest on your $5000 borrowed. If you had been assessed a single simple interest charge at an annual rate of 6%, you would have paid a total off $5000 \times 0.06 \times 4 = \$1200$ in interest, quite a bit more! The reason for this large difference lies in how interest is charged on your loan and how your periodic payments affect the repayment of the actual amount borrowed. Your car loan is an example of a typical **installment loan**. The distinguishing characteristic here is that *a periodic interest charge is assessed at the time a payment is made; each payment you make pays all periodic interest charged to the loan balance since the last payment was made. The remainder of the payment goes towards reducing the balance owed on the original loan amount.* We say here that you are **amortizing** the loan.

In this case, you have an annual interest rate of 6%. This corresponds to a monthly interest rate of 0.5%, so each month the interest charged on your loan is 0.5% of the account balance as of the time your last payment made is credited to the loan. We are assuming here that your first payment is due one month after you take out the loan and that your payments are equally spaced for the entire 48-month term of the loan. We summarize what is happening with your loan balance and payments for the first six payments you make in the following table. When necessary, rounding is to the nearest cent.

| Payment | Beginning Loan Balance ($B$) | Periodic Interest Charged to Balance ($I$) | Payment ($PMT$) | Ending Loan Balance ($B + I - PMT$) |
|---|---|---|---|---|
| 1 | 5000 | $5000 \times \frac{0.06}{12} = 25$ | 117.43 | 4907.57 |
| 2 | 4907.57 | $4907.57 \times \frac{0.06}{12} \approx 24.54$ | 117.43 | 4814.68 |
| 3 | 4814.68 | $4814.68 \times \frac{0.06}{12} \approx 24.07$ | 117.43 | 4721.32 |
| 4 | 4721.32 | $4721.32 \times \frac{0.06}{12} \approx 23.61$ | 117.43 | 4627.50 |
| 5 | 4627.50 | $4627.50 \times \frac{0.06}{12} \approx 23.14$ | 117.43 | 4533.21 |
| 6 | 4533.21 | $4533.21 \times \frac{0.06}{12} \approx 22.67$ | 117.43 | 4438.45 |

Notice that with each successive payment, the amount of interest you're charged decreases so the amount of your payment that goes toward reducing the unpaid balance of the loan increases. The next table summarizes the last six payments on the loan. Again, rounding is to the nearest cent.

| Payment | Beginning Loan Balance ($B$) | Periodic Interest Charged to Balance ($I$) | Payment ($PMT$) | Ending Loan Balance ($B + I - PMT$) |
|---|---|---|---|---|
| 43 | 692.16 | $692.16 \times \frac{0.06}{12} \approx 3.47$ | 117.43 | 578.20 |
| 44 | 578.20 | $578.20 \times \frac{0.06}{12} \approx 2.89$ | 117.43 | 463.66 |
| 45 | 463.66 | $463.66 \times \frac{0.06}{12} \approx 2.32$ | 117.43 | 348.55 |
| 46 | 348.55 | $348.55 \times \frac{0.06}{12} \approx 1.74$ | 117.43 | 232.86 |
| 47 | 232.86 | $232.86 \times \frac{0.06}{12} \approx 1.16$ | 117.43 | 116.59 |
| 48 | 116.59 | $116.59 \times \frac{0.06}{12} \approx 0.58$ | 117.17 | 0 |

It should be noted that amount of the last payment differs from the previous 47 payments because of rounding that was done when computing the monthly payment (which will be discussed presently) and rounding that occurred in the subsequent calculations of interest charged on the loan balance between successive payments.

Now that it has been demonstrated exactly how your monthly payments will be applied to the loan, we turn to another issue: how the monthly payment amount was determined. Rather than dealing with the current scenario involving 48 payments, we will investigate a similar situation that involves only four payments.

Suppose instead that the terms of your car loan are such that after each year (since the loan was taken out) you will make a payment of $$PMT$, and that you are charged 6% interest each year on the unpaid loan balance. We immediately observe that after the first year, you will be charged $5000 \times 0.06 = $300$ in interest. If $B_1$ denotes the unpaid balance of your car loan after the first payment, then

$$B_1 = 5000 + 5000(0.06) - PMT$$

$$B_1 = 5000(1.06) - PMT$$

Similarly, if $B_2$ is the unpaid loan balance after the second payment, then

$$B_2 = B_1 + B_1(0.06) - PMT$$

$$B_2 = B_1(1.06) - PMT$$

$$B_2 = [5000(1.06) - PMT](1.06) - PMT$$

$$B_2 = 5000(1.06)^2 - PMT(1.06) - PMT$$

After the third payment, the unpaid loan balance is given by

$$B_3 = B_2 + B_2(0.06) - PMT$$

$$B_3 = B_2(1.06) - PMT$$

$$B_3 = [5000(1.06)^2 - PMT(1.06) - PMT](1.06) - PMT$$

$$B_3 = 5000(1.06)^3 - PMT(1.06)^2 - PMT(1.06) - PMT$$

Finally, after the fourth payment, the unpaid loan balance is given by

$$B_4 = B_3 + B_3(0.06) - PMT$$

$$B_4 = B_3(1.06) - PMT$$

$$B_4 = [5000(1.06)^3 - PMT(1.06)^2 - PMT(1.06) - PMT](1.06) - PMT$$

$$B_4 = 5000(1.06)^4 - PMT(1.06)^3 - PMT(1.06)^2 - PMT(1.06) - PMT$$

Notice, however, that after the fourth payment your debt has been completely repaid, so that $B_4 = 0$ and we have

$$5000(1.06)^4 - PMT(1.06)^3 - PMT(1.06)^2 - PMT(1.06) - PMT = 0$$

Equivalently,

$$5000(1.06)^4 = PMT(1.06)^3 + PMT(1.06)^2 + PMT(1.06) + PMT$$

The right-hand side of the above equation should look familiar, as it is exactly the same sort of sum we encountered when studying future value of an ordinary annuity. Using the same technique as in Section 2.2, we see that

$$5000(1.06)^4 = PMT\left[\frac{(1.06)^4 - 1}{0.06}\right]$$

Multiplying each side to the above equation by $(1.06)^{-4}$ yields

$$5000 = PMT\left[\frac{(1.06)^4 - 1}{0.06}\right](1.06)^{-4}$$

$$5000 = PMT\left[\frac{1 - (1.06)^{-4}}{0.06}\right]$$

Interestingly enough, this last equation has the form of the formula for present value of an ordinary annuity $(PV = 5000, i = 0.06, n = 4, PMT$ unknown). Solving for $PMT$ gives

$$PMT = 5000\left[\frac{0.06}{1 - (1.06)^{-4}}\right]$$

$$PMT \approx \$1442.96$$

Rounded to the nearest cent, this gives an annual payment of \$1442.96 in order to repay the loan over the course of four years.

What this investigation points to, then, is that we can interpret a standard installment loan as an ordinary annuity where the present value is just the loan balance given the number of payments yet to be made, the corresponding periodic interest rate, and periodic payment amount.

Returning to the original scenario, your original loan of \$5000 is to be repaid over the course of 48 months given an annual interest rate of 6% compounded monthly. Thus we see that

$$5000 = PMT\left[\frac{1 - \left(1 + \dfrac{0.06}{12}\right)^{-48}}{\left(\dfrac{0.06}{12}\right)}\right]$$

$$5000\left[\frac{\left(\dfrac{0.06}{12}\right)}{1 - \left(1 + \dfrac{0.06}{12}\right)^{-48}}\right] = PMT$$

$$PMT \approx \$117.43$$

Rounded to the nearest cent, this calculation gives a monthly payment of $117.43, just as proposed earlier.

To summarize, then, an installment loan can be viewed as an ordinary annuity, and the present value formula for ordinary annuities gives the relationship between loan balance, periodic payment amount, periodic interest rate, and number of payments:

---

### Installment Loans

$$PV = PMT\left[\frac{1 - (1 + i)^{-n}}{i}\right],$$

where $PMT$ = amount of periodic payments

$i$ = periodic interest rate (written as a decimal)

$n$ = total number of payments

$PV$ = unpaid loan balance with $n$ payments

---

Typically, installment loans are the types of loans offered for car loans and home mortgages, among other things.

---

## EXAMPLE 5 ◆ Conventional Home Mortgage

Newlyweds Lori and Jerome are looking into purchasing their first home. Right now, they can probably obtain a 30-year mortgage on a home with an interest rate of 5.75%. Lori and Jerome figure that they can probably afford to pay about $600 each month on loan payments. Given this, how much can they afford to borrow when they get set to purchase a home?

**Solution**

We use $PV = PMT\left[\frac{1 - (1 + i)^{-n}}{i}\right]$.

$$PV = 600\left[\frac{1 - \left(1 + \dfrac{0.0575}{12}\right)^{-360}}{\left(\dfrac{0.0575}{12}\right)}\right]$$

$$PV \approx \$102{,}814.93$$

Rounded to the nearest cent, the newlyweds can borrow up to $102,814.93.

---

It is worth remarking here that even if Lori and Jerome do wind up with a 30-year mortgage as described above, it is most likely that their monthly payments will exceed $600 as it is common practice to include in the monthly home payment money for real estate taxes, homeowners insurance, and things of that nature.

## EXAMPLE 6 ◆ Investigating a Mortgage

Suppose that Lori and Jerome find a house they wish to purchase and borrow $90,000, with the terms of the loan being the same as in Example 5. Given this,

(a) What will their monthly loan payments be?

(b) How much of their first payment is interest?

(c) How much interest will be paid over the course of 30 years?

(d) What is the unpaid balance of their loan after 10 years of payments?

**Solution**

All of the questions here can be answered using tools already developed.

(a) Here we use $PV = PMT\left[\frac{1 - (1 + i)^{-n}}{i}\right]$, where $PMT$ is what we seek.

$$90000 = PMT\left[\frac{1 - \left(1 + \frac{0.0575}{12}\right)^{-360}}{\left(\frac{0.0575}{12}\right)}\right]$$

$$90000\left[\frac{\left(\frac{0.0575}{12}\right)}{1 - \left(1 + \frac{0.0575}{12}\right)^{-360}}\right] = PMT$$

$$PMT \approx \$526.94$$

Rounded to the nearest cent, the couple will have monthly loan payments of $526.94.

(b) At the time of their first loan payment, the original loan balance of $90,000 will be charged interest at the periodic (monthly) rate. Thus the first payment must account for

$$90000 \times \frac{0.0575}{12} = \$431.25$$

in interest. Notice that this means $526.94 - 431.25 = \$95.69$ of their first payment goes towards reducing the unpaid balance on the loan.

(c) Since the mortgage calls for 360 equal payments of $526.94, it is expected that Lori and Jerome will pay a total of $526.94 \times 360 = \$189{,}698.40$ in loan payments. However, the loan was only for $90,000, so this means that there will be a total of $189{,}698.40 - 90{,}000 = \$99{,}698.40$ in interest paid on the loan.

(d) We again use $PV = PMT\left[\frac{1 - (1 + i)^{-n}}{i}\right]$, where the unpaid loan balance is $PV$, but the number of payments ($n$) is the number of payments **remaining** on the loan. After 10 years of payments, Lori and Jerome have 240 payments remaining. Consequently,

$$PV = 526.94\left[\frac{1 - \left(1 + \frac{0.0575}{12}\right)^{-240}}{\left(\frac{0.0575}{12}\right)}\right]$$

$$PV \approx \$75{,}053.75$$

Rounded to the nearest cent, the unpaid balance of the home mortgage after 120 payments is $75,053.75.

---

Perhaps the results of the calculations above seem somewhat surprising. In particular, the fact that the interest charged on the loan is *more* than was originally borrowed may be disturbing. However, given the nature of the amount of money that the newlyweds needed to borrow, it is not at all realistic to expect to be able to pay off a loan of that size over a short period of time. Imagine what their monthly payments would be if they needed to pay off the same amount of money in, say, 10 years! The large amount of interest that is paid is the trade-off for not needing to come up with relatively large amounts of money to either purchase the home outright or pay off a shorter-term loan.

You may on occasion hear advertisements for "home equity loans." Typically these are installment loans, but the loans are secured by the **equity** that lies in a home. We define the equity in some object as the difference between the value of the object and how much is owed on the object, if money has been borrowed to purchase the object. So, for instance, a home equity loan is a loan based on the difference between the value of the home and the unpaid balance(s) of any loan(s) taken out on the home. In the previous example, we found that after 10 years, Lori and Jerome would have an unpaid balance of about $75,053.75 on their home. If, at that point in time, the value of their home (based on market prices) was $110,000, then they would have $110,000 - 75,053.75 = \$34,496.25$ in equity in their home.

In our final example, we construct a type of **amortization schedule** for an installment loan. This is merely a table that shows how payments are applied to an unpaid loan balance, much like the tables shown in the discussion preceding Example 5 do.

## EXAMPLE 6 ◆ Constructing an Amortization Schedule

Johnny Rotten has just borrowed $12,000 for the installation of stained glass windows on his house. The loan is to be repaid over the course of two years with equal quarterly payments. The interest rate on the loan is 14.5%. The following amortization table shows the anticipated activity for his loan over the course of two years. Rounding is to the nearest cent. You should check the calculations, in particular the quarterly payment amount.

| Payment | Beginning Loan Balance ($B$) | Periodic Interest Charged to Balance ($I$) | Payment ($PMT$) | Loan Reduction ($PMT - I$) | Ending Loan Balance ($B + I - PMT$) |
|---|---|---|---|---|---|
| 1 | 12,000 | $12000 \times \frac{0.145}{4} = 435$ | 1754.84 | 1319.84 | 10,680.16 |
| 2 | 10,680.16 | $10680.16 \times \frac{0.145}{4} \approx 387.16$ | 1754.84 | 1367.68 | 9312.48 |
| 3 | 9312.48 | $9312.48 \times \frac{0.145}{4} \approx 337.58$ | 1754.84 | 1417.26 | 7895.22 |
| 4 | 7895.22 | $7895.22 \times \frac{0.145}{4} \approx 286.20$ | 1754.84 | 1468.64 | 6426.58 |
| 5 | 6426.58 | $6426.58 \times \frac{0.145}{4} \approx 232.96$ | 1754.84 | 1521.88 | 4904.70 |
| 6 | 4904.70 | $4904.70 \times \frac{0.145}{4} \approx 177.80$ | 1754.84 | 1577.04 | 3327.66 |
| 7 | 3327.66 | $3327.66 \times \frac{0.145}{4} \approx 120.63$ | 1754.84 | 1634.21 | 1693.45 |
| 8 | 1693.45 | $1693.45 \times \frac{0.145}{4} \approx 61.39$ | 1754.84 | 1693.45 | 0 |

The fact that the final payment was exactly the same as the previous payments is merely good fortune, as this is often not the case. Over the course of a large number of payments and many instances of rounding calculations, it is not likely that the final payment of an installment loan will match the preceding payments, as the final payment needs to pay both the amount of interest due and the unpaid loan balance as of the next-to-last payment.

# EXERCISES

*For the following exercises, assume a 365-day year unless directed otherwise. Also, when calculating interest rates, give your result as a percentage, correct to one hundredth of a percent. Also, unless specified otherwise, stated interest rates are annual rates.*

1. Calculate the amount that can be borrowed under an installment loan with payments of $600 made

   (a) annually at 6.5% for 12 years
   (b) quarterly at 9.75% for 10 years
   (c) monthly at 7% for 5.5 years
   (d) weekly at 5.99% for 3 years

2. Calculate the amount that can be borrowed under an add-on loan with payments of $500 made

   (a) semiannually at 4% for 8.5 years
   (b) monthly at 8.85% for 12 years
   (c) quarterly at 6.3% for 9 years
   (d) weekly at 4% for 1 year

3. For each of the following you are given the amount of an installment loan, an annual interest rate, payment frequency, and an amount of time. Find the corresponding periodic loan payments.

   (a) $5000; 5%; monthly; 4 years
   (b) $26,000; 8.99%; quarterly; 10 years
   (c) $31,500; 6.49%; annually; 15 years
   (d) $852,456.30; 9.2%; weekly; 18 years

4. For each of the following you are given the amount of an add-on loan, an annual interest rate, an amount of time, and a payment frequency. Assuming that the loans and all interest will be paid in equal periodic payments, find the corresponding periodic loan payments.

   (a) $8800; 5.4%; 8 years; semiannual payments
   (b) $10,000; 2.8%; 4 years; monthly payments
   (c) $13,200; 3%; 6 years; quarterly payments
   (d) $6500; 18%; 1 year; weekly payments

5. For each of the following you are given the total amount to be repaid for a discounted loan, a discount rate, and an amount of time. Calculate the amount of money loaned in each case.

   (a) $7500; 3%; 5 years
   (b) $6000; 2.99%; 10 years
   (c) $20,000; 6.5%; 3 years
   (d) $500; 10%; 8 months

6. Jack loaned Jill $1000 to buy some furniture for her apartment. She repaid him with a single payment of $1100 after nine months. What annual interest rate did Jack charge on the loan?

7. In order to cover a bad night at the casino, Caroline had to borrow some money from a less-than-reputable source. She has three months to come up with a payment of $8000. Interest on her loan is compounding daily at a rate of 1%

per day for that three months (91 days). How much money did she need to borrow?

8. Some types of student education loans allow for payments to be deferred until after the borrower leaves school. However, interest is still charged on the loan. Suppose that Timothy takes out such a student loan upon entering college. The loan is for $5000 and interest on the loan is 8% compounded monthly. He graduates from college and exactly five years after taking out the loan, he needs to begin making monthly payments. The loan and all interest will be repaid over the course of 10 years. Calculate his monthly payments and how much interest is ultimately charged on the student loan.

9. A car dealer offers you a 5% add-on loan of $8000, repayable in 48 monthly payments. Suppose you also know you can obtain a 4% discounted loan with the same repayment terms.

   (a) Calculate the amount of the discounted loan so that you have $8000 available to purchase a vehicle.
   (b) Calculate the monthly payment amount for the discounted loan.
   (c) Calculate the total amount of interest that would be paid under each loan.

10. Elizabeth can amortize a $2000 debt with equal monthly payments at 0.5% **monthly** interest over the course of two years. Find the amount of her monthly payments and the total amount of interest she pays on the loan.

11. A family borrowed $72,000 to buy a house. The home loan was a standard 30-year mortgage with an annual interest rate of 7% and monthly payments. Determine the monthly mortgage payments and for each of the first three payments determine how much of each payment goes for interest and how much goes for loan balance reduction.

12. Charlotte borrows $3500 for 20 months and repays the installment loan with monthly payments of $195. Estimate the annual interest rate charged on the loan.

13. Ichabod repays an add-on loan of $4000 over the course of three years with quarterly payments of $350. Estimate the annual interest rate on the loan.

14. Which is a better 40-month loan for a consumer: a $3500 add-on loan at 6%, or a $3500 installment loan at an annual rate of 8% with monthly payments? Explain your claim.

15. Zach wants to borrow money to purchase some music equipment. He can afford to make quarterly payments of $400 for three years to finance his debt. Find the amount of money he can borrow for each of the following.

    (a) an add-on loan at 8%
    (b) a 5% discounted loan
    (c) an installment loan at 9.6%

16. A home was purchased eight years ago with a $20,000 down payment and a 30-year mortgage of $108,000 at 6.25% with monthly payments. Now, the house has a market value of $130,000. Calculate the equity in the home right now and calculate the amount of interest paid on the loan to date.

17. Casey purchased a tractor five years ago using an installment loan carrying an interest rate of 9.9% with semiannual payments. The loan was for $86,000 and was to be repaid over a 12-year span. Now, he can refinance the existing balance on the loan by taking advantage of a special offer from a rival bank that is offering an installment loan for whatever he currently owes on the tractor at an interest rate of 6.8%, repayable with monthly payments over the course of six years. If Casey decides to take this offer, how much will he ultimately save in interest charges?

18. A house is purchased with a 25-year mortgage of $84,000. Interest on the loan is 7.8% with monthly payments.

    (a) Calculate the monthly payments.

    (b) If the borrowers decided to add $100 to the payment amount, then how long would it take to repay the loan? How much would this save in interest? (Be careful with the final payment.)

19. Jason takes out a $12,000 add-on loan at 6.5% payable in equal monthly payments over three years. At the same time, Heather takes out an installment loan for $10,000 at 7.9% payable over four years with monthly payments. Determine the number of payments made by each before Jason has a lower unpaid loan balance than Heather (consider unpaid principal only).

20. The only loan you could get for your Harley was a 6% discounted loan with a value of $20,000 payable with monthly payments over the course of seven years. If you could have gotten an installment loan with identical monthly payments for the same amount of time, what interest rate would the loan have needed to carry so that you borrowed exactly the same amount for your new hog?

# Interlude: Practice Problems 2.4, Part 1

1. Mick has been paying a rent-to-buy company $76.50 each month for 20 months for his laptop computer, which was worth $1200 when he entered into this purchase agreement, and is now paid for. Assuming the company has been treating this purchase as an add-on loan, what annual interest rate has Mick been charged?

2. How much should be borrowed on a discounted loan with a discount rate of 7.5% for 16 months so that the borrower obtains $4000?

3. Martin just made his last monthly payment on a car loan he took out six years ago. The loan was for $38,000 and the annual interest rate on this installment loan was 6.6%

   (a) What were his monthly payments?

   (b) How much interest was due on his first payment?

   (c) How much total interest did he pay?

   (d) How much of the loan had he repaid after four years of payments?

4. Fred just took out an installment loan to purchase a used car. He will repay the loan over the course of the next 38 months. If his monthly payments are $298 and the interest rate on the loan is 5.25%, then how much was this loan?

5. An installment loan for $6200 with a term of 60 months has monthly payments of $157. What is the annual interest rate on that loan? Give your result as a percentage, accurate to two decimal places.

6. Suppose that you need to borrow $1000 and have available to you either an add-on loan or a discounted loan, both for the same period of time and at the same interest rate. Which will have the lower payment, if each is to be repaid with equal periodic payments?

# Interlude: Practice Problems 2.4, Part 2

Consider an installment loan of $18,000 carrying an annual interest rate of 6% to be repaid with quarterly payments. The loan is to be amortized over a period of two years. Fill out the following amortization schedule for the loan.

| Payment # | Payment Amount | Interest Due | Loan Reduction | Ending Loan Balance |
|-----------|----------------|--------------|----------------|---------------------|
| 1 | | | | |
| 2 | | | | |
| 3 | | | | |
| 4 | | | | |
| 5 | | | | |
| 6 | | | | |
| 7 | | | | |
| 8 | | | | |

What was the total amount of money paid on this loan?

How much interest was paid on this loan?

Return to the loan of $18,000. The interest rate and proposed term remain the same as before, but suppose the borrower wishes to make quarterly payments of $3000. Fill out a new amortization schedule for the loan. Keep in mind that the final payment must pay all interest due and the remaining loan balance.

| Payment # | Payment Amount | Interest Due | Loan Reduction | Loan Balance |
|---|---|---|---|---|
|  |  |  |  |  |
|  |  |  |  |  |
|  |  |  |  |  |
|  |  |  |  |  |
|  |  |  |  |  |
|  |  |  |  |  |
|  |  |  |  |  |
|  |  |  |  |  |

How much interest was paid on this loan?

Calculate the savings in interest the borrower realizes by making $3000 payments instead of the payments indicated by the terms of the loan.

# Interlude: Practice Problems 2.4, Part 3

1. An *adjustable rate mortgage* (ARM) is a type of installment loan that allows for changing interest rates over the repayment term. Suppose that you take out an ARM for $60,000 and a repayment term of 15 years with monthly payments. For the first three years, the interest rate is 3.5%. That rate goes up to 4.75% for the next five years, and then to 8.25% for the final seven years. Assume that each time the interest rate changes, a new monthly payment is calculated based on the current interest rate and the remaining term of the loan.

   (a) Calculate the monthly payments for the first three years.

   (b) Calculate the monthly payments for the second five years.

   (c) Calculate the monthly payments for the final seven years.

   (d) How much interest is paid over the entire 15 year repayment period?

2. Suppose that you take out a home improvement loan for $20,000 and the lender charges you a processing fee of $500. The loan is an installment loan with an interest rate of 8% and a repayment term of six years, with monthly payments. The processing fee can be repaid, interest-free, with monthly payments over the term of the loan. If you consider the processing fee as additional interest, what actual interest rate are you being charged on the loan? Give your percentage accurate to two decimal places.

3. Some types of student loans allow the option of making interest payments while you are still in school (without actually repaying the principal borrowed). Suppose you have one such loan for $3500 that carries an interest rate of 6.8%. You elect to make quarterly interest-only payments to the lender for the five years you take to complete college, and then the loan becomes a standard installment loan requiring monthly payments over the course of 10 years.

   (a) How much interest will you pay while in school?

   (b) By the time you have repaid the loan, how much total interest will you have paid?

# Chapter 2  Progress Check I

1. A municipal bond with a face value of $10,000 matured over a period of eight years. Including all interest payments made over that period of time, the purchaser of the bond was paid a sum of money amounting to $13,424. What annual interest rate was the bond carrying?

2. Margot recently received a lump sum payment of $200,000 from winning a lottery. She wishes to deposit all of her winnings in an investment account, leaving it to earn interest for 15 years, at which time she hopes to begin making monthly withdrawals of $8000 for the ensuing 10 years, at which time the account will be empty. Given this, what annual interest rate, compounded monthly, does the investment account need to pay? Give your result as a percentage, accurate to two decimal places.

3. A single deposit of $6500 that earned interest compounded quarterly doubled in value over the course of 14 years. Determine the annual interest rate that was earned, accurate to two decimal places.

4. Paul and Janet are arguing the merits of two different schemes for putting money into a retirement account. The retirement account earns 6.9% interest compounded monthly, and it is assumed that in 30 years the money in the account will be available.

   *Paul's Plan:* Make monthly deposits of $200 for the first five years, make monthly deposits of $150 for the following five years, and then make monthly deposits of $100 for the remaining 20 years.

   *Janet's Plan:* Make monthly deposits of $125 for 30 years.

   Which plan do you recommend, and why?

# Chapter 2 Progress Check II

1. Suppose that you borrow $175,000 to purchase a home. The loan is a standard mortgage that will be repaid over the course of 30 years with equal monthly payments. The annual interest rate on the loan is 4.8%, giving monthly payments of $918.16.

   (a) How much interest is due on the first payment? How much interest is due on the second payment?

   (b) What will the unpaid loan balance be after 20 years of payments?

   (c) If you can afford monthly payments of $1200, then how much could you have borrowed under the same loan terms (same interest rate and repayment period)?

   (d) Suppose that you can only afford monthly payments of $850 on the mortgage. What annual interest rate would the loan have to carry (with the same repayment term) so that you could afford the payments? Give the percentage accurate to two decimal places.

2. Montrose, Inc. is starting a savings fund that will be used in 10 years to replace aging equipment. The goal is to have $50,000 saved up, and to do that the company will deposit money semiannually into an account that earns 6% compounded semiannually.

   (a) How much does the company need to deposit semiannually?

   (b) Suppose that after six years, the savings fund interest rate increases to 7% compounded semi-annually. If the company wishes to adjust its semiannual deposit amount so that the fund still has $50,000 at the end of the 10 year period, what should the new deposit amount be?

3. The annual interest rate on my savings account is 0.25% compounded quarterly. If I could get the same annual interest rate compounded daily, then how much would I need to have in my account so that I would earn $0.01 more in interest over the course of one year with daily compounding compared to quarterly compounding?

# Chapter 2 Progress Check III

1. Crystal recently purchased a condominium for $229,000. She obtained an installment loan with a term of 26 years carrying an annual interest rate of 4% to make the purchase.

    (a) Determine the monthly payments on her loan.

    (b) Suppose that Crystal wants to add $100 to each of her loan payments. If she does this, starting with her first payment, then how long will it take her to repay the loan?

    (c) Suppose that after making monthly payments for 10 years, Crystal refinances the unpaid loan balance at an annual interest rate of 3.25%, to be repaid with equal monthly payments over the course of another 15 years. How much will she save in interest by doing this compared to sticking with the original 26-year loan?

2. A credit card carries an annual interest rate of 18%, but compounds interest daily on any unpaid balance. Suppose that right now you are carrying a card balance of $4000. You can afford to make $500 payments every 30 days, with your first payment made 30 days from now. Assume that you will not be using the credit card to make any purchases while you carry a balance.

   (a) What will your statement balance be when you make your first payment?

   (b) What will your statement balance be when you make your second payment?

   (c) How much is your unpaid statement balance **after** you have made your fourth payment?

# Chapter 2 Progress Check IV

1. Fast Freddy's Financial is known for offering somewhat nonstandard investment options for customers. For instance, a certain savings account option available offers an annual interest rate of 4.2109% with interest compounded "frequently" (according to advertisements) and boasts an APY of 4.2999%. Given this, how many compounding periods per year does this account offer?

2. Another interesting product from Fast Freddy's Financial is a type of installment loan with changing payments. Suppose you are offered such a loan, which carries an annual interest rate of 8% and a total term of four years, with the proposal that for the second half of the repayment term, the periodic payments are twice as much as the periodic payments the first half of the payment term. You decide that you can afford to make semiannual payments of $1000 for the first four years and then semiannual payments of $2000 for the remaining four years. Given this, how much money can you borrow right now?

3.  Some lenders offer an "interest only" repayment plan, where the borrower makes payments that only pay the periodic interest charges over part of the loan term. Suppose you agree to such a plan on a 20-year mortgage of $80,000 with an interest rate of 6%. For the first five years of the repayment term, you will make monthly payments that only pay the monthly interest charges. After that, you will make monthly payments for the remaining 15 years sufficient to pay off the entirety of the mortgage.

    (a)  What are your scheduled monthly payments for the first five years?

    (b)  How much will you pay in interest charges over the 20 year term?

    (c)  How much more do you wind up paying in total interest charges by going with this "interest only" plan as compared to just paying the $80,000 loan off with equal monthly payments over the course of 20 years?

    (d)  Suppose that you want to make payments that are $50 more than what you calculated in (a) for the first five years. After that, your monthly payment for the remaining 15 years will be set based on the unpaid loan balance. How much do you save in interest charges by doing this rather than sticking with the "interest only" payments for the first 5 years?

# CHAPTER 2 ANSWERS TO ODD-NUMBERED EXERCISES

## Section 2.1

**1.** **(a)** $56   **(b)** $63   **(c)** $72.61
   **(d)** $170.10   **(e)** $441   **(f)** $4.79

**3.** 12%

**5.** 12.5 yrs

**7.** **(a)** A = $9528.13, Interest: $1528.13
   **(b)** A = $9403.46, Interest: $1403.46
   **(c)** A = $8979.42, Interest: $979.42
   **(d)** A = $9823.78, Interest: $1823.78
   **(e)** A = $9081.77, Interest: $1081.77
   **(f)** A = $10,162.90, Interest: $2162.90

**9.** 2.97%

**11.** **(a)** 5%   **(b)** 5.06%   **(c)** 5.09%
   **(d)** 5.12%   **(e)** 5.13%   **(f)** 5.12%

**13.** Weekly rate of 0.2865%

**15.** 5.39%

**17.** 12.91%

**19** 10,502 days

**21.** $0.010625

**23.** **(a)** 2.75%
   **(b)** $25,500
   **(c)** 3.29% and 3.32%

**25.** **(a)** 5.88%
   **(b)** 5.73%
   **(c)** 5.72%
   **(d)** 5.72%

## Section 2.2

**1.** **(a)** $16,911.28   **(b)** $123,418.03
   **(c)** $1,595,050.07   **(d)** $2,421,100.07

**3.** **(a)** $163.32   **(b)** $1532.02
   **(c)** $844.19   **(d)** $1155.64

**5.** $4486.23

**7.** $169,705.02

**9.** $314,919.03

**11.** $4714.07

**13.** $37,208.14

**15.** **(a)** Emily: $113,140.54, Interest: $93,140.54,
   Nick: $109,729.02, Interest: $59,729.02

## Section 2.3

**1.** **(a)** $5276.44   **(b)** $37,542.11
   **(c)** $68,475.51   **(d)** $241,374.98

**3.** **(a)** $1519.19   **(b)** $1128.20
   **(c)** $2143.32   **(d)** $1999.15

**5.** 4.74%

**7.** *PMT* = $34.13, Total amount: $2047.80

**9.** $2710.97

**11.** **(a)** $16,960.79   **(b)** $358.84
   **(c)** $1,736,621.14   **(d)** $338,324.39

## Section 2.4

**1.** **(a)** $4895.24   **(b)** $15,221.44
   **(c)** $32,789.60   **(d)** $85,626.94

**3.** **(a)** $115.15   **(b)** $992.19
   **(c)** $3347.96   **(d)** $1864.68

**5.** **(a)** $6375   **(b)** $4206
   **(c)** $16,100   **(d)** $466.67

**7.** $3234.78

**9.** **(a)** $9523.81   **(b)** $198.41
   **(c)** Discounted loan interest: $1523.81, Add-on interest:
   $1600

**11.** *PMT* = $479.02, 1st payment: $420 interest, $59.02 bal.
   red.; 2nd payment: $419.66 interest, $59.36 bal. red.; 3rd
   payment: $419.31 interest, $58.71 bal. red.

**13.** 1.67%

**15.** **(a)** $3870.97   **(b)** $4080   **(c)** $4128.06

**17.** Yes, $11,652.28

**19.** 14 payments

# Linear Systems and Matrices

In the previous chapter, we discussed modeling a situation involving two related quantities where a function reasonably represented the situation. We now turn to other situations that are too complex to be modeled by a single function. In particular, we investigate situations where a collection of linear structures are used to model the relationships between some number of quantities.

## 3.1  SYSTEMS OF LINEAR EQUATIONS: A FIRST LOOK

Consider the following problem.

> A student organization held a fundraiser dinner one evening. Adult dinners cost $6 each, while children's dinners each cost $3. The group raised a total of $771 serving 189 dinners. How many adult dinners were served, and how many children's dinners were served?

We see here that there are two quantities that need to be identified. We will denote these quantities by $a$ and $c$, where

$$a = \text{the number of adult dinners served}$$

$$c = \text{the number of children's dinners served}$$

We also have two explicit relationships between $a$ and $c$ directly stated. First, we know that the total number of dinners served is 189, so

$$a + c = 189$$

In addition, we know that the total money raised from serving the 189 dinners is $771. Since adult dinners cost $6 and children's dinners cost $3, we have

$$6a + 3c = 771$$

This gives us a pair of linear equations, which we now write together.

$$a + c = 189$$

$$6a + 3c = 771$$

We say here that we have a **system of linear equations.** Specifically, we have a system of two linear equations in two unknowns. Solving the problem that has been posed requires finding appropriate values of $a$ and $c$ that simultaneously satisfy *both* equations. Any pair $(a, c)$ that does this is called a **solution to the system.** Notice that the pair $(63, 126)$ is *not* a solution to the system since when $a = 63$ and $c = 126$ we see that

$$63 + 126 = 189$$

but

$$6(63) + 3(126) = 756 \neq 771$$

That pair satisfies the first equation in the system but not the second.

We will set the problem at hand aside and begin looking at common methods for finding solutions to systems of linear equations in two unknowns. Later, we will investigate methods that can be applied to systems of linear equations in any number of unknowns.

## Graphical Solutions

The first example we consider involves a simple system and illustrates a geometric method for finding solutions to a system of linear equations in two unknowns.

**EXAMPLE 1 ◆ Solving a System Graphically**

Solve the system

$$3x - y = 11$$
$$x + 2y = -8$$

**Solution**

The graph of each equation is a line. Points on a line represent pairs of values $(x, y)$ that satisfy the equation for the line. So, when solving a system of equations graphically, we seek all points that the graphs of the equations have in common.

It is convenient to rewrite each of the equations in the system and use a graphing calculator to investigate their graphs. Working with the first equation, solving for $y$, we have

$$3x - y = 11$$
$$3x - 11 = y$$

Similarly, we rewrite the second equation.

$$x + 2y = -8$$
$$y = (-8 - x)/2$$

We simultaneously graph the equations

$$y_1 = 3x - 11$$
$$y_2 = (-8 - x)/2$$

on a graphing utility to obtain a graph as shown in Figure 3.1:

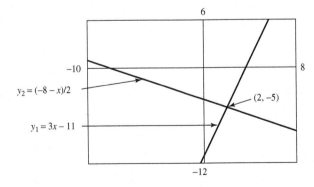

*FIGURE 3.1*

Using an intersection routine on the calculator, we find that the lines intersect at the point $(2, -5)$. Thus the pair $x = 2$ and $y = -5$ satisfies both of the equations in the system (check this) and hence forms a solution to the system.

---

Take note that in the previous example, the pair of numbers $x = 2$ and $y = -5$ forms *one* solution, not two solutions.

Even when using a graphing calculator, the graphical method of finding solutions to a system of linear equations in two variables sometimes results in **approximate solutions** rather than **exact solutions.** We can use algebraic methods to obtain exact solutions.

## The Substitution Method

The basic approach to the method of substitution is easily summarized:

1. Solve for a variable in one of the equations in a system of equations.
2. Substitute for that variable in the other equation(s).

We illustrate this method in the following example.

## EXAMPLE 2 ◆ Solving a System by Substitution

Solve the system

$$3x - y = 11$$
$$x + 2y = -8$$

**Solution**

We may choose either variable to solve for and use either equation. In this case it is easy to solve for $x$ in the second equation.

$$x = -8 - 2y$$

We substitute this expression for $x$ in the first equation, $3x - y = 11$, as follows.

$$3(-8 - 2y) - y = 11$$
$$-24 - 6y - y = 11$$
$$-24 - 7y = 11$$
$$-7y = 35$$
$$y = -5$$

We now **back-substitute** to determine a value for $x$. Since $x = -8 - 2y$,

$$x = -8 - 2(-5)$$
$$x = -8 + 10$$
$$x = 2$$

We could have substituted $y = -5$ in either of the original equations. In any event, we find that the pair $x = 2$ and $y = -5$ is a solution to the given system, as we found in Example 1.

---

The next algebraic method we look at is particularly important because later we will expand upon it to develop a relatively efficient approach to finding solutions to systems of linear equations in more than two variables.

## The Method of Elimination

The method of elimination is a systematic approach to finding solutions to a system of linear equations by virtue of transforming the given system into a "simpler" but "equivalent" system. When we say that two systems of linear equations are **equivalent,** we mean that the systems have exactly the same solution(s). The objective in the method of elimination is to modify one (or more) of the equations in a given system to produce an equivalent system where one (or more) of the equations contains just one unknown, making it relatively easy to solve for that unknown.

There are only three legitimate operations that can be used to modify a system of linear equations to produce an equivalent system. However, these operations can be performed in any sequence for as many steps as one wishes. These operations are:

---

### Operations on Equations

- Interchange any two equations.

- Multiply (all terms of) an equation by any nonzero constant and replace that equation with the result.

- Multiply (all terms of) an equation by any constant and add (corresponding terms) to any other equation, replacing the second equation in this sum with the result.

Performing any sequence of these operations results in an equivalent system of equations.

---

We will use a type of shorthand notation for these operations. For a given system of linear equations, we let:

---

### Notation for Operations on Equations

- $E_k \leftrightarrow E_j$ denote the operation "interchange equations $k$ and $j$"

- $cE_k \rightarrow E_k$ denote the operation "multiply equation $k$ by $c$ and let the result replace equation $k$"

- $cE_k + E_j \rightarrow E_j$ denote the operation "multiply equation $k$ by $c$, add to equation $j$ and let the result replace equation $j$"

---

We illustrate the method of elimination in the next two examples.

---

**EXAMPLE 3  ◆  Solving a System by Elimination**

Solve the system

$$3x - y = 11$$
$$x + 2y = -8$$

**Solution**

There are several reasonable ways to begin here. We start by rearranging the order in which the system is written; that is, we perform the operation $E_1 \leftrightarrow E_2$ to produce the equivalent system

$$x + 2y = -8$$
$$3x - y = 11$$

Since the coefficient of the $x$ variable in the first equation is 1, we will "eliminate" the $x$ variable from the second equation by performing the operation $-3E_1 + E_2 \to E_2$ on the most recent system.

$$-3x - 6y = 24 \quad (-3E_1)$$

$$\underline{\phantom{-3x} 3x - y = 11}$$

$$-7y = 35 \quad \text{(replaces second equation)}$$

This yields the equivalent system

$$x + 2y = -8$$

$$-7y = 35$$

We now perform the operation $\frac{-1}{7} E_2 \to E_2$ on this system to obtain

$$x + 2y = -8$$

$$y = -5$$

We immediately see that $y = -5$, and back-substitution into $x + 2y = -8$ gives

$$x + 2(-5) = -8$$

$$x = 2$$

So, the pair $x = 2$ and $y = -5$ is the solution to the original system, as expected.

---

## EXAMPLE 4 ◆ Solving a System by Elimination

Solve the system

$$3x + 4y = 9$$

$$2x - 5y = 15$$

**Solution**

Although it is apparent that the arithmetic involved will be somewhat more difficult here compared to the last example, we proceed in a similar manner. We will eliminate the $x$ variable from the second equation via the operation $\frac{-2}{3} E_1 + E_2 \to E_2$, so that

$$-2x - \frac{8}{3}y = -6 \quad \left(\frac{-2}{3}E_1\right)$$

$$\underline{\phantom{-2x} 2x - 5y = 15}$$

$$-\frac{23}{3}y = 9 \quad \text{(replaces second equation)}$$

This gives the equivalent system

$$3x + 4y = 9$$

$$\frac{-23}{3}y = 9$$

We modify the second equation in this system by the operation $\frac{-3}{23} E_2 \to E_2$, which yields the equivalent system

$$3x + 4y = 9$$

$$y = \frac{-27}{23}$$

Since $y = \frac{-27}{23}$, we immediately back-substitute into $3x + 4y = 9$ to obtain

$$3x + 4\left(\frac{-27}{23}\right) = 9$$

$$3x - \frac{108}{23} = 9$$

$$3x = \frac{315}{23}$$

$$x = \frac{105}{23}$$

Thus the pair $x = \frac{105}{23}$ and $y = \frac{-27}{23}$ is the solution to the original system.

---

Example 4 helps illustrate the difference between exact solutions and approximate solutions. Had we opted to look at the given system graphically, even using an intersection routine on a graphing calculator would have only given a decimal approximation to the actual solution. In practice, a *good* approximation is often as useful as an exact answer, but it is useful at times to retain the option of finding exact solutions.

Thus far, we have seen systems of two linear equations in two unknowns where there was exactly one solution to the system. It is natural to wonder if there are other possible outcomes when we attempt to solve systems of equations. This is where we now turn our attention.

## When Strange Things Happen

Consider the system

$$x + y = 2$$

$$3x + 3y = 9$$

We may elect to attempt to solve this system using the method of substitution, as it is easy to solve the first equation for $x$:

$$x = 2 - y$$

Substituting into the second equation finds

$$3(2 - y) + 3y = 9$$

$$6 - 3y + 3y = 9$$

$$6 = 9$$

The equation $6 = 9$ is certainly untrue (a **contradiction**). How should we interpret this? Recalling that solving a system of linear equations in two unknowns graphically boils down to finding all points where the graphs of the equations intersect, we observe that the nature of this situation is somewhat different than what was encountered earlier. If we rewrite the given linear equations in slope-intercept form, then the first equation becomes

$$y = -x + 2$$

while the second equation becomes

$$3y = -3x + 9$$

$$y = -x + 3$$

Graphically, this means that we have two parallel lines, as illustrated in Figure 3.2.

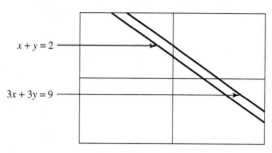

*FIGURE 3.2*

Since the lines corresponding to the equations in the system have no point of intersection, there is no pair $(x, y)$ that simultaneously satisfies each equation—the system of equations does not have a solution. Any system of equations having no solution is called an **inconsistent system.**

It is relatively easy to identify an inconsistent system of linear equations, no matter how many unknowns, since when using algebraic methods to solve such a system, eventually a contradictory equation will appear, just as happened above. The next situation we consider, however, is much more subtle in nature.

Consider the system

$$2x - y = 4$$
$$-4x + 2y = -8$$

We will attempt to find a solution for this system using the method of elimination. In this case, the operation $2E_1 + E_2 \rightarrow E_2$ seems reasonable, as this will eliminate x from the second equation.

$$4x - 2y = 8 \qquad (2E_1)$$
$$\underline{-4x + 2y = -8}$$
$$0 = 0 \qquad \text{(replaces second equation)}$$

This gives the equivalent system

$$2x - y = 4$$
$$0 = 0$$

The second equation in this system ($0 = 0$) is most certainly a true statement. However, it is not at all enlightening, nor does it appear to be helpful, as we are not able to solve for a single unknown. We again look at the original system graphically in an attempt to ascertain what is happening. Notice that when we rewrite each equation in slope-intercept form, we have

$$2x - y = 4$$
$$y = 2x - 4$$

and

$$-4x + 2y = -8$$
$$2y = 4x - 8$$
$$y = 2x - 4$$

The two equations in the original system have exactly the same slope-intercept form, so the graphs of each equation are the same line. As such, any point $(x, y)$ on that line represents a solution to the system. Since there are an infinite number of points on a line, this means that the given system has an infinite number of solutions! See Figure 3.3.

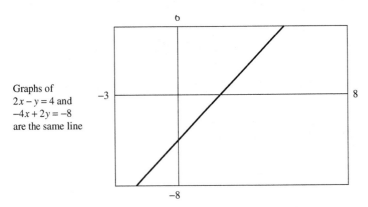

Graphs of
$2x - y = 4$ and
$-4x + 2y = -8$
are the same line

*FIGURE 3.3*

Thus far we have encountered three scenarios in terms of the number of solutions to a system of linear equations—no solutions, exactly one solution, or an infinite number of solutions. In the case of a system of equations in two unknowns, it seems clear that these are the only possibilities (a pair of lines cannot have exactly two points of intersection, for example). As it turns out, *this is the case for systems of linear equations in any number of variables.*

We conclude this section with an application involving a system of linear equations where a noteworthy occurrence within the solving process will be investigated.

## EXAMPLE 5 ◆ Application

Cybil has a combination of nickels and dimes, 51 in all, in her pocket. The number of dimes is twice the number of nickels, and the total value of the change is $4.25. How many of each kind of coin does she have in her pocket?

**Solution**

We begin by declaring our unknowns. Let

$$d = \text{the number of dimes Cybil has in her pocket}$$

$$n = \text{the number of nickels Cybil has in her pocket}$$

We have several relationships between $d$ and $n$. We know that Cybil has twice as many dimes as nickels, so

$$d = 2n$$

We also know that she has 51 coins in all, so

$$d + n = 51$$

Finally, the total value of the collection of coins is $4.25, so

$$0.10d + 0.05n = 4.25$$

Rewriting $d = 2n$ as $d - 2n = 0$, we have modeled the problem with a system of linear equations:

$$d - 2n = 0$$

$$d + n = 51$$

$$0.10d + 0.05n = 4.25$$

We can take any approach to solving this system, but for the purpose of illustration we opt to use the method of elimination. We will eliminate $d$ from the second and third equations by performing the operations $-1E_1 + E_2 \rightarrow E_2$ and $-0.10E_1 + E_3 \rightarrow E_3$. We see that

$$-d + 2n = 0 \qquad (-1E_1)$$

$$\underline{d + n = 51}$$

$$3n = 51 \qquad \text{(replaces second equation)}$$

and

$$-0.10d + 0.2n = 0 \qquad (-0.10E_1)$$

$$\underline{0.10d + 0.05n = 4.25}$$

$$0.25n = 4.25 \qquad \text{(replaces third equation)}$$

This gives us the equivalent system

$$d - 2n = 0$$

$$3n = 51$$

$$0.25n = 4.25$$

Although this may seem superfluous, we now perform the operation $\frac{1}{3} E_2 \rightarrow E_2$ to obtain the system

$$d - 2n = 0$$

$$n = 17$$

$$0.25n = 4.25$$

And finally, we perform the operation $-0.25E_2 + E_3 \rightarrow E_3$

$$-0.25n = -4.25 \qquad (-0.25E_2)$$

$$\underline{0.25n = 4.25}$$

$$0 = 0 \qquad \text{(replaces third equation)}$$

This results in the equivalent system

$$d - 2n = 0$$

$$n = 17$$

$$0 = 0$$

The third equation in this system ($0 = 0$) reminds us of the situation encountered earlier where a system had an infinite number of solutions. It may be tempting here to jump to that conclusion, but common sense suggests that that is not a reasonable outcome. After all, are there really an infinite number of ways to have a collection of a finite number of two different types of coins? Notice, though, that the second equation already gives us a value for $n$, and substituting that value of $n$ into the first equation gives

$$d - 2(17) = 0$$

$$d - 34 = 0$$

$$d = 34$$

Thus the pair $d = 34$ and $n = 17$ satisfies all three of the equations in the original system (check), and we conclude that Cybil has 34 dimes and 17 nickels in her pocket.

## EXERCISES

1. For each of the following systems, determine if the pair $x = 3, y = -2$ is a solution.

   (a) $2x - y = 8$
   $x + y = 1$

   (b) $5x + 27y = 10$
   $-3x + 27y = 7$

   (c) $x - 4y = 11$
   $4x - 3y = -12$

   (d) $x - y = 5$
   $3x + 4y = 1$

2. Suppose that $(-3, 1)$ and $(2, 4)$ are both solutions to a system of linear equations in two unknowns. What can you conclude about the number of solutions to that system?

*In problems 3–8, solve the system **graphically**.*

3. $3x + 2y = 3$
   $x - 3y = -10$

4. $2y - 10x = -14$
   $3y - 15x = 3$

5. $10s - 5t = 25$
   $-8s + 4t = -20$

6. $7v - 4w = -1$
   $14v + 8w = 10$

7. $4x_1 - 3x_2 = -4$
   $2x_1 + x_2 = 3$

8. $x - 3y = 18$
   $4x + 2y = 5$

*In problems 9–12, solve the system by **substitution**. Give exact answers.*

9. $-5x + 2y = 10$
   $3x - y = 4$

10. $3k - 5l = -27$
    $k + 2l = 13$

11. $4m - 3n = 7$
    $2m + 5n = 14$

12. $-3x + 2y = 6$
    $-x - y = 10$

*In problems 13–18, solve the system by **elimination**. Give exact answers.*

13. $3x - 2y = 12$
    $4x + 2y = 2$

14. $3x - 4y = -10$
    $5x + y = 14$

15. $-\frac{1}{2}s + t = 5$
    $2s - 4t = -20$

16. $\frac{2}{3}x - 4y = 6$
    $2x + \frac{1}{3}y = 4$

17. $-7x_1 + 5x_2 = 16$
    $2x_1 + 3x_2 = 10$

18. $-6x + 2y = 10$
    $9x - 3y = 12$

19. The Completely Cookoo Chocolate Company makes two varieties of chocolate milk. Regular chocolate milk uses 1 gallon of syrup to 12 gallons of milk. Extra-thick chocolate milk uses 2 gallons of syrup to 19 gallons of milk. The company has 260 gallons of syrup and 2720 gallons of milk. Determine the number of gallons of each type of chocolate milk the company should produce in order to use up all of the syrup and all of the milk.

20. Randolph has invested $1000 in two different certificates of deposit (CDs). One gives a return of 6% and the other gives a return of 7%. The total return on his investments is $64. How much did he invest in each CD?

21. A year ago, Kris had twice as much money in his savings account as Shelley. Kris deposited an additional $100 in his account this last year. However, Shelley deposited $200 in her account in that same period of time, and now she has $50 more saved than Kris. How much does each of them have in their savings accounts right now?

22. A batch of Sinfully Silky gourmet ice cream calls for 200 gallons of a milk-cream mixture. Milk costs the producer $1.50 per gallon and cream costs $2.75 per gallon. Profit studies suggest that the company should spend only $450 for the milk-cream mixture for a batch of ice cream. How many gallons each of milk and cream should be mixed together to make a mixture that costs $450?

23. At a recent rodeo event, a total of 4200 adult and children's tickets were sold. The gate revenue was $39,900. Adult tickets sell for $10.50 apiece and children's tickets sell for $7.50 apiece. Determine how many adult tickets were sold and how many children's tickets were sold.

# Interlude: Practice Problems 3.1, Part 1

1. For an experiment a chemist needs 500 liters of a 25% acid solution. On hand are solutions that are 18% and 30% acid, respectively. You need to determine how many liters of each of the available solutions should be mixed together to achieve the desired result.

   (a) Set up a system of linear equations that can be used to solve the problem. Be sure to indicate what your variables represent.

   (b) Solve the system of equations from (a) graphically. Show the graphs you used, label the point of intersection appropriately, and indicate how much of each solution is used.

2.  Consider the system $-5x + 2y = 3$
    $$20x + ky = -27.$$

    (a)  Determine all possible values of $k$ so that the system has exactly one solution. Show how you arrived at your conclusion.

    (b)  Determine all possible values of $k$ so that the system has an infinite number of solutions. Show how you arrived at your conclusion.

    (c)  Determine all possible values of $k$ so that the system has no solutions. Show how you arrived at your conclusion.

# Interlude: Practice Problems 3.1, Part 2

1. A jar contains a combination of 60 nickels and dimes worth a total of $4.30. Suppose that it is deemed desirable to determine how many of each kind of coin are in the jar.

   (a) Set up a system of linear equations that can be used to solve the problem. Be sure to indicate what your variables represent.

   (b) Solve the system of equations from (a) graphically. Show the graphs you used, label the point of intersection appropriately, and indicate how many of each coin are in the jar.

   (c) Solve the system of equations from (a) algebraically. Show your method, and indicate how many of each coin are in the jar.

2. A bake shop makes two different sizes of blueberry muffins using prepackaged dough and blueberries. A large muffin requires 5 ounces of dough and 2 ounces of blueberries. A small muffin requires 2 ounces of dough and 1 ounce of blueberries. Each day, the shop takes delivery of 450 ounces of dough and 200 ounces of blueberries. The bake shop would like to know how many of each size muffin to bake each day to use up all of the dough and blueberries.

(a) Set up a system of linear equations that can be used to solve the problem. Be sure to indicate what your variables represent.

(b) Solve the system of equations from (a) graphically. Show the graphs you used, label the point of intersection appropriately, and indicate how many of each size of muffin to bake.

(c) Solve the system of equations from (a) algebraically. Show your method, and indicate how many of each size of muffin to bake.

# 3.2 LARGER SYSTEMS: MATRIX REPRESENTATION AND GAUSS-JORDAN ELIMINATION

Many situations can be modeled quite naturally by a system of linear equations involving more than one unknown. Consider the following problem.

> A testing lab in a pharmaceutical company has a group of young mice, adult female mice, and adult male mice. There are 33 mice in all. On average, the mice consume 194 mg of food and 53.7 ml of water daily. Typically, young mice will consume 2 mg of food and 1.0 ml of water in a day. Similarly, adult female mice and adult male mice will consume 6 mg of food and 1.75 ml of water, and 8 mg of food and 1.85 ml of water, respectively. Given this, how many young mice, adult female mice, and adult male mice are there?

We see that the problem that has been posed has three unknown but sought quantities. Denote these quantities by $x$, $y$, and $z$, where

$x =$ the number of young mice in the group

$y =$ the number of adult female mice in the group

$z =$ the number of adult male mice in the group

We have several pieces of information about these quantities, which we summarize in a table.

|            | Young | Female | Male | Totals |
|------------|-------|--------|------|--------|
| Number     | $x$   | $y$    | $z$  | 33     |
| Food (mg)  | 2     | 6      | 8    | 194    |
| Water (ml) | 1.0   | 1.75   | 1.85 | 53.7   |

This information readily leads to equations relating $x$, $y$, and $z$:

$$x + y + z = 33 \qquad \text{(total number of mice)}$$
$$2x + 6y + 8z = 194 \qquad \text{(total mg of food consumed)}$$
$$1.0x + 1.75y + 1.85z = 53.7 \qquad \text{(total ml of water consumed)}$$

We now turn to the task of finding a solution to a system such as this, where there are more than two unknowns. In general, with systems of this nature, a graphical approach is not feasible, so we will look at algebraic methods.

## The Method of Elimination, Revisited

Although it looks to be a labor-intensive process, we can certainly use the method of elimination to attempt to find a solution to the system of equations we just generated. To start, we will eliminate $x$ from the second and third equations. Since the coefficient of $x$ is 1 in the first equation, it is natural to use the first equation in the subsequent eliminations. We perform the operations $-2E_1 + E_2 \rightarrow E_2$ and $-1.0E_1 + E_3 \rightarrow E_3$.

$$-2x - 2y - 2z = -66 \qquad (-2E_1)$$
$$\underline{2x + 6y + 8z = 194}$$
$$4y + 6z = 128 \qquad \text{(replaces second equation)}$$

$$-1.0x - 1.0y - 1.0z = -33 \qquad (-1.0E_1)$$
$$\underline{1.0x + 1.75y + 1.85z = 53.7}$$
$$0.75y + 0.85z = 20.7 \qquad \text{(replaces third equation)}$$

This results in the equivalent system

$$x + y + z = 33$$
$$4y + 6z = 128$$
$$0.75y + 0.85z = 20.7$$

Notice that the last two equations only involve the variables $y$ and $z$. We focus on those two equations. We first perform the operation $\frac{1}{4} E_2 \to E_2$ to obtain

$$x + y + z = 33$$
$$y + 1.5z = 32$$
$$0.75y + 0.85z = 20.7$$

and then use the new second equation in that system to eliminate $y$ from the third equation via the operation $-0.75E_2 + E_3 \to E_3$.

$$-0.75y - 1.125z = -24 \qquad (-0.75E_2)$$
$$\underline{0.75y + 0.85z = 20.7}$$
$$-0.275z = -3.3 \qquad \text{(replaces third equation)}$$

This gives the equivalent system

$$x + y + z = 33$$
$$y + 1.5z = 32$$
$$-0.275z = -3.3$$

The third equation in this system is readily solved for $z$: $z = 12$. Back-substitution into the equation $y + 1.5z = 32$ yields

$$y + 1.5(12) = 32$$
$$y + 18 = 32$$
$$y = 14$$

Finally, back-substituting the pair $y = 14$ and $z = 12$ into $x + y + z = 33$ gives

$$x + 14 + 12 = 33$$
$$x = 7$$

So, the triple $x = 7$, $y = 14$, and $z = 12$ is a solution to the original system (check), indicating that the group of lab mice consisted of 7 young mice, 14 adult female mice, and 12 adult male mice.

While the method of elimination can be used on systems of linear equations in any number of variables, it can be rather tedious. Notice, though, that when using this method, the coefficients of the variables in the equations determined the choice of multipliers chosen for the subsequent operations. We now turn our attention to a method of solving a system of linear equations that is equivalent to the method of elimination, but uses an object that will allow us to work only with coefficients and constant terms in the system and not the variables.

## Matrices

A **matrix** is a rectangular array of numbers arranged in rows and columns. Each number in this array is called an **element** of the matrix. When we write a matrix, we typically enclose the array in brackets.

Matrices (plural) come in many sizes, as determined by the number of rows and the number of columns. If a matrix has $m$ rows and $n$ columns, then we say that the **size of the matrix** is $m \times n$, read "$m$ by $n$." The following are examples of matrices of various sizes.

$$\begin{bmatrix} 1 & 5 & -6 \\ 2 & 3 & 0 \end{bmatrix} \qquad \begin{bmatrix} 0.5 & -3 & 4 \\ 0 & 1 & 9.4 \\ 6 & 0.25 & 99 \end{bmatrix} \qquad \begin{bmatrix} 7 \\ 13 \\ -24 \end{bmatrix} \qquad \begin{bmatrix} -8 & 2 & 0 & 0 & 4 \\ 9 & 6 & 3 & -5 & -1 \end{bmatrix}$$

$$2 \times 3 \qquad\qquad 3 \times 3 \qquad\qquad 3 \times 1 \qquad\qquad 2 \times 5$$

When we enumerate rows and columns of a given matrix, we count rows from top to bottom and count columns from left to right.

Since a matrix is just an array of numbers, we often see matrices used to record information, especially if rows and columns of a matrix can be understood to represent categories. As such, we can certainly use matrices to record pertinent information about a system of linear equations—coefficients of the variables as well as constants on the right-hand side of equations in the system.

Consider the system

$$2x_1 + 6x_2 = 9$$

$$-3x_1 + 5x_2 = -12$$

We adopt a convention here of using subscripted variables rather than individual letter variables to avoid possible difficulties in the number of available letters. We construct a $2 \times 3$ matrix, called the **augmented matrix for the system,** where each row represents information for a particular equation and each column represents either coefficients of a variable or the constants on the right-hand side of the equations. This matrix is written as follows.

$$\begin{bmatrix} 2 & 6 & 9 \\ -3 & 5 & -12 \end{bmatrix}$$

Notice the correspondence between the rows of this matrix and the equations in the system as well as the correspondence between the columns of the matrix and the coefficients and constant terms in the equations. The vertical line has no real purpose except to serve as a visual reminder of the location of the equal signs in the system and hence a separation between the coefficients of the variables and the constants on the right-hand side of the equations.

Before delving into use of these augmented matrices representing systems, we pause to put forth some terminology and notation. Recall that in the method of elimination we had three operations that could be used to produce equivalent systems of linear equations. We have a similar collection of **row operations** that we perform on matrices. Two matrices are said to be **row-equivalent** if one is obtained from the other by some sequence of row operations. These operations are as follows:

---

### Row Operations

- Interchange any two rows.
- Multiply (all elements in) a row by any nonzero constant and replace that row with the result.
- Multiply (all elements in) a row by any constant and add (corresponding elements) to any other row, replacing the second row in this sum with the result.

Performing any sequence of these operations results in a row-equivalent matrix.

---

Notice the similarity between these operations and the operations employed in the method of elimination. We use a similar shorthand notation to indicate performing a particular row operation as well.

> ## Notation for Row Operations
> - $R_k \leftrightarrow R_j$ denotes the operation "interchange rows $k$ and $j$"
> - $cR_k \rightarrow R_k$ denotes the operation "multiply row $k$ by $c$ and let the result replace row $k$"
> - $cR_k + R_j \rightarrow R_j$ denotes the operation "multiply row $k$ by $c$ and add to row $j$ and let the result replace row $j$"

In the case where a matrix is the augmented matrix representing a system of linear equations, performing a row operation on the matrix is equivalent to performing the corresponding operation on a system of equations. So, *row-equivalent matrices represent equivalent systems of linear equations*. To demonstrate how to use augmented matrices to find solutions to systems of linear equations, we will show in parallel operations in the method of elimination and the corresponding row operations.

## EXAMPLE 1 ◆ Using Augmented Matrices to Solve a System

Solve the system

$$x_1 + 2x_2 - x_3 = 3$$
$$x_1 + 3x_2 - x_3 = 4$$
$$2x_1 - 2x_2 + 2x_3 = 8$$

### Solution

Consider the following sequences of equivalent systems and row-equivalent matrices. Pay close attention to the correspondence between elements in the matrices and terms in the systems of equations.

| Sequence of Equivalent Systems | Subsequent Operations on Equations | Sequence of Row-Equivalent Matrices | Subsequent Row Operations |
|---|---|---|---|
| *Original System* <br> $x_1 + 2x_2 - x_3 = 3$ <br> $x_1 + 3x_2 - x_3 = 4$ <br> $2x_1 - 2x_2 + 2x_3 = 8$ | $-1E_1 + E_2 \rightarrow E_2$ <br> $-2E_1 + E_3 \rightarrow E_3$ <br> (eliminate $x_1$ from the last two equations) | *Original Augmented Matrix* <br> $\begin{bmatrix} 1 & 2 & -1 & \vert & 3 \\ 1 & 3 & -1 & \vert & 4 \\ 2 & -2 & 2 & \vert & 8 \end{bmatrix}$ | $-1R_1 + R_2 \rightarrow R_2$ <br> $-2R_1 + R_3 \rightarrow R_3$ <br> (get 0s in row 2, col 1 and row 3, col 1) |
| $x_1 + 2x_2 - x_3 = 3$ <br> $x_2 = 1$ <br> $-6x_2 + 4x_3 = 2$ | $-2E_2 + E_1 \rightarrow E_1$ <br> $6E_2 + E_3 \rightarrow E_3$ <br> (eliminate $x_2$ from first and third equations) | $\begin{bmatrix} 1 & 2 & -1 & \vert & 3 \\ 0 & 1 & 0 & \vert & 1 \\ 0 & -6 & 4 & \vert & 2 \end{bmatrix}$ | $-2R_2 + R_1 \rightarrow R_1$ <br> $6R_2 + R_3 \rightarrow R_3$ <br> (get 0s in row 1, col 2 and row 3, col 2) |
| $x_1 - x_3 = 1$ <br> $x_2 = 1$ <br> $4x_3 = 8$ | $\frac{1}{4}E_3 \rightarrow E_3$ <br> (modify third equation to get $x_3$ coefficient 1) | $\begin{bmatrix} 1 & 0 & -1 & \vert & 1 \\ 0 & 1 & 0 & \vert & 1 \\ 0 & 0 & 4 & \vert & 8 \end{bmatrix}$ | $\frac{1}{4}R_3 \rightarrow R_3$ <br> (get 1 in row 3, col 3) |
| $x_1 - x_3 = 1$ <br> $x_2 = 1$ <br> $x_3 = 2$ | $E_3 + E_1 \rightarrow E_1$ <br> (eliminate $x_3$ from first equation) | $\begin{bmatrix} 1 & 0 & -1 & \vert & 1 \\ 0 & 1 & 0 & \vert & 1 \\ 0 & 0 & 1 & \vert & 2 \end{bmatrix}$ | $R_3 + R_1 \rightarrow R_1$ <br> (get 0 in row 1, col 3) |
| $x_1 = 3$ <br> $x_2 = 1$ <br> $x_3 = 2$ | | $\begin{bmatrix} 1 & 0 & 0 & \vert & 3 \\ 0 & 1 & 0 & \vert & 1 \\ 0 & 0 & 1 & \vert & 2 \end{bmatrix}$ | |

We see, then, that the solution to the system is expressly given in the last system. Notice how the solution can also be read expressly from the last matrix as well. We performed more operations in this case than we had done previously to illustrate that, at least in this case, the extra operations didn't result in significantly more work than a series of back-substitutions and to illustrate how, in the final augmented matrix, the solution to the system is easily read.

After arriving at the end of the sequence of row operations and resulting matrices in the previous example, we would say that we have **reduced** the augmented matrix that represented the system of equations being investigated to a simpler matrix where the solution to the system was evidently displayed. This reduction process is called **Gauss-Jordan elimination.** When we engage in this process in an organized manner, it is somewhat more efficient than the method of elimination, at least relative to systems of equations in more than two variables. It may seem that the process will still be very tedious, but many graphing calculators and computer programs work very well for such a task, as the process is really just arithmetic on an array of numbers.

Bear in mind that when using this method, every matrix that is generated really represents a system of linear equations, and any two matrices that are row-equivalent represent equivalent systems.

Before attempting to summarize the method of Gauss-Jordan elimination, we introduce the notion of **reduced echelon form** of a matrix. A matrix is said to be in reduced echelon form if all of the following are true:

---

### Reduced Echelon Form

1.  All rows consisting entirely of zeros are grouped at the bottom of the matrix.

2.  The leftmost nonzero number in each row is 1 (this element is called the **leading 1 of the row**).

3.  The leading 1 of a row is to the right of the leading 1 of the previous row.

4.  All other elements in the same column as a leading 1 are zeros.

---

The following matrices are in reduced echelon form. Considering them as augmented matrices representing a system of linear equations, the corresponding system is also listed.

| Reduced Echelon Form Matrix | Corresponding System of Linear Equations |
|:---:|:---:|
| $\begin{bmatrix} 1 & 0 & 0 & 6 \\ 0 & 1 & 0 & -4 \\ 0 & 0 & 1 & 0 \end{bmatrix}$ | $x_1 = 6$<br>$x_2 = -4$<br>$x_3 = 0$ |
| $\begin{bmatrix} 1 & 0 & \frac{3}{2} \\ 0 & 1 & -10 \\ 0 & 0 & 0 \end{bmatrix}$ | $x_1 = \frac{3}{2}$<br>$x_2 = -10$<br>$0 = 0$ |
| $\begin{bmatrix} 1 & 2 & 0 & 5 & 0 \\ 0 & 0 & 1 & -4 & 0 \\ 0 & 0 & 0 & 0 & 1 \\ 0 & 0 & 0 & 0 & 0 \end{bmatrix}$ | $x_1 + 2x_2 + 5x_4 = 0$<br>$x_3 - 4x_4 = 0$<br>$0 = 1$<br>$0 = 0$ |
| $\begin{bmatrix} 1 & -5 & 0 & 8 \\ 0 & 0 & 1 & -13 \\ 0 & 0 & 0 & 0 \end{bmatrix}$ | $x_1 - 5x_2 = 8$<br>$x_3 = -13$<br>$0 = 0$ |

It may seem superfluous to list equations like $0 = 0$, but it is important when interpreting what exactly is happening with solutions to linear systems to clearly list all information rather than run the risk of jumping to a hasty conclusion or ignoring a piece that is vital.

We can actually summarize the method of Gauss-Jordan elimination in three statements:

---

**Gauss-Jordan Elimination**

1. Represent a system of linear equations with an augmented matrix.
2. Reduce the original augmented matrix to reduced echelon form.
3. Interpret the reduced echelon form matrix.

---

Naturally, this summary does not indicate the nature of detail in the process. For instance, leading 1s are an important feature of reduced echelon form. During the reduction process, it is beneficial to obtain leading 1s "quickly" and to use these in the process of reducing other elements in the matrix to zeros. Also, we need to deal with the issue of interpreting the information presented in the resulting reduced echelon form matrix. We explore some of this in the next examples.

## EXAMPLE 2 ◆ Gauss-Jordan Elimination

Solve the following system using matrices, if possible.

$$x_2 - 3x_3 = 2$$
$$2x_1 + 4x_2 + 6x_3 = -4$$
$$3x_1 + 5x_2 + 2x_3 = 2$$

**Solution**

We begin by writing the augmented matrix associated with the system and proceed to reduce that matrix to reduced echelon form. Notice that the coefficient of the $x_1$ variable in the first equation is 0.

| Sequence of Matrices | Subsequent Row Operations | Comments |
|---|---|---|
| $\begin{bmatrix} 0 & 1 & -3 & 2 \\ 2 & 4 & 6 & -4 \\ 3 & 5 & 2 & 2 \end{bmatrix}$ | $R_1 \leftrightarrow R_2$ | We will be able to get a leading 1 in row 1 |
| $\begin{bmatrix} 2 & 4 & 6 & -4 \\ 0 & 1 & -3 & 2 \\ 3 & 5 & 2 & 2 \end{bmatrix}$ | $\frac{1}{2} R_1 \rightarrow R_2$ | Get leading 1 in row 1 |
| $\begin{bmatrix} 1 & 2 & 3 & -2 \\ 0 & 1 & -3 & 2 \\ 3 & 5 & 2 & 2 \end{bmatrix}$ | $-3R_1 + R_3 \rightarrow R_3$ | Use the leading 1 in row 1 to reduce the rest of column 3 |
| $\begin{bmatrix} 1 & 2 & 3 & -2 \\ 0 & 1 & -3 & 2 \\ 0 & -1 & -7 & 8 \end{bmatrix}$ | $-2 R_2 + R_1 \rightarrow R_1$ <br> $R_2 + R_3 \rightarrow R_3$ | Row 2 already has a leading 1 <br> Use the leading 1 in row 2 to reduce the rest of column 2 |
| $\begin{bmatrix} 1 & 0 & 9 & -6 \\ 0 & 1 & -3 & 2 \\ 0 & 0 & -10 & 10 \end{bmatrix}$ | $-\frac{1}{10} R_3 \rightarrow R_3$ | Get leading 1 in row 3 |

$$\begin{bmatrix} 1 & 0 & 9 & | & -6 \\ 0 & 1 & -3 & | & 2 \\ 0 & 0 & 1 & | & -1 \end{bmatrix}$$

$-9R_3 + R_1 \rightarrow R_1$

$3R_3 + R_2 \rightarrow R_2$

Use the leading 1 in row 3 to reduce the rest of column 3

$$\begin{bmatrix} 1 & 0 & 0 & | & 3 \\ 0 & 1 & 0 & | & -1 \\ 0 & 0 & 1 & | & -1 \end{bmatrix}$$

Reduction Complete

The final matrix in the sequence indicates that the system of equations has a single solution, the triple $x_1 = 3, x_2 = -1$, and $x_3 = -1$ (check).

## EXAMPLE 3 ◆ Gauss-Jordan Elimination

Solve the following system using matrices, if possible.

$$x_1 + x_2 - 2x_3 = 3$$

$$-2x_1 - x_2 + 5x_3 = -2$$

$$-3x_1 + 2x_2 + 11x_3 = 11$$

**Solution**

Again, we begin by writing the augmented matrix associated with the system and proceed to reduce that matrix to reduced echelon form.

| Sequence of Matrices | Subsequent Row Operations | Comments |
|---|---|---|
| $\begin{bmatrix} 1 & 1 & -2 & \vert & 3 \\ -2 & -1 & 5 & \vert & -2 \\ -3 & 2 & 11 & \vert & 11 \end{bmatrix}$ | $2R_1 + R_2 \rightarrow R_2$ <br> $3R_1 + R_3 \rightarrow R_3$ | Row 1 already has a leading 1 <br> Use the leading 1 in row 1 to reduce the rest of column 1 |
| $\begin{bmatrix} 1 & 1 & -2 & \vert & 3 \\ 0 & 1 & 1 & \vert & 4 \\ 0 & 5 & 5 & \vert & 20 \end{bmatrix}$ | $-1R_2 + R_1 \rightarrow R_1$ <br> $-5R_2 + R_3 \rightarrow R_3$ | Row 2 already has a leading 1 <br> Use the leading 1 in row 2 to reduce the rest of column 2 |
| $\begin{bmatrix} 1 & 0 & -3 & \vert & -1 \\ 0 & 1 & 1 & \vert & 4 \\ 0 & 0 & 0 & \vert & 0 \end{bmatrix}$ | | Reduction is complete |

At this point, we see that the final matrix represents the system

$$x_1 - 3x_3 = -1$$

$$x_2 + x_3 = 4$$

$$0 = 0$$

There are no contradictions present, but we cannot solve explicitly (find a real number result) for any of the variables. Consequently, the system of equations has an infinite number of solutions. Techniques for describing these solutions are not going to be examined in this text.

## **EXAMPLE 4** ♦ Gauss-Jordan Elimination

Solve the following system using matrices, if possible.

$$2x_1 + 7x_2 + 12x_3 = 1$$

$$-x_1 + 4x_2 + 9x_3 = 8$$

$$x_1 + 2x_2 + 3x_3 = -1$$

### Solution

Again, we begin by writing the augmented matrix associated with the system and proceed to reduce that matrix to reduced echelon form.

| Sequence of Matrices | Subsequent Row Operations | Comments |
|---|---|---|
| $\begin{bmatrix} 2 & 7 & 12 & \bigm| & 1 \\ -1 & 4 & 9 & \bigm| & 8 \\ 1 & 2 & 3 & \bigm| & -1 \end{bmatrix}$ | $R_1 \leftrightarrow R_3$ | Get a leading 1 in row 1 |
| $\begin{bmatrix} 1 & 2 & 3 & \bigm| & -1 \\ -1 & 4 & 9 & \bigm| & 8 \\ 2 & 7 & 12 & \bigm| & 1 \end{bmatrix}$ | $R_1 + R_2 \to R_2$ <br> $-2R_1 + R_3 \to R_3$ | Use the leading 1 in row 1 to reduce the rest of column 1 |
| $\begin{bmatrix} 1 & 2 & 3 & \bigm| & -1 \\ 0 & 6 & 12 & \bigm| & 7 \\ 0 & 3 & 6 & \bigm| & 3 \end{bmatrix}$ | $\frac{1}{6} R_2 \to R_2$ | Get a leading 1 in row 2 |
| $\begin{bmatrix} 1 & 2 & 3 & \bigm| & -1 \\ 0 & 1 & 2 & \bigm| & \frac{7}{6} \\ 0 & 3 & 6 & \bigm| & 3 \end{bmatrix}$ | $-2R_2 + R_1 \to R_1$ <br> $-3R_2 + R_3 \to R_3$ | Use the leading 1 in row 2 to reduce the rest of column 2 |
| $\begin{bmatrix} 1 & 0 & -1 & \bigm| & -\frac{10}{3} \\ 0 & 1 & 2 & \bigm| & \frac{7}{6} \\ 0 & 0 & 0 & \bigm| & -\frac{1}{2} \end{bmatrix}$ | | There is no need to continue |

The system represented by the last matrix shown is

$$x_1 - x_3 = -\frac{10}{3}$$

$$x_2 + 2x_3 = \frac{7}{6}$$

$$0 = -\frac{1}{2}$$

The last equation in this system is a contradiction. Consequently, the original system of equations is inconsistent and has no solution.

In summary, the nonzero rows of the reduced echelon form of the augmented matrix for a system of linear equations give all of the information needed to ascertain the nature of solutions to the system. There are only three possibilities.

1. **The system has no solution.** One row of the reduced echelon form matrix will indicate a contradiction—all zeros in the coefficient portion of the matrix and a nonzero constant in the last column position of that row.

   In the event that the system has solutions (no contradictions), there are two possibilities.

2. **The solution is unique.** The number of nonzero rows in the reduced echelon form matrix is exactly the same as the number of variables in the system (this *does not* mean that the system must have the same number of equations as variables).

3. **The system has an infinite number of solutions.** The number of nonzero rows in the reduced echelon form matrix is less than the number of variables in the system (this *does not* mean that the system must have fewer equations than variables).

# EXERCISES

1. For each of the following augmented matrices, indicate whether or not it is in reduced echelon form. If it is not in reduced echelon form, then state why it is not.

   (a) $\begin{bmatrix} 1 & 0 & 0 & | & 2 \\ 0 & 1 & 0 & | & 3 \\ 0 & 0 & 1 & | & 0 \end{bmatrix}$

   (b) $\begin{bmatrix} 1 & 2 & 0 & | & 0 \\ 0 & 0 & 1 & | & 0 \\ 0 & 0 & 0 & | & 1 \end{bmatrix}$

   (c) $\begin{bmatrix} 1 & 0 & 0 & 3 & | & -2 \\ 0 & 1 & 0 & 1 & | & 1 \\ 0 & 0 & 0 & 1 & | & -4 \\ 0 & 0 & 1 & 0 & | & 2 \end{bmatrix}$

   (d) $\begin{bmatrix} 1 & 0 & | & -2 \\ 0 & 1 & | & 3 \\ 0 & 0 & | & 0 \end{bmatrix}$

*Let* $M = \begin{bmatrix} 1 & 0 & 1 & 1 \\ 2 & 4 & 6 & 7 \\ 0 & 2 & -1 & 0 \end{bmatrix}$. *In problems 2–5, specify row operations that transform M into the given matrix.*

2. $\begin{bmatrix} -2 & 0 & -2 & -2 \\ 2 & 4 & 6 & 7 \\ 0 & 2 & -1 & 0 \end{bmatrix}$

3. $\begin{bmatrix} 1 & 0 & 1 & 1 \\ 3 & 4 & 7 & 8 \\ 0 & 2 & -1 & 0 \end{bmatrix}$

4. $\begin{bmatrix} 1 & 0 & 1 & 1 \\ 0 & 2 & -1 & 0 \\ 2 & 4 & 6 & 7 \end{bmatrix}$

5. $\begin{bmatrix} 1 & 0 & 1 & 1 \\ 0 & 1 & 1 & 1.25 \\ 0 & 0 & -3 & -2.5 \end{bmatrix}$

6. For each system of linear equations, write the corresponding augmented matrix.

   (a) $\begin{aligned} x &+ 2y - z &= 5 \\ 3x &- y + z &= 6 \\ x &\phantom{- y} + z &= 3 \end{aligned}$

   (b) $\begin{aligned} -x_1 + 3x_2 &- x_3 + x_4 &= -2 \\ 2x_2 &+ 3x_3 - x_4 &= 7 \\ 3x_1 \phantom{+ 3x_2} &- x_3 + 5x_4 &= 1 \end{aligned}$

   (c) $\begin{aligned} 2x_1 - x_2 &= 4 \\ x_1 + x_2 &= 6 \end{aligned}$

   (d) $\begin{aligned} 4a + 3b - c &= -10 \\ 2a - b &= 5 \end{aligned}$

7. For each of the following systems of linear equations, indicate whether or not the triple $x_1 = 3, x_2 = -1, x_3 = 4$ is a solution.

   (a) $\begin{aligned} x_1 + x_2 + x_3 &= 6 \\ 2x_1 - x_2 - 2x_3 &= -1 \\ -3x_1 + 3x_2 - x_3 &= 5 \end{aligned}$

   (b) $\begin{aligned} 2x_1 - 3x_2 + 5x_3 &= 29 \\ x_1 - x_2 - x_3 &= 0 \\ 4x_1 + x_2 + 3x_3 &= 23 \end{aligned}$

   (c) $\begin{aligned} x_1 - 5x_2 + x_3 &= 9 \\ 2x_1 \phantom{- 5x_2} + 3x_3 &= 18 \end{aligned}$

   (d) $\begin{aligned} -3x_1 + 2x_2 - x_3 &= -15 \\ x_2 + 4x_3 &= 15 \end{aligned}$

*In problems 8–14, solve the system using Gauss-Jordan Elimination.*

8. $\begin{aligned} x - 2y &= 9 \\ 2x + y &= -2 \end{aligned}$

9. $\begin{aligned} 3x_1 - 2x_2 + x_3 &= 5 \\ x_1 + x_2 - x_3 &= 3 \\ 2x_1 - x_2 &= 2 \end{aligned}$

**10.** $-x_1 + 2x_2 + 3x_3 = 4$
$-x_1 - 3x_2 + 2x_3 = -1$
$3x_1 - 9x_2 + 6x_3 = 1$

**11.** $-6x_1 + 18x_2 - 12x_3 = -3$
$3x_1 - 9x_2 + 6x_3 = 1.5$
$4x_1 - 12x_2 + 8x_3 = 2$

**12.** $2x_1 - x_2 + 4x_3 + x_4 = 13$
$-x_1 + x_2 - 2x_3 + 4x_4 = -15$
$3x_1 + 3x_2 + x_3 - x_4 = 14$
$-2x_1 + x_2 - 7x_3 + 2x_4 = -28$

**13.** $3x_1 - 2x_2 = -5$
$6x_1 + x_2 = 0$
$-2x_1 - \dfrac{1}{3}x_2 = 0$

**14.** $3x_1 + 5x_2 - x_3 + 2x_4 = 2$
$2x_1 - x_3 = 0$
$2x_2 - x_4 = -3$
$x_1 - 3x_2 - x_3 = 1$

**15.** Verify that the following system has an infinite number of solutions.

$$x_1 + 2x_2 - x_3 = 5$$
$$2x_1 + 3x_2 - 3x_3 = 8$$
$$x_2 + x_3 = 2$$

Find three specific solutions to this system. (Hint: Specify a value for $x_3$, then solve for $x_1$ and $x_2$.)

# Interlude: Practice Problems 3.2, Part 1

1. Given $\begin{bmatrix} -3 & -1 & 2 & 4 \\ 5 & 0 & -2 & 7 \\ 6 & -2 & 1 & 9 \end{bmatrix}$, I performed this row operation: $-3R_2 + R_3 \rightarrow R_3$.

    (a) Show the resulting matrix.

    (b) What row operation will transform the matrix from (a) back into the original matrix?

2. Write a system of three linear equations in three variables that has $(-3, 1, 4)$ as one of an infinite number of solutions. Your system should have all nonzero coefficients. Show clearly that your system actually satisfies the given requirement.

3.  Consider the matrix $A = \begin{bmatrix} 1 & 1 & 1 & 1 & | & 20 \\ 0 & 1 & 1 & 1 & | & 0 \\ 0 & 0 & 1 & 1 & | & 13 \\ 0 & 2 & 0 & -2 & | & -10 \end{bmatrix}$.

(a)  Write the system of equations represented by this matrix.

(b)  Use row operations to completely row-reduce $A$ and show the resulting matrix after each row operation.

(c)  Give the solution to the system represented by $A$.

# Interlude: Practice Problems 3.2, Part 2

1. Consider the following matrices. For each, write the system of equations it represents, and determine how many solutions there are to that system.

$$A = \begin{bmatrix} 1 & 0 & 6 & | & 4 \\ 0 & 1 & -3 & | & 8 \\ 0 & 0 & 1 & | & 0 \end{bmatrix} \qquad B = \begin{bmatrix} 1 & 5 & -2 & | & 0 \\ 0 & 0 & 1 & | & -1 \\ 0 & 0 & 5 & | & -5 \end{bmatrix}$$

2. Consider the matrices $A = \begin{bmatrix} 3 & -4 & -5 & 8 \\ 1 & -2 & 1 & 0 \\ -4 & 5 & 9 & -9 \end{bmatrix}$ and $B = \begin{bmatrix} 3 & -24 & 12 & -27 \\ 0 & 1 & -2 & -42 \\ 0 & 0 & 2 & 58 \end{bmatrix}$.

Are $A$ and $B$ row-equivalent? Explain why or why not.

3. Verify that the system $\begin{aligned} 5x_1 - x_2 - 2x_3 + 5x_4 &= -9 \\ -4x_1 + x_2 + 2x_3 - 4x_4 &= 11 \\ 2x_1 - x_2 - x_3 &= -10 \end{aligned}$ has an infinite number of solutions, and find three distinct solutions to the system.

# 3.3   MORE MODELING AND APPLICATIONS

Thus far, a few problems have been presented where the situation at hand was modeled by a system of linear equations. We now concentrate on more types of problems that are naturally modeled by systems of linear equations.

Prior to Chapter 1, a five-step problem-solving approach was set forth. Familiarizing yourself with the problem situation and setting up the subsequent model is, typically, the most challenging part of a problem. In the types of problems we will be considering here, it is often helpful to summarize and organize the information, perhaps using a table, before starting the modeling process.

## EXAMPLE 1 ♦ A Dietician Problem

A dietician must make a meal consisting of 700 calories, 450 grams of protein, and 220 milligrams of vitamin A from three types of food: ground beef, baked potato, and mixed vegetables. One serving of ground beef contains 200 calories, 150 grams of protein, and 10 milligrams of vitamin A. One serving of baked potato contains 150 calories, 100 grams of protein, and 30 milligrams of vitamin A. One serving of mixed vegetables contains 100 calories, 50 grams of protein, and 75 milligrams of vitamin A. How many servings of each type of food should the meal contain?

**Solution**

Since we seek the number of servings of the three types of food needed to satisfy various dietary requirements, we begin by summarizing information about servings of ground beef, baked potato, and mixed vegetables in a table. We also list the dietary requirements.

|  | **Ground Beef** | **Baked Potato** | **Mixed Vegetables** | **Requirement** |
|---|---|---|---|---|
| calories | 200 | 150 | 100 | 700 |
| protein (g) | 150 | 100 | 50 | 450 |
| vitamin A (mg) | 10 | 30 | 75 | 220 |

We now declare our unknowns. Let

$$x_1 = \text{the number of servings of ground beef needed}$$

$$x_2 = \text{the number of servings of baked potato needed}$$

$$x_3 = \text{the number of servings of mixed vegetables needed}$$

We can see from the table constructed above that $x_1, x_2$ and $x_3$ must satisfy three equations, one for each dietary requirement.

$$200x_1 + 150x_2 + 100x_3 = 700 \quad \text{(calorie requirement)}$$

$$150x_1 + 100x_2 + 50x_3 = 450 \quad \text{(protein requirement)}$$

$$10x_1 + 30x_2 + 75x_3 = 220 \quad \text{(vitamin A requirement)}$$

Thus the situation is modeled by a system of three linear equations in three unknowns. We will solve the system using the associated augmented matrix and Gauss-Jordan elimination. The augmented matrix for the system is

$$\begin{bmatrix} 200 & 150 & 100 & | & 700 \\ 150 & 100 & 50 & | & 450 \\ 10 & 30 & 75 & | & 220 \end{bmatrix}$$

It can be shown that the reduced echelon form of this matrix is

$$\begin{bmatrix} 1 & 0 & 0 & | & 1 \\ 0 & 1 & 0 & | & 2 \\ 0 & 0 & 1 & | & 2 \end{bmatrix}$$

This yields the corresponding system

$$x_1 = 1$$
$$x_2 = 2$$
$$x_3 = 2$$

We conclude, then, that the meal should consist of one serving of ground beef, two servings of baked potato, and two servings of mixed vegetables (check this in the original system).

## EXAMPLE 2 ◆ Portfolio Analysis

The following table describes the expected annual appreciation and potential loss of three types of investments.

|  | Growth Stocks | Dividend Stocks | Tax-Free Bonds |
|---|---|---|---|
| Expected Appreciation | 12% | 8% | 5% |
| Potential Loss | 10% | 8% | 3% |

A total of $250,000 is to be invested among the three types of investments listed above. The desired overall annual appreciation for the portfolio is 7%, with a potential loss of 5%. How should the $250,000 be allocated among the three types of investments to meet these goals?

### Solution

Given a $250,000 portfolio, an annual appreciation of 7% would be growth of $250,000(0.07) = \$17,500$. A loss of 5% would be a loss of $250,000(0.05) = \$12,500$. Since we need to decide how much money to allocate to each of the three types of investments, we have three unknowns, which we declare now. Let

$x_1$ = the number of dollars allocated to growth stocks

$x_2$ = the number of dollars allocated to dividend stocks

$x_3$ = the number of dollars allocated to tax-free bonds

We summarize the situation in a table.

|  | Growth Stocks | Dividend Stocks | Tax-Free Bonds | Total |
|---|---|---|---|---|
| Amount Allocated | $x_1$ | $x_2$ | $x_3$ | $250,000 |
| Expected Appreciation | 12% | 8% | 5% | $17,500 |
| Potential Loss | 10% | 8% | 3% | $12,500 |

This naturally leads to a system of three equations:

$$x_1 + x_2 + x_3 = 250000 \quad \text{(total allocation)}$$
$$0.12x_1 + 0.08x_2 + 0.05x_3 = 17500 \quad \text{(appreciation goal)}$$
$$0.1x_1 + 0.08x_2 + 0.03x_3 = 12500 \quad \text{(loss maximum)}$$

Again, we reduce the augmented matrix for the system. The matrix

$$\begin{bmatrix} 1 & 1 & 1 & 250{,}000 \\ 0.12 & 0.08 & 0.05 & 17{,}500 \\ 0.1 & 0.08 & 0.03 & 12{,}500 \end{bmatrix}$$

reduces to

$$\begin{bmatrix} 1 & 0 & 0 & 71{,}428.57 \\ 0 & 1 & 0 & 0 \\ 0 & 0 & 1 & 178{,}571.43 \end{bmatrix}$$

The entries in the last column have been rounded to two decimal places. This gives the system

$$x_1 = 71{,}428.57$$

$$x_2 = 0$$

$$x_3 = 178{,}571.43$$

We conclude that \$71,428.57 should be allocated to growth stocks, \$178,571.43 should be allocated to tax-free bonds, and no money should be allocated to dividend stocks in order to meet the established objectives (check this in the original system).

## EXAMPLE 3 ♦ Agriculture

According to data from a 1984 Texas agricultural report the amount of nitrogen (in lb/acre), phosphate (in lb/acre) and labor (in hr/acre) needed to grow honeydews, yellow onions, and lettuce is given in the following table.

| | **Honeydews** | **Onions** | **Lettuce** |
|---|---|---|---|
| Nitrogen (lb/acre) | 120 | 150 | 180 |
| Phosphate (lb/acre) | 180 | 80 | 80 |
| Labor (hr/acre) | 4.97 | 4.45 | 4.65 |

If a farmer has 220 acres of land, 29,100 lbs. of nitrogen, 32,600 lbs. of phosphate, and money for 480 hours of labor, how many acres should be allotted for each crop in order to use all of the available resources?

**Solution**

Once again we must draw a conclusion about how to divide a quantity among three possible options. Let

$$x_1 = \text{the number of acres allotted to honeydews}$$

$$x_2 = \text{the number of acres allotted to yellow onions}$$

$$x_3 = \text{the number of acres allotted to lettuce}$$

We summarize the situation in a table.

| | **Honeydews** | **Onions** | **Lettuce** | **Total** |
|---|---|---|---|---|
| Acres Allotted | $x_1$ | $x_2$ | $x_3$ | 220 |
| Nitrogen (lb/acre) | 120 | 150 | 180 | 29,100 |
| Phosphate (lb/acre) | 180 | 80 | 80 | 32,600 |
| Labor (hr/acre) | 4.97 | 4.45 | 4.65 | 480 |

This naturally leads to a system of linear equations:

$$x_1 + x_2 + x_3 = 220$$

$$120x_1 + 150x_2 + 180x_3 = 29{,}100$$

$$180x_1 + 80x_2 + 80x_3 = 32{,}600$$

$$4.97x_1 + 4.45x_2 + 4.65x_3 = 480$$

The augmented matrix for the system is

$$\left[\begin{array}{ccc|c} 1 & 1 & 1 & 220 \\ 120 & 150 & 180 & 29{,}100 \\ 180 & 80 & 80 & 32{,}600 \\ 4.97 & 4.45 & 4.65 & 480 \end{array}\right]$$

which reduces to

$$\left[\begin{array}{ccc|c} 1 & 0 & 0 & 0 \\ 0 & 1 & 0 & 0 \\ 0 & 0 & 1 & 0 \\ 0 & 0 & 0 & 1 \end{array}\right]$$

Notice that the last row of the reduced echelon form matrix corresponds to the equation $0 = 1$, which is a contradiction. Consequently, we conclude that it is **not** possible for the farmer to use all of the available resources, no matter how many acres are allotted to each crop.

## EXAMPLE 4 ◆ Agriculture

Repeat Example 3, except this time assume that the farmer has resources available for 1061 hours of labor.

### Solution

The same analysis as done above will lead to a system of four linear equations in four unknowns. The only difference in the systems will be in the last equation. Here we have

$$4.97x_1 + 4.45x_2 + 4.65x_3 = 1061$$

The associated augmented matrix becomes

$$\left[\begin{array}{ccc|c} 1 & 1 & 1 & 220 \\ 120 & 150 & 180 & 29{,}100 \\ 180 & 80 & 80 & 32{,}600 \\ 4.97 & 4.45 & 4.65 & 1061 \end{array}\right]$$

which reduces to

$$\left[\begin{array}{ccc|c} 1 & 0 & 0 & 150 \\ 0 & 1 & 0 & 50 \\ 0 & 0 & 1 & 20 \\ 0 & 0 & 0 & 0 \end{array}\right]$$

This gives the system

$$x_1 = 150$$

$$x_2 = 50$$

$$x_3 = 20$$

$$0 = 0$$

The last equation in this system does not give any useful information, but we do have explicit values for the three unknowns. Consequently, the farmer should plant 150 acres of honeydews, 50 acres of yellow onions, and 20 acres of lettuce to use all of the available resources (check this in the original system).

---

### EXAMPLE 5 ◆ Traffic Control

During rush hour, substantial traffic congestion is encountered at the traffic intersections shown in the figure below. The arrows indicate that the streets are one-way streets.

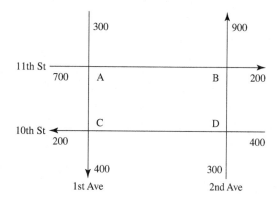

City engineers wish to speed the flow of traffic through these intersections by improving traffic signals. What the diagram shows is that 700 cars per hour come down 11th Street to intersection $A$ and 300 cars per hour come down 1st Avenue to intersection $A$. Let $x_1$ denote the number of cars that leave intersection $A$ along 11th Street and let $x_2$ denote the number of cars that leave intersection $A$ along 1st Avenue. Since a total of 1000 cars per hour enter intersection $A$, 1000 cars per hour must leave, and so we note that

$$x_1 + x_2 = 1000$$

Similarly, let $x_3$ denote the number of cars that leave intersection $D$ along 2nd Avenue and let $x_4$ denote the number of cars that leave intersection $D$ along 10th Street. We see that

$$x_3 + x_4 = 700$$

You should also verify that we must have

$$x_1 + x_3 = 1100 \quad \text{(intersection } B\text{)}$$

$$x_2 + x_4 = 600 \quad \text{(intersection } C\text{)}$$

Putting all of these linear equations together gives a system with augmented matrix

$$\left[\begin{array}{cccc|c} 1 & 1 & 0 & 0 & 1000 \\ 0 & 0 & 1 & 1 & 700 \\ 1 & 0 & 1 & 0 & 1100 \\ 0 & 1 & 0 & 1 & 600 \end{array}\right]$$

The reduced echelon form of this matrix is

$$\left[\begin{array}{cccc|c} 1 & 0 & 0 & -1 & 400 \\ 0 & 1 & 0 & 1 & 600 \\ 0 & 0 & 1 & 1 & 700 \\ 0 & 0 & 0 & 0 & 0 \end{array}\right]$$

This yields the system

$$x_1 - x_4 = 400$$
$$x_2 + x_4 = 600$$
$$x_3 + x_4 = 700$$
$$0 = 0$$

This system has an infinite number of solutions. The question, then, is what information these solutions provide. Although many things can be investigated here, we focus only on intersection $A$ and what solutions to the system of equations can indicate there.

For instance, $x_1$ gives the number of cars leaving intersection $A$ along 11th Street. From above we see that $x_1 - x_4 = 400$, and so

$$x_1 = 400 + x_4$$

Since $x_4 \geq 0$, the smallest possible value for $x_1$ is 400, meaning that the least number of cars that can leave intersection $A$ along 11th Street is 400 per hour. The largest number of cars that can leave intersection $A$ along 11th Street is 1100 per hour since $x_4 \leq 700$. Also, we see that $x_2 + x_4 = 600$, and so

$$x_2 = 600 - x_4$$

Since $x_4 \geq 0$, the largest number of cars that can leave intersection $A$ along 1st Avenue is 600 per hour (and the least, of course, is zero).

Similar analysis of the other intersections gives the city engineers some bounds on the number of cars that can be expected to be approaching or leaving other intersections as well. Based on these bounds, decisions can be made on how to adjust traffic signals in order to speed the flow of traffic through the intersections.

# EXERCISES

1. Jack spends 22 hours per week doing 3 types of exercises: running, biking, and swimming. He spends twice as much time biking as he does running and he spends 3 more hours biking than swimming. How much time per week does Jack spend on each activity?

2. Sue was doing homework for two classes, math and chemistry. She can do 10 math problems per hour and 15 chemistry problems per hour. One night she worked for 4 hours and completed 49 problems. How many math problems and how many chemistry problems did she do?

3. A company that rents small moving trucks wants to purchase 25 trucks with a combined capacity of 28,000 cubic feet. Three different types of trucks are available: a 10-foot truck with a capacity of 350 cubic feet, a 14-foot truck with a capacity of 700 cubic feet and a 24-foot truck with a capacity of 1,400 cubic feet. How many of each type of truck should the company purchase? (Hint: There may be more than one solution. Find restrictions on the variables by using the fact that the number of each type of truck purchased must be a nonnegative, whole number.)

4. Every left-handed widget uses 5 ounces of plastic and 10 ounces of steel, while every right-handed widget uses 4 ounces of plastic and 11 ounces of steel. What is the total number of widgets that can be made from 300 ounces of plastic and 750 ounces of steel?

5. A manufacturer of women's blouses makes three types of blouses: sleeveless, short-sleeve, and long-sleeve. The time required by each department to produce a dozen blouses of each type is shown in the accompanying table.

| Department | Sleeveless | Short-Sleeve | Long-Sleeve |
|---|---|---|---|
| Cutting | 9 min. | 12 min. | 15 min. |
| Sewing | 22 min. | 24 min. | 28 min. |
| Packaging | 6 min. | 8 min. | 8 min. |

The cutting, sewing, and packaging departments have available a maximum of 80, 160, and 48 hours, respectively, per day. How many dozens of each type of blouse can be produced each day if the plant is operated at full capacity?

6. A man has $260,000 invested at 4.7 percent, 5.5 percent, 6 percent, and 7 percent. His annual income from the investments is $16,058.40. His investments at 4.7 percent and 5.5 percent total $64,800 while his investment at 6 percent is $5800 more than twice his investment at 4.7 percent. How much does he have invested at each rate?

7. A company produces three combinations of mixed vegetables that sell in 1-kg packages. Italian style combines 0.3 kg of zucchini, 0.3 kg of broccoli, and 0.4 kg of carrots. French style combines 0.6 kg of broccoli and 0.4 kg of carrots. Oriental style combines 0.2 kg of zucchini, 0.5 kg of broccoli, and 0.3 kg of carrots. The company has a stock of 16,200 kg of zucchini, 41,400 kg of broccoli, and 29,400 kg of carrots. How many packages of each style should it prepare to use up existing supplies?

8. A dietician is preparing a meal consisting of foods A, B, and C. Each ounce of food A contains 2 units of protein, 3 units of fat, and 4 units of carbohydrate. Each ounce of food B contains 3 units of protein, 2 units of fat, and 1 unit of carbohydrate. Each ounce of food C contains 3 units of protein, 3 units of fat, and 2 units of carbohydrate. If the meal must provide exactly 25 units of protein, 24 units of fat, and 21 units of carbohydrate, how many ounces of each type of food should be used?

9. The Hiker and Biker Outfitting Shop makes packages of snacks to order. One customer likes peanuts, raisins, and chocolate chips but wants only two ingredients per snack. In peanut-raisin mixes she likes twice as many peanuts as raisins, in peanut-chocolate chip mixes she likes twice as many chocolate chips as peanuts, and in raisin-chocolate chip mixes she likes equal amounts of the two foods. Her total purchase contains 4 ounces of peanuts, 6 ounces of raisins, and 12 ounces of chocolate chips. How many ounces of each mix did she buy?

10. The head dietician at Gigantic State University has 330 pounds of fresh citrus fruit and 230 pounds of mixed fruit which must be used today. She has recipes for salads which use both. Citrus salad requires 3 pounds of fresh citrus and 1 pound of mixed fruit to produce 4 pounds of salad, and fruit salad requires 1.5 pounds of citrus fruit and 2.5 pounds of mixed fruit to produce 4 pounds of salad. How many pounds of each type of salad must be made to use all of the fruit?

11. In 20 oz of one alloy there are 6 oz of copper, 4 oz of zinc, and 10 oz of lead. In 20 oz of a second alloy there are 12 oz of copper, 5 oz of zinc, and 3 oz of lead, while in 20 oz of a third alloy there are 8 oz of copper, 6 oz of zinc, and 6 oz of lead. How many ounces of each alloy should be combined to make a new alloy containing 34 oz of copper, 17 oz of zinc, and 19 oz of lead?

12. At a fast food restaurant, one group of customers bought 8 deluxe hamburgers, 6 orders of large fries, and 6 large colas for $26.10. A second group ordered 10 deluxe hamburgers, 6 large fries, and 8 large colas and paid $31.60. Is there sufficient information to determine the price of each food item? If not, construct a table showing the various possibilities. Assume that the hamburgers cost between $1.75 and $2.25, the fries between $0.75 and $1.00, and the colas between $0.60 and $0.90.

13. A Florida juice company completes the preparation of its products by sterilizing, filling, and labeling bottles. Each case of orange juice requires 9 minutes for sterilizing, 6 minutes for filling, and 1 minute for labeling. Each case of grapefruit juice requires 10 minutes for sterilizing, 4 minutes for filling, and 2 minutes for labeling. Each case of tomato juice requires 12 minutes for sterilizing, 4 minutes for filling, and 1 minute for labeling. If the company runs the sterilizing machine for 398 minutes, the filling machine for 164 minutes, and the labeling machine for 58 minutes, how many cases of each type of juice are prepared?

# Interlude: Practice Problems 3.3, Part 1

1. A farmer in Minnesota has a choice of three varieties of rye (call them varieties $R_1$, $R_2$, and $R_3$) to grow and wishes to devote part of his available acreage to each variety. The farmer needs to harvest 9,000 bushels of grain at the end of the growing season to ensure a reasonable profit. He has 45 acres available for rye. Variety $R_1$ requires 3 person-hours of labor per bushel during the growing season and is expected to yield 400 bushels per acre. Variety $R_2$ requires 2 person-hours of labor per bushel during the growing season but is only expected to yield 200 bushels per acre. Variety $R_3$ requires only 1 person-hour of labor per bushel during the growing season but is also the lowest-yielding variety, only expected to yield 100 bushels per acre. The farmer has enough of a work force to provide 20,000 person-hours of labor for the growing season. Determine how many bushels of each variety the farmer should plan to sow in order to use up all available labor-hours. Also determine how many acres will be devoted to each variety of rye. Set up an appropriate model for the problem, declaring what all variables represent.

2. In preparation for Halloween, Fantastically Freakish Figurines is scheduling production of its three specialty figurines: Astrocreep, Bubonican, and Cthuluazon. Each figurine requires some processing time using three different machines. Call them Machine I, II, and III, respectively. Manufacture of Astrocreep requires 2 minutes on Machine I, 1 minute on Machine II, and 2 minutes on Machine III. Manufacture of Bubonican requires 1 minute on Machine I, 3 minutes on Machine II, and 1 minute on Machine III. Manufacture of Cthuluazon requires 1 minute on Machine I, 2 minutes on Machine II, and 2 minutes on Machine III. Machine I is available for 3 hours a day, Machine II is available for 5 hours a day, and Machine III is available for 4 hours a day. Determine how many of each type of figurine can be produced each day by using all of the available machine time. Set up an appropriate model for the problem, declaring what all variables represent.

# Interlude: Practice Problems 3.3, Part 2

1. A large farming operation is in need of 35,000 pounds of potash, 68,000 pounds of nitrogen, and 25,000 pounds of phosphoric acid. The local distributor carries three different brands of fertilizer, Growex, Wyesure, and Zeeplot. Each contains some amount of potash, nitrogen, and phosphoric acid, as given in the following table. The operation needs to determine how many tons of each brand of fertilizer to order to meet its needs.

| (pounds/ton) | potash | nitrogen | phosphoric acid |
|---|---|---|---|
| Growex | 400 | 500 | 300 |
| Wyesure | 600 | 1000 | 400 |
| Zeeplot | 700 | 1600 | 500 |

(a) Write a system of equations modeling the farming operation's problem. Declare what the variables represent.

(b) Solve this problem.

2. Percy has invested $40,000 in three stocks. The first year, stock A paid 6% dividends and increased 3% in value; stock B paid 7% dividends and increased 4% in value; stock C paid 8% dividends and increased 2% in value. If the total dividends were $2730 and the total increase in value was $1080, how much was invested in each stock? Set up an appropriate model declaring what all variables represent, and solve the problem. Clearly state the conclusion.

# Interlude: Practice Problems 3.3, Part 3

1. A chemist mixes three different solutions with concentrations of 20%, 30%, and 45% glucose to obtain 10 liters of a 38% glucose solution. If the amount of 30% solution used is one liter more than twice the amount of 20% solution used, then find the amount of each solution used. Set up an appropriate model, declare what all variables represent, and solve the problem. Clearly state the conclusion.

2.  As part of a promotion, a local bank invites its customers to view a large sack full of $5, $10, and $20 gold pieces, promising to give the sack to the first person able to state the number of coins for each denomination. Customers are told there are exactly 250 coins and with a total face value of $1875. If there are seven times as many $5 gold pieces as $20 gold pieces, then how many of each denomination are there? Set up an appropriate model, declare what all variables represent, and solve the problem. Clearly state the conclusion.

# 3.4 OTHER APPLICATIONS INVOLVING MATRICES

Recall that a matrix is just a rectangular array of numbers. Previously, we have used matrices to store information about systems of linear equations. We now look at other applications involving matrices that have little to do with systems of equations.

It is common practice to "name" matrices with letters (*A*, *B*, *C*, etc.). Recall that the size of a matrix is a descriptor of the number of rows and columns that the matrix has. As such, if *A* and *B* are two matrices of the same size, we say that *A* and *B* are **equal,** and write *A* = *B*, if the corresponding elements of *A* and *B* are equal.

We now describe arithmetic involving matrices. As with real number arithmetic, we will discuss *addition (sum)*, *subtraction (difference)*, and *multiplication (product)*.

## Sums and Differences

Consider the following problem.

> A drug company is testing two groups of 100 people each to see if a new headache medication is effective. Half of the patients in each group receive the medicine, while half receive a placebo. The results are summarized in a pair of labeled matrices.

| | | *Pain Relief Obtained* | | | | | *Pain Relief Obtained* | |
|---|---|---|---|---|---|---|---|---|
| | | Yes | No | | | | Yes | No |
| Group I: | Medicine | 42 | 8 | | Group II: | Medicine | 38 | 12 |
| | Placebo | 12 | 38 | | | Placebo | 20 | 30 |

> Of those that took the placebo, how many experienced pain relief? Of those that took the medicine, how many experienced no relief? On the basis of these results, does it appear that the medicine is effective?

Two matrices have been used to summarize data from the test, and the rows and columns have been labeled to indicate what the elements in the arrays represent. It is apparent that in order to answer the first two questions posed above, we merely need to find the appropriate entries in each matrix and add the results. For instance, in Group I, 12 people taking the placebo experienced pain relief, while in Group II 20 people taking the placebo experienced pain relief, giving a total of 32 people taking the placebo that experienced pain relief. In fact, we could easily form a third 2 × 2 matrix that gives the relevant data for the entire test population

| | | *Pain Relief Obtained* | |
|---|---|---|---|
| | | Yes | No |
| Entire Test Population: | Medicine | 80 | 20 |
| | Placebo | 32 | 68 |

Notice how the entries in this latest matrix are simply the sums of corresponding elements from the first two data matrices. In essence, the last matrix was formed by "adding" the first two matrices. From this, it appears that the medicine is effective, as 80% of the test group that took the medicine actually experienced pain relief.

So, it appears that there is a natural way to "add" two matrices. In fact, if *A* and *B* are two matrices of the same size, say *m* × *n*, then the **sum of *A* and *B*,** denoted by *A* + *B*, is the *m* × *n* matrix whose elements are just the sums of corresponding elements from *A* and *B*. Analogously, the **difference of *A* and *B*,** denoted by *A* − *B*, is the *m* × *n* matrix whose elements are just the differences between corresponding elements from *A* and *B*.

## EXAMPLE 1 ◆ Sum and Difference

The matrices $M$ and $F$ give the death rates, per million person trips, for male and female drivers for various ages and number of passengers, according to one source.

**Male ($M$)**  Number of Passengers

|     |     | 0 | 1 | 2 | ≥3 |
|-----|-----|------|------|------|------|
| Age | 16 | 2.61 | 4.39 | 6.29 | 9.08 |
|     | 17 | 1.63 | 2.77 | 4.61 | 6.92 |
|     | 30–59 | 0.92 | 0.75 | 0.62 | 0.54 |

**Female ($F$)**  Number of Passengers

|     |     | 0 | 1 | 2 | ≥3 |
|-----|-----|------|------|------|------|
| Age | 16 | 1.38 | 1.72 | 1.94 | 3.31 |
|     | 17 | 1.26 | 1.48 | 2.82 | 2.28 |
|     | 30–59 | 0.41 | 0.33 | 0.27 | 0.4 |

Write a matrix showing the combined death rates for both male and female drivers as well as a matrix showing the difference between the death rates of male and female drivers.

### Solution

The matrix showing the combined death rates is $M + F$, while the matrix showing the difference between the death rates is $M - F$. Here

Number of Passengers

|     |     | 0 | 1 | 2 | ≥3 |     |
|-----|-----|------|------|------|------|-----|
| Age | 16 | 3.99 | 6.11 | 8.23 | 12.39 |     |
|     | 17 | 2.89 | 4.25 | 7.43 | 9.2 | $= M + F$ |
|     | 30–59 | 1.33 | 1.08 | 0.89 | 0.94 |     |

and

Number of Passengers

|     |     | 0 | 1 | 2 | ≥3 |     |
|-----|-----|------|------|------|------|-----|
| Age | 16 | 1.23 | 2.67 | 4.35 | 5.77 |     |
|     | 17 | 0.37 | 1.29 | 1.79 | 4.64 | $= M - F$ |
|     | 30–59 | 0.51 | 0.42 | 0.35 | 0.14 |     |

Notice that all of the elements of $M - F$ are positive. What does that suggest, if the source of this information is correct?

## Matrices and Multiplication

While the way sums and differences of matrices are defined is quite natural, the concept of multiplication poses some challenges. In fact, we actually discuss *two* types of multiplication with respect to matrices.

A **scalar** is just a real number. Given any scalar $k$ and a matrix $A$, the **scalar product of $k$ and $A$,** denoted by $kA$, is the matrix whose elements are exactly the elements of $A$ multiplied by $k$. For instance, if

$$A = \begin{bmatrix} -6 & 0 \\ 3 & 4 \\ 8 & 9 \end{bmatrix}$$

then the scalar product $-3A$ is

$$-3A = -3 \begin{bmatrix} -6 & 0 \\ 3 & 4 \\ 8 & 9 \end{bmatrix} = \begin{bmatrix} 18 & 0 \\ -9 & -12 \\ -24 & -27 \end{bmatrix}$$

Before we formally consider the product of two matrices, we pose an example and develop a calculation that will serve as a necessary part of the idea of multiplication of two matrices.

## EXAMPLE 2 ◆ Calculating Weighted Totals

Deb, Mandy, and Mike are all taking the same Sociology class. Their course grades are determined according to performance on three 100-point exams. The instructor of the course **weights** the exam scores, though, so that the first exam counts as 20% of the final grade, the second exam counts as 30% of the final grade, and the third exam counts as 50% of the final grade. The instructor computes a weighted point total for each student, so that at the end of the semester each student obtains a (weighted) point total that cannot exceed 100 points. The matrix

$$
\begin{array}{c} \\ \text{Deb} \\ \text{Mandy} \\ \text{Mike} \end{array}
\begin{array}{ccc} \text{Ex. 1} & \text{Ex. 2} & \text{Ex. 3} \\ \left[\begin{array}{ccc} 69 & 78 & 70 \\ 95 & 67 & 88 \\ 52 & 83 & 89 \end{array}\right] \end{array} = P
$$

gives the scores the three students received on each exam. The matrix

$$
\begin{array}{c} \\ \text{Ex. 1} \\ \text{Ex. 2} \\ \text{Ex. 3} \end{array}
\begin{array}{c} \text{Weight} \\ \left[\begin{array}{c} 0.20 \\ 0.30 \\ 0.50 \end{array}\right] \end{array} = W
$$

gives the weighting information. What is the weighted point total for each student?

**Solution**

We will calculate Deb's weighted point total. The calculation of the weighted point totals for the other students is done similarly.

| Deb's Scores | Weighting Factor | Weighted Point Total |
|---|---|---|
| Exam 1: 69 | 0.2 | $69 \times 0.2 = 13.8$ |
| Exam 2: 78 | 0.3 | $78 \times 0.3 = 23.4$ |
| Exam 3: 70 | 0.5 | $70 \times 0.5 = 35$ |
| | **Total** | $13.8 + 23.4 + 35 = 72.2$ |

By the same type of calculations, we find weighted point totals for Mandy and Mike of 83.1 and 79.8, respectively. We record these results in a $3 \times 1$ matrix, $T$, where

$$
\begin{array}{c} \\ \text{Deb} \\ \text{Mandy} \\ \text{Mike} \end{array}
\begin{array}{c} \text{Weighted} \\ \text{Score} \\ \left[\begin{array}{c} 72.2 \\ 83.1 \\ 79.8 \end{array}\right] \end{array} = T
$$

For convenience, we record the matrices $P$, $W$, and $T$ from the last example together:

$$
P = \left[\begin{array}{ccc} 69 & 78 & 70 \\ 95 & 67 & 88 \\ 52 & 83 & 89 \end{array}\right] \quad W = \left[\begin{array}{c} 0.20 \\ 0.30 \\ 0.50 \end{array}\right] \quad T = \left[\begin{array}{c} 72.2 \\ 83.1 \\ 79.8 \end{array}\right]
$$

Looking at the elements of $P$ and $W$ and considering the calculations done in Example 2, it is clear that multiplying the first, second, and third elements in the first row of $P$ by the first, second, and third elements, respectively, of the first (only) column of $W$ and adding the results gives the element in row 1, column 1 of $T$. A similar calculation involving row 2 of $P$ and column 1 of $W$ gives the element in row 2, column 1 of $T$. Likewise considering row 3 of $P$ and column 1 of $W$. We say here that the calculation giving Deb's weighted point total is the **inner product** of the first row of $P$ and the first column of $W$. Similarly, Mandy's weighted point total is the inner product of the second row of $P$ and the first column of $W$, while Mike's weighted point total is the inner product of row 3 of $P$ and column 1 of $W$. Of course, these weighted point totals are the elements of $T$.

What this suggests is a type of "multiplication" of $P$ and $W$ that results in $T$. As such we might write

$$PW = \begin{bmatrix} 69 & 78 & 70 \\ 95 & 67 & 88 \\ 52 & 83 & 89 \end{bmatrix} \begin{bmatrix} 0.20 \\ 0.30 \\ 0.50 \end{bmatrix} = \begin{bmatrix} 72.2 \\ 83.1 \\ 79.8 \end{bmatrix} = T$$

More generally, suppose that $A$ and $B$ are two matrices such that the number of columns of $A$ and the number of rows of $B$ are equal, say $A$ has $n$ columns and $B$ has $n$ rows. Then for any integers $j$ and $k$ between 1 and $n$, inclusive, let $R_j$ denote the $j$th row of $A$ and let $C_k$ denote the $k$th column of $B$. We write

$$R_j = [a_1 \ a_2 \ \ldots \ a_n] \text{ and } C_k = \begin{bmatrix} b_1 \\ b_2 \\ \vdots \\ b_n \end{bmatrix}$$

and define the **inner product of $R_j$ and $C_k$,** denoted $Rj \bullet C_k$, by

$$R_j \bullet C_k = [a_1 \ a_2 \ \ldots \ a_n] \bullet \begin{bmatrix} b_1 \\ b_2 \\ \vdots \\ b_n \end{bmatrix} = a_1 b_1 + a_2 b_2 + \ldots + a_n b_n$$

### EXAMPLE 3 ◆ Calculating Inner Products

Given the matrices $A$ and $B$ below, calculate the inner products, if possible.

(a) row 1 of $A$ and column 1 of $B$

(b) row 2 of $A$ and column 3 of $B$

(c) row 1 of $B$ and column 2 of $A$

$$A = \begin{bmatrix} -4 & 2 & 0 \\ 9 & -1 & 5 \end{bmatrix} \quad B = \begin{bmatrix} 4 & 8 & -2 \\ 6 & 0 & 1 \\ 3 & -5 & 1 \end{bmatrix}$$

**Solution**

(a) $[-4 \ 2 \ 0] \bullet \begin{bmatrix} 4 \\ 6 \\ 3 \end{bmatrix} = (-4)(4) + (2)(6) + (0)(3) = -4$

(b) $[9 \ -1 \ 5] \bullet \begin{bmatrix} -2 \\ 1 \\ 1 \end{bmatrix} = (9)(-2) + (-1)(1) + (5)(1) = -14$

(c) Since rows of $B$ have three elements but columns of $A$ have two elements, that inner product is not defined.

We are now prepared to define the product of a pair of matrices. Let $A$ be an $m \times n$ matrix and let $B$ be an $n \times p$ matrix. The **product of $A$ and $B$**, denoted by $AB$, is an $m \times p$ matrix. The element in row $i$, column $j$ of $AB$ is the inner product of row $i$ from $A$ and column $j$ from $B$.

This definition of matrix multiplication may seem very confusing, but with a little practice it is not difficult. We illustrate matrix multiplication in the next example.

---

**EXAMPLE 4** ◆ **Product of Matrices**

Calculate the products $AB$ and $BA$, if possible, given the matrices $A$ and $B$ below.

$$A = \begin{bmatrix} -1 & 0 & 4 \\ 3 & 6 & 8 \end{bmatrix} \quad B = \begin{bmatrix} 10 & 0 \\ -4 & 2 \\ 7 & 8 \end{bmatrix}$$

**Solution**

Referencing rows from $A$ and columns from $B$, the product $AB$ is a $2 \times 2$ matrix given by

$$AB = \begin{bmatrix} -1 & 0 & 4 \\ 3 & 6 & 8 \end{bmatrix} \begin{bmatrix} 10 & 0 \\ -4 & 2 \\ 7 & 8 \end{bmatrix} = \begin{bmatrix} R_1 \bullet C_1 & R_1 \bullet C_2 \\ R_2 \bullet C_1 & R_2 \bullet C_2 \end{bmatrix}$$

$$= \begin{bmatrix} (-1)(10) + (0)(-4) + (4)(7) & (-1)(0) + (0)(2) + (4)(8) \\ (3)(10) + (6)(-4) + (8)(7) & (3)(0) + (6)(2) + (8)(8) \end{bmatrix}$$

$$= \begin{bmatrix} 18 & 32 \\ 62 & 76 \end{bmatrix}$$

Similarly, referencing rows from $B$ and columns from $A$, the product $BA$ is a $3 \times 3$ matrix given by

$$BA = \begin{bmatrix} 10 & 0 \\ -4 & 2 \\ 7 & 8 \end{bmatrix} \begin{bmatrix} -1 & 0 & 4 \\ 3 & 6 & 8 \end{bmatrix} = \begin{bmatrix} R_1 \bullet C_1 & R_1 \bullet C_2 & R_1 \bullet C_3 \\ R_2 \bullet C_1 & R_2 \bullet C_2 & R_2 \bullet C_3 \\ R_3 \bullet C_1 & R_3 \bullet C_2 & R_3 \bullet C_3 \end{bmatrix}$$

$$= \begin{bmatrix} (10)(-1) + (0)(3) & (10)(0) + (0)(6) & (10)(4) + (0)(8) \\ (-4)(-1) + (2)(3) & (-4)(0) + (2)(6) & (-4)(4) + (2)(8) \\ (7)(-1) + (8)(3) & (7)(0) + (8)(6) & (7)(4) + (8)(8) \end{bmatrix}$$

$$= \begin{bmatrix} -10 & 0 & 40 \\ 10 & 12 & 0 \\ 17 & 48 & 92 \end{bmatrix}$$

---

In the previous example, note that if $B$ had been a $3 \times 3$ matrix, then the product $AB$ would still be defined, but the product $BA$ would no longer be defined as it would not be possible to calculate inner products of rows of $B$ and columns of $A$.

---

**EXAMPLE 5** ◆ **Product of Matrices**

Calculate the products $AB$ and $BA$, if possible, given the matrices $A$ and $B$ below.

$$A = \begin{bmatrix} -5 & -2 \\ 3 & 8 \end{bmatrix} \quad B = \begin{bmatrix} 4 & -6 \\ 9 & 13 \end{bmatrix}$$

**Solution**

Notice that both $AB$ and $BA$ are defined.

$$AB = \begin{bmatrix} -5 & -2 \\ 3 & 8 \end{bmatrix}\begin{bmatrix} 4 & -6 \\ 9 & 13 \end{bmatrix} = \begin{bmatrix} (-5)(4) + (-2)(9) & (-5)(-6) + (-2)(13) \\ (3)(4) + (8)(9) & (3)(-6) + (8)(13) \end{bmatrix} = \begin{bmatrix} -38 & 4 \\ 84 & 86 \end{bmatrix}$$

$$AB = \begin{bmatrix} 4 & -6 \\ 9 & 13 \end{bmatrix}\begin{bmatrix} -5 & -2 \\ 3 & 8 \end{bmatrix} = \begin{bmatrix} (4)(-5) + (-6)(3) & (4)(-2) + (-6)(8) \\ (9)(-5) + (13)(3) & (9)(-2) + (13)(8) \end{bmatrix} = \begin{bmatrix} -38 & -56 \\ -6 & 86 \end{bmatrix}$$

One thing illustrated by Example 5 is that even if the matrix products $AB$ and $BA$ are defined, it need not be the case that $AB$ and $BA$ are equal. In other words, *in general, matrix multiplication is not commutative*.

We conclude with an example using repeated multiplication with a single matrix. Recall that if $r$ is a real number, then $r^2 = r \times r$, $r^3 = r \times r \times r$, and so on. We use similar "power" notation with matrices.

### EXAMPLE 6 ◆ Flight Options

The matrix $I$ below is an **incidence matrix**—it describes availability of direct flights between select cities, as offered by a regional airline. The cities are simply denoted $A$, $B$, $C$, and $D$. An entry of 1 in the matrix indicates that the two cities of that row and column are connected by a direct flight, while an entry of 0 indicates that no direct flight is offered. See the accompanying diagram (Figure 3.1).

$$\text{Origin} \begin{array}{c} \\ A \\ B \\ C \\ D \end{array}\begin{matrix} \overset{\text{Destination}}{A\ B\ C\ D} \\ \begin{bmatrix} 0 & 1 & 0 & 1 \\ 1 & 0 & 1 & 1 \\ 0 & 1 & 0 & 0 \\ 1 & 1 & 0 & 0 \end{bmatrix} \end{matrix} = I$$

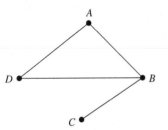

*FIGURE 3.1*

Now, $I^2 = II$, so

$$I^2 = \begin{bmatrix} 0 & 1 & 0 & 1 \\ 1 & 0 & 1 & 1 \\ 0 & 1 & 0 & 0 \\ 1 & 1 & 0 & 0 \end{bmatrix}\begin{bmatrix} 0 & 1 & 0 & 1 \\ 1 & 0 & 1 & 1 \\ 0 & 1 & 0 & 0 \\ 1 & 1 & 0 & 0 \end{bmatrix} = \begin{array}{c} A \\ B \\ C \\ D \end{array}\begin{matrix} \overset{A\ B\ C\ D}{} \\ \begin{bmatrix} 2 & 1 & 1 & 1 \\ 1 & 3 & 0 & 1 \\ 1 & 0 & 1 & 1 \\ 1 & 1 & 1 & 2 \end{bmatrix} \end{matrix}$$

The question, then, is how to interpret $I^2$. Consider the first row of $I^2$. Notice that to start and end a trip at city $A$, two flights must be taken—either from $A$ to $B$ and back or from $A$ to $D$ and back. So, there are two ways to start and end in city $A$ by taking two flights. Similarly, there is only one way to start in $A$ and end in $B$ by taking two flights, there is only one way to start in $A$ and end in $C$ by taking two flights, and only one way to start in $A$ and end in $D$ by taking two flights. This suggests, then, that $I^2$ records the number of flight options between cities consisting of two flights. You should check that this interpretation holds for the remaining elements of $I^2$.

# EXERCISES

*For problems 1–12, refer to the following matrices and perform the indicated calculations, if possible.*

$$A = \begin{bmatrix} 0 & 2 & 6 \\ 1 & -1 & 4 \\ -5 & 3 & 3 \\ 2 & 7 & 1 \end{bmatrix} \qquad B = \begin{bmatrix} 1 & 0 \\ 0 & 1 \end{bmatrix} \qquad C = \begin{bmatrix} 4 & 0 \\ 3 & -2 \end{bmatrix}$$

$$D = \begin{bmatrix} 3 & -2 & 4 \\ 2 & -7 & 6 \end{bmatrix} \qquad E = \begin{bmatrix} 0 & 5 & 8 \\ -4 & 2 & 3 \end{bmatrix} \qquad F = \begin{bmatrix} 2 & -9 & 1 \\ 6 & 12 & 4 \\ 10 & 5 & 6 \end{bmatrix}$$

1. $B + C$

2. Inner product of row 1 of $D$ and column 2 of $F$

3. Inner product of row 2 of $F$ and column 3 of $A$

4. $2A$

5. $CD$

6. $DC$

7. $AF$

8. $\frac{1}{2}D - 2E$

9. $B + E$

10. $-4F$

11. $BE$

12. $D - E$

13. If $A^2 = \begin{bmatrix} -2 & -1 \\ 2 & -1 \end{bmatrix}$ and $A^3 = \begin{bmatrix} -2 & 1 \\ -2 & -3 \end{bmatrix}$, what is $A$?

14. Ally, Marta, and Joe are taking a math class, and the instructor gives three exams, each consisting of two parts, an in-class part worth 80 points, and a take-home part worth 20 points. Each of the three exams, then, is worth 100 total points. On the in-class parts, Ally got scores of 64, 72, and 75, Marta got 70, 59, and 62, and Joe got 79, 68, and 60. On the take-home parts, the scores were as follows: Ally got 18, 16 and 16, and Marta and Joe (who worked together) each got 14, 17, and 18. Write two matrices, $I$ and $T$, whose sum gives the scores out of 100 for each of the three exams for Ally, Marta, and Joe. Label the rows and columns of the matrices.

15. A company makes radios and TV sets. Each radio requires 3 hours of assembly and 0.5 hour of packaging, while each TV set requires 5 hours of assembly and 1 hour of packaging.

    (a) Write a matrix $T$ representing the required time for assembly and packaging of radios and TV sets.

    (b) The company receives an order from a retail outlet for 30 radios and 20 TV sets. Find a matrix $S$ so that either $ST$ or $TS$ gives the total assembly time and the total packaging time required to fill the order. What is the total assembly time? What is the total packaging time?

16. The matrix

    $$A = \begin{bmatrix} \overset{\text{Baseballs}}{20} & \overset{\text{Footballs}}{50} & \overset{\text{Basketballs}}{15} \\ 40 & 10 & 80 \end{bmatrix} \begin{matrix} \text{Uptown store} \\ \text{Downtown store} \end{matrix}$$

    gives the number of baseballs, footballs, and basketballs in a sporting goods store's two branches. The store sells both leather and synthetic models for these three products. The prices charged (dollars) for these products are given in the matrix

    $$P = \begin{bmatrix} \overset{\text{Leather}}{\$10} & \overset{\text{Synthetic}}{\$5} \\ \$45 & \$30 \\ \$75 & \$20 \end{bmatrix} \begin{matrix} \text{Baseballs} \\ \text{Footballs} \\ \text{Basketballs} \end{matrix}$$

    Compute the matrix product $AP$ and interpret the elements in this matrix.

17. A certain sheep rancher owns the following numbers of a given breed of sheep.

| Age | Males | Females |
|---|---|---|
| 0–1 yr | 300 | 450 |
| 1–2 yr | 350 | 400 |
| 2 yr and up | 500 | 500 |

A neighboring ranch contains the following numbers of sheep.

| Age | Males | Females |
|---|---|---|
| 0–1 yr | 350 | 400 |
| 1–2 yr | 450 | 400 |
| 2 yr and up | 650 | 550 |

(a) Represent the sheep population of each ranch as a matrix.

(b) Use matrix arithmetic to determine by how many sheep in each category the second ranch exceeds the first ranch.

(c) The second rancher is thinking of buying the first ranch. If he does, what will his total sheep holdings be in each of the categories? Use matrix arithmetic.

**18.** The following matrix, *I*, is an incidence matrix which describes the availability of direct flights between five cities, *A*, *B*, *C*, *D*, and *E*, as described in Example 6.

$$
\begin{array}{c}
\phantom{A}\ \ A\ B\ C\ D\ E \\
\begin{array}{c} A \\ B \\ C \\ D \\ E \end{array}
\begin{bmatrix}
0 & 1 & 1 & 1 & 0 \\
1 & 0 & 1 & 0 & 0 \\
1 & 1 & 0 & 1 & 0 \\
1 & 0 & 1 & 0 & 1 \\
0 & 0 & 0 & 1 & 0
\end{bmatrix}
\end{array}
$$

**(a)** Compute $I^2$ and interpret.

**(b)** How many ways are there to begin in city *A* and end in city *B* by taking two flights?

# Interlude: Practice Problems 3.4, Part 1

1. The table below summarizes the closing price per share for Baracus (BA), Hammersmith Labs (HL), and Suncorp (SUN) common stock at the end of the indicated year, as well as the number of shares of the stock owned by Nigel, Bonnie, and Jamie at those times.

| $/share | BA | HL | SUN |
|---------|-------|-------|-------|
| 2011 | 89.92 | 47.26 | 18.55 |
| 2012 | 94.33 | 46.81 | 22.14 |
| 2013 | 92.65 | 49.42 | 26.77 |

| shares | BA | HL | SUN |
|--------|-----|-----|-----|
| Nigel | 420 | 180 | 300 |
| Bonnie | 225 | 175 | 450 |
| Jamie | 190 | 360 | 95 |

(a) Create a $3 \times 3$ matrix $P$ to represent the stock price information, labeling the rows and columns of $P$ appropriately.

(b) Create a $3 \times 3$ matrix $S$ to represent the holdings of the three people, labeling the rows and columns of $S$ appropriately.

(c) Show and calculate a matrix product that gives the value of the stock holdings of the three people at the end of the indicated years. Label the rows and columns of the product matrix appropriately.

2. SpecialTees sells specialty T-shirts and sweaters, with production plants in Verndale and Menahga. The company offers this apparel in three quality levels: standard, deluxe, and premium. Last fall the Verndale plant produced 2980 standard, 1950 deluxe, and 1220 premium T-shirts, along with 1840 standard, 1020 deluxe, and 760 premium sweatshirts. The Menahga plant produced 3980 standard, 2450 deluxe, and 1330 premium T-shirts, along with 2840 standard, 3650 deluxe, and 990 premium sweatshirts in the same time period.

   (a) Write "production matrices" of the same size for each plant that show how many of each different type of apparel were manufactured. Label your rows and columns appropriately.

   (b) Use the matrices from (a) to determine how many more or less articles of each type were produced in Menahga compared to in Verndale. Be clear about what matrix operation you are using.

   (c) Suppose that both plants expect a 3% increase in business next fall. Using an appropriate matrix operation on the matrices from (a), construct new matrices that show how many items of each type are expected to be produced next fall. Be clear about what matrix operation you are using.

# Interlude:  Practice Problems 3.4, Part 2

The $n \times n$ **identity matrix** (denoted $I_n$) is the square matrix whose entries along the *main diagonal* are 1 and all other entries are 0. For instance, $I_2 = \begin{bmatrix} 1 & 0 \\ 0 & 1 \end{bmatrix}$, $I_3 = \begin{bmatrix} 1 & 0 & 0 \\ 0 & 1 & 0 \\ 0 & 0 & 1 \end{bmatrix}$, and so on.

Given a pair of $n \times n$ matrices $A$ and $B$, we say that the two are **inverses** of each other if both $AB = I_n$ and $BA = I_n$.

1.  Let $A = \begin{bmatrix} 2 & -6 \\ 9 & -1 \end{bmatrix}$ and $B = \begin{bmatrix} a & b \\ c & d \end{bmatrix}$. Find the appropriate values of $a, b, c,$ and $d$ so that $A$ and $B$ are inverses of each other.

Given a matrix $A$, the ***transpose of*** $A$, denoted $A^T$ is the matrix whose rows are formed from the columns of $A$.

2. Let $A = \begin{bmatrix} 2 & -3 & 5 \\ 1 & 0 & -7 \end{bmatrix}$ and let $B = \begin{bmatrix} 8 & 2 \\ -3 & 4 \\ 6 & 9 \end{bmatrix}$.

   (a) Calculate $A^T$ and $B^T$.

   (b) Calculate $(AB)^T$.

   (c) Calculate $B^T A^T$.

# Chapter 3 Progress Check I

1. A survey conducted in Plasterton, WI revealed some interesting information about eligible voters and political affiliation. For voters under the age of 30, 50% identified themselves as Republican, 30% identified themselves as Democrat, and 20% identified themselves as Independent. For voters ages 30 to 50, the corresponding numbers were 45%, 40%, and 15%, respectively. For voters over the age of 50, the corresponding numbers were 40%, 50%, and 10%, respectively.

   (a) Write a matrix $A$ representing the reported political affiliations according to age group. Label rows and columns appropriately.

   (b) In Plasterton, there are 1500 voters under the age of 30, 1800 voters between the ages of 30 and 50, and 1050 voters over the age of 50. Find a matrix $V$ so that either $VA$ or $AV$ gives total number of eligible voters in the city that identify themselves as Democrat, Republican, or Independent. Label the rows and columns and the appropriately.

   (c) Calculate the appropriate product ($VA$ or $AV$) and show the resulting matrix. Label the rows and columns appropriately.

2. Dewey, Skrughem, and Howe is a real estate development firm that plans to build a new apartment complex close to campus consisting of one-bedroom apartments and two- and three-bedroom townhouses. A total of 192 units is planned. The total number of two- and three-bedroom townhouses will be the same as the number of one-bedroom apartments. The number of one-bedroom apartments will be three times the numer of three-bedroom townhouses. Determine how many of each type of housing unit will be in the complex. Set up an appropriate model, declare what all variables represent, and solve the problem. Clearly state the conclusion.

# Chapter 3  Progress Check II

1. Centerville Community College is planning to offer courses in Finite Math, Applied Calculus, and Web Design. Each section of Finite Math holds 40 students and generates $40,000 in tuition revenue. Each section of Applied Calculus holds 40 students and generates $60,000 in tuition revenue. Each section of Web Design holds 10 students and generates $20,000 in tuition revenue. The school will offer a total of six sections of the courses, in order to accommodate 210 students and will need to total tuition revenue of $260,000 to run the courses. How many of sections of each of the courses need to be offered? Set up an appropriate model, declare what all variables represent, and solve the problem. Clearly state the conclusion.

2. Quinn's Quickmart is open from 8:00 in the morning until midnight, every day of the year. The store has three shifts: 8 a.m. to 4 p.m., noon to 8 p.m., and 4 p.m. to midnight. These shifts must be apportioned among 60 workers. Management wishes to have 45 workers on hand during the overlap between the first two shifts and 40 workers on hand during the overlap between the last two shifts. Determine how the 60 workers should be allotted to the three shifts. Set up an appropriate model and declare what all variables represent.

# Chapter 3 Progress Check III

1. Let $A = \begin{bmatrix} 3 & -6 \\ 1 & -2 \end{bmatrix}$. Show that there is no matrix $B$ so that $A$ and $B$ are inverses (see p. 165).

2. Let $A = \begin{bmatrix} 4 & 7 \\ 1 & 2 \end{bmatrix}$, $B = \begin{bmatrix} 2 & -7 \\ -1 & 4 \end{bmatrix}$, $C = \begin{bmatrix} 6 & 2 \\ 8 & 3 \end{bmatrix}$, and $D = \begin{bmatrix} 1.5 & -1 \\ -4 & 3 \end{bmatrix}$.

   (a) Show that $A$ and $B$ are inverses, and that $C$ and $D$ are inverses.

   (b) Calculate $AC$ and $DB$.

3.  If $A = \begin{bmatrix} 1 & -2 \\ -2 & 5 \end{bmatrix}$ and $AB = \begin{bmatrix} -1 & 2 & -1 \\ 6 & -9 & 3 \end{bmatrix}$, then find $B$.

4.  Two farmers, John and Patty, have a wager going. Patty has challenged John to find a way to purchase exactly 100 chickens for $100. Roosters cost $5 each, hens cost $3 each, and baby chicks cost $0.05 each. Patty has agreed to pay for the cost of the chickens if John can figure out how to accomplish the purchase goal, provided he purchase at least one rooster, at least one hen, and at least one baby chick. If he cannot figure out how to do this (and hence lose the wager), then he must clean Patty's chicken coop. Can John actually win this wager? If so, determine how. If not, explain why not. Your analysis must include an appropriate mathematical model of the situation and justification for your conclusion.

# ANSWERS TO SELECTED EXERCISES

## Section 3.1

1. (a) Yes (b) No (c) No (d) Yes
3. $x = -1, y = 3$
5. Infinite number of solutions
7. $x_1 = .5, x_2 = 2$
9. $x = 18, y = 50$
11. $m = \dfrac{77}{26}, n = \dfrac{21}{13}$
13. $x = 2, y = -3$
15. Infinite number of solutions
17. $x_1 = \dfrac{2}{31}, x_2 = \dfrac{102}{31}$
19. 1300 gallons of regular chocolate milk, 1680 gallons of extra-thick chocolate milk
21. Kris has $200, Shelley has $250
23. 2800 adult and 1400 children's tickets

## Section 3.2

1. (a) Yes
   (b) Yes
   (c) No, doesn't satisfy condition 2 of reduced echelon form, for example.
   (d) Yes
3. $R_1 + R_2 \to R_2$
5. $-2R_1 + R_2 \to R_2, \dfrac{1}{4}R_2 \to R_2, -2R_2 + R_3 \to R_3$
7. (a) No (b) Yes (c) No (d) Yes
9. $x_1 = 3, x_2 = 4, x_3 = 4$
11. Infinite number of solutions
13. $x_1 = -\dfrac{1}{3}, x_2 = 2$
15. $x_1 = 1, x_2 = 2, x_3 = 0; x_1 = 4, x_2 = 1, x_3 = 1; x_1 = 7, x_2 = 0, x_3 = 2$

## Section 3.3

1. 5 hrs. running, 10 hrs. biking, 7 hrs. swimming
3. 0 10-ft. trucks, 10 14-ft. trucks, and 15 24-ft. trucks; 2 10-ft. trucks, 7 14-ft. trucks, and 16 24-ft. trucks; 4 10-ft., 4 14-ft., and 17 24-ft.; 6 10-ft., 1 14-ft., and 18 24-ft.
5. 80 doz. sleeveless, 140 doz. short-sleeve, 160 doz. long-sleeve
7. 18,000 packages Italian style, 15,000 packages French style, 54,000 packages Oriental style

9. 1.2 oz. peanut-raisin, 9.6 oz. peanut-chocolate chip, 11.2 oz. raisin-chocolate chip
11. 20 oz. of the first alloy, 40 oz. of the second alloy, 10 oz. of the third alloy
13. 6 cases orange juice, 20 cases grapefruit juice, 12 cases tomato juice

## Section 3.4

1. $\begin{bmatrix} 5 & 0 \\ 3 & -1 \end{bmatrix}$
3. Not defined
5. $\begin{bmatrix} 12 & -8 & 16 \\ 5 & 8 & 0 \end{bmatrix}$
7. $\begin{bmatrix} 72 & 54 & 44 \\ 36 & -1 & 21 \\ 38 & 96 & 25 \\ 56 & 71 & 36 \end{bmatrix}$
9. Not defined
11. $\begin{bmatrix} 0 & 5 & 8 \\ -4 & 2 & 3 \end{bmatrix}$
13. $\begin{bmatrix} 0 & -1 \\ 2 & 1 \end{bmatrix}$

15. (a)

|  | Radios | TV sets |  |
|---|---|---|---|
| $T = $ | 3 | 5 | Assembly |
|  | .5 | 1 | Packaging |

(b)

|  | number |  |
|---|---|---|
| $S = $ | 30 | Radios |
|  | 20 | TV sets |

Assembly time: 190 hours
Packaging time: 35 hours

17. (a)

|  | Males | Females |  |
|---|---|---|---|
| $R = $ | 300 | 450 | 0–1 yr |
|  | 350 | 400 | 1–2 yr |
|  | 500 | 500 | 2+ yr |

|  | Males | Females |  |
|---|---|---|---|
| $N = $ | 350 | 400 | 0–1 yr |
|  | 450 | 400 | 1–2 yr |
|  | 650 | 550 | 2+ yr |

**(b)**

$$N - R = \begin{bmatrix} 50 & -50 \\ 100 & 0 \\ 150 & 50 \end{bmatrix} \begin{matrix} \text{0–1 yr} \\ \text{1–2 yr} \\ \text{2+ yr} \end{matrix}$$

Males   Females

**(c)**

$$N + R = \begin{bmatrix} 650 & 850 \\ 800 & 800 \\ 1150 & 1050 \end{bmatrix} \begin{matrix} \text{0–1 yr} \\ \text{1–2 yr} \\ \text{2+ yr} \end{matrix}$$

Males   Females

# An Introduction to Linear Programming

As was discovered in the previous chapter, sometimes a problem has only one possible solution, or perhaps no solution at all. However, some problems have many solutions. In a situation where there are many reasonable solutions to a problem, how do we decide which solution is "best"? Linear programming came about in the 1940s as a result of a military project concerned with computing the most efficient and economical way to distribute men, weapons, and supplies during World War II. Since then linear programming has become a very important part of modern decision making on issues like resource allocation and scheduling. "Real" linear programming problems often involve thousands of variables and extremely large models, so many computer applications have been developed to deal with these types of problems. We will discuss more modest problems.

## 4.1  INEQUALITIES, LINEAR INEQUALITIES, AND GRAPHS

### Inequalities

Inequalities come in a variety of forms. Any statement involving one of the symbols $>$ (greater than), $\geq$ (greater than or equal to), $<$ (less than), or $\leq$ (less than or equal to) is termed an **inequality.** Inequalities of the $<$ type and $>$ type are sometimes called **strict inequalities.** Two kinds of inequalities, absolute inequalities and conditional inequalities, are to be distinguished. An inequality that involves numbers only or holds for all permissible values of the variables involved is called an **absolute inequality.**

$$-3 < -2, \quad 8 > \frac{1}{2}, \quad \text{and} \quad x^2 \geq 0$$

are absolute inequalities. An inequality that does not hold for all permissible values of the variables involved is called a **conditional inequality,** or more simply, an **inequality.**

$$2x + 3 < 7, \quad x^2 \leq 16, \quad \text{and} \quad 2x + 3y \geq 6$$

are conditional inequalities.

When 1 is substituted for $x$ in the inequality

$$2x + 3 < 7$$

we obtain $2(1) + 3 < 7$, or $5 < 7$, which is a correct statement. The number 1 or any other number that yields a correct statement when substituted for $x$ in $2x + 3 < 7$ (such as $\frac{3}{2}, \frac{1}{2}, 0, -4$) is said to be a solution of $2x + 3 < 7$. The number 1 and the other solutions of $2x + 3 < 7$ are said to satisfy this inequality. The number 2 is not a solution of (does not satisfy) $2x + 3 < 7$ since substitution of 2 in $2x + 3 < 7$ yields $2(2) + 3 < 7$, or $7 < 7$, which is an incorrect statement. More generally, a number is said to be a **solution** of a one-variable inequality if a correct statement is obtained when the number is substituted for the variable of the inequality.

A **solution** of an inequality in two variables, $x$ and $y$, let us say, is any ordered pair of numbers that yields a correct statement when substituted for $x$ and $y$, respectively. For example, $(3, 0)$ is a solution of

$$2x + 3y \geq 6$$

since substitution of 3 for $x$ and 0 for $y$ yields $2(3) + 3(0) \geq 6$, or $6 \geq 6$, which is a correct statement. $(3, 0)$ is said to satisfy $2x + 3y \geq 6$. $(0, 1)$ is not a solution of (does not satisfy) $2x + 3y \geq 6$, since substitution of 0 for $x$ and 1 for $y$ yields $2(0) + 3(1) \geq 6$, or $3 \geq 6$, which is an incorrect statement.

A **solution** of an inequality in three variables, $x, y$, and $z$, is any ordered triple of numbers that yields a correct statement when substituted for $x, y$, and $z$, respectively. Thus $(-1, 3, 4)$ is a solution of

$$x^2 y + z < 10$$

since substitution of $-1$ for $x$, 3 for $y$, and 4 for $z$ yields $(-1)^2(3) + 4 < 10$, or $7 < 10$, which is a correct statement. Here too we say that $(-1, 3, 4)$ satisfies $x^2 y + z < 10$. The ordered triple of numbers $(2, 4, -1)$ does not satisfy (is not a solution of) $x^2 y + z < 10$, since substitution of 2 for $x$, 4 for $y$, and $-1$ for $z$ yields $(2)^2(4) - 1 < 10$, or $15 < 10$, which is an incorrect statement. The notion of solution of an inequality in any number of variables is defined in an analogous way.

# EXERCISES

*Determine which of $(0, 2)$, $(3, 22)$, $(-4, 0)$, $(\frac{1}{2}, 5)$, $(2, 3)$, and $(6, 2)$ satisfy the following inequalities.*

1. $3x + 2y \leq 6$
2. $2x - y^2 < 4$
3. $2x + xy \geq 10$
4. $2x - 3y < 4$
5. $5x^2 - y > 6$
6. $2x^2 - y^2 \leq 4$

*Determine which of $(2, 3, 1)$, $(1, 3, -1)$, $(4, 2, 1)$, $(3, -1, 4)$, $(\frac{1}{2}, \frac{1}{3}, -3)$, and $(-2, 3, 1)$ satisfy the following inequalities.*

7. $3x + 2y - z \leq 8$
8. $2x + 5y - 3z \leq 10$
9. $3x^2 y - 2z > 6$
10. $x^2 + 2y^2 - z \leq 12$
11. $\dfrac{2x^2 - y^2}{z} \leq 6$
12. $4x - 2y + 3z \leq 8$

## Operations of Equivalence

Two inequalities are said to be **equivalent** if they have the same solutions. The following operations of equivalence lead to equivalent inequalities:

> 1. The same constant may be added to, or subtracted from, both sides of an inequality; the direction of the inequality is maintained. That is, if $a \leq b$ and $c$ is any number, then $a + c \leq b + c$ (and $a - c \leq b - c$).

For example, adding 5 to both sides of $2x - 5 \leq 3$ yields the equivalent inequality $2x \leq 8$.

> 2. A term may be transferred from one side of an inequality to the other side, provided that its sign is changed; the direction of the inequality is maintained. This operation is called **transposition.**

For example, transposing $2x$ from the left side to the right side of $2x + y \leq 8$ yields the equivalent inequality $y \leq 8 - 2x$.

> 3. Both sides of an inequality may be multiplied or divided by the same nonzero constant; multiplying or dividing by a positive constant maintains the direction of the inequality; multiplying or dividing by a negative constant reverses the direction of the inequality. That is, if $a \leq b$ and $c$ is positive, then $ac \leq bc$ and $a/c \leq b/c$. If $a \leq b$ and $c$ is negative, then $ac \geq bc$ and $a/c \geq b/c$.

For example, multiplying both sides of $2 < 3$ by $-1$ yields $-2 > -3$. The direction of the inequality is reversed since $-1$ is negative. Dividing both sides of $2x \le 8$ by 2 yields the equivalent inequality $x \le 4$. The direction of the original inequality is maintained since 2 is positive.

These operations of equivalence for inequalities are analogous to the ones cited for equations, with one difference. In multiplying or dividing both sides of an inequality by a value one must pay particular attention to the sign of the value. Positive values maintain the direction of an inequality; negative values reverse the direction of an inequality.

## EXAMPLE 1

Determine the solutions of $10 - 3x \ge 7$.

Subtracting 10 from both sides of $10 - 3x \ge 7$ yields

$$-3x \ge -3$$

By dividing both sides of $-3x \ge -3$ by $-3$ (which is negative), we obtain

$$x \le 1$$

Since the operations employed lead to equivalent inequalities, $10 - 3x \ge 7$ is equivalent to $x \le 1$. Every number that is less than or equal to 1 is a solution of $10 - 3x \ge 7$, and every solution of $10 - 3x \ge 7$ is less than or equal to 1.

## EXAMPLE 2

Show that $y \ge \frac{1}{5}(x + y)$ is equivalent to $-x + 4y \ge 0$.

Multiplying both sides of $y \ge \frac{1}{5}(x + y)$ by 5 yields

$$5y \ge x + y$$

By transposing $x + y$ we obtain

$$-x + 4y \ge 0$$

## EXAMPLE 3

Show that $200 - x - y \ge 0$ is equivalent to $x + y \le 200$.

Subtracting 200 from both sides of $200 - x - y \ge 0$ yields

$$-x - y \ge -200$$

Multiplying both sides of $-x - y \ge -200$ by $-1$ (which is negative) gives us

$$x + y \le 200$$

## EXAMPLE 4

Show that $2(x - 4) > 3x + 2$ is equivalent to $x < -10$.

$$2(x - 4) > 3x + 2$$

Since $2(x - 4) = 2x - 8$, we have

$$2x - 8 > 3x + 2$$

Transposing $3x$ yields

$$-x - 8 > 2$$

Adding 8 to both sides gives us

$$-x > 10$$

Multiplying both sides by $-1$ yields

$$x < -10$$

# EXERCISES

13. Show that $100 - x \geq 0$ is equivalent to $x \leq 100$.

14. Show that $300 - y \geq 0$ is equivalent to $y \leq 300$.

15. Show that $1000 - x - y \geq 0$ is equivalent to $x + y \leq 1000$.

16. Show that $2x + y < 6$ is equivalent to $y < 6 - 2x$.

17. Show that $x \leq 2$ is equivalent to $3x - 4 \leq 2$.

18. Show that $2(x - 3y) < 8$ is equivalent to $x < 4 + 3y$.

19. Show that $3(x + 2y) < 2x + 4$ is equivalent to $x < 4 - 6y$.

20. Show that $-2(x + 3) < 4x + 12$ is equivalent to $x > -3$.

*Find the solutions of the following inequalities in one variable. Sketch their graphs.*

21. $3x + 6 > 4$

22. $-2x \geq 5$

23. $5x + 3 \leq 13$

24. $3y + 2 < 11$

25. $3 - 4y > -5$

26. $2 - 3y < 17$

27. $2(x + 4) < x - 7$

28. $-3(2x + 2) < 6x + 5$

29. $2(x - 4) < 3(2x - 6)$

30. $4(3 - 2x) \geq 7 - 3x$

31. $3(4 - 2y) \leq y - 2$

32. $-2(3 - y) \geq y + 1$

## Graphs of Two-Variable Inequalities

The **graph** of an inequality in two variables is the collection of all those points, and only those points, whose coordinates satisfy the inequality. In a number of situations the geometric view of an inequality provided by its graph is most useful. We shall find graphs of particular importance in our discussion of basic linear programming in Chapter 4. In this connection especially, graphs of linear inequalities occupy the spotlight.

A two-variable linear equation is of the form

$$Ax + By = C$$

where $A$ and $B$ are not both zero. A **two-variable linear inequality** is an inequality that can be obtained from $Ax + By = C$ by replacing the equality condition ($=$) by any one of the inequality conditions $>$, $\geq$, $<$, or $\leq$. $2x - 3y \leq 6$, for example, is a two-variable linear inequality; it can be obtained from the linear equation $2x - 3y = 6$ by replacing $=$ by $\leq$.

The graph of $2x - 3y = 6$ is a line that divides the rest of the coordinate plane into two components (see Figure 4.1). All points on one side of this boundary line satisfy $2x - 3y < 6$, while all points on the other side satisfy $2x - 3y > 6$. To find which side corresponds to which inequality, it suffices to choose a test point on one side of $2x - 3y = 6$ and substitute it into $2x - 3y < 6$. If this test point satisfies $2x - 3y < 6$, then all points on the same side as this test point satisfy $2x - 3y < 6$, and all points on the other side of the boundary line satisfy $2x - 3y > 6$. If the test point does not satisfy $2x - 3y < 6$, then it satisfies $2x - 3y > 6$, as do all other points on the same side as the test point. The points on the other side of the boundary line satisfy $2x - 3y < 6$.

To illustrate, let us take $(0, 0)$ as our test point. (Any other point not on the boundary line $2x - 3y = 6$ would do just as well.) Substituting $(0, 0)$ into $2x - 3y < 6$ yields $0 < 6$, which is a correct statement. Thus $(0, 0)$ satisfies $2x - 3y < 6$, and all points on the same side as $(0, 0)$ (above $2x - 3y = 6$) satisfy $2x - 3y < 6$. The inequality $2x - 3y > 6$ is satisfied by all points below $2x - 3y = 6$ (see Figure 4.2).

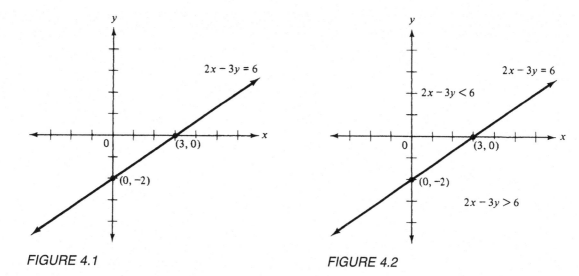

FIGURE 4.1                                    FIGURE 4.2

More generally, the graph of $Ax + By = C$ is a line that divides the rest of the coordinate plane into two components. All points on one side of $Ax + By = C$ satisfy $Ax + By < C$, while all points on the other side satisfy $Ax + By > C$. Examples 5 and 6 further illustrate the graphing of linear inequalities.

## EXAMPLE 5

Sketch the graph of $5x + 2y \geq 10$.

We first graph the boundary line $5x + 2y = 10$. For $x = 0, y = 5$; for $y = 0, x = 2$. Thus $5x + 2y = 10$ passes through $(0, 5)$ and $(2, 0)$ (see Figure 4.3). We next choose a test point not on $5x + 2y = 10$, $(0, 0)$, for example, and determine if it satisfies $5x + 2y > 10$. Substituting 0 for $x$ and 0 for $y$ in $5x + 2y > 10$ yields $0 > 10$, which is an incorrect statement. Since $(0, 0)$ does not satisfy $5x + 2y > 10$, the points that do satisfy $5x + 2y > 10$ are on the other side of (above) the boundary line $5x + 2y = 10$. The graphs of $5x + 2y \geq 10$ consists of all points on and above the boundary line $5x + 2y = 10$. This is indicated visually by drawing $5x + 2y = 10$ as a solid line and shading the region above it (see Figure 4.4).

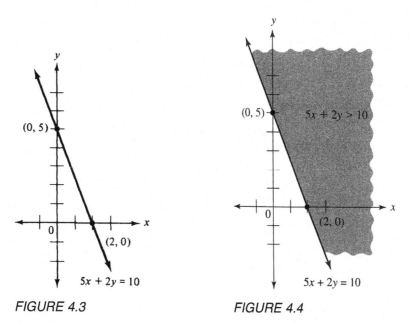

FIGURE 4.3                          FIGURE 4.4

## EXAMPLE 6

Sketch the graph of $x - 2y < 4$. We first graph the boundary line $x - 2y = 4$. For $x = 0, y = -2$; for $y = 0, x = 4$. Thus $x - 2y = 4$ passes through $(0, -2)$ and $(4, 0)$. Since this boundary line is not part of the graph of $x - 2y < 4$ (this inequality is a strict inequality), we draw a dashed line as shown in Figure 4.5. We next choose a test point not on $x - 2y = 4$, $(0, 0)$, for example, and determine if it satisfies $x - 2y < 4$. Substituting 0 for $x$ and 0 for $y$ in $x - 2y < 4$ yields $0 < 4$, which is a correct statement. Since $(0, 0)$ satisfies our inequality and is above the boundary line $x - 2y = 4$, the graph of $x - 2y < 4$ consists of all points above this boundary line (see Figure 4.6).

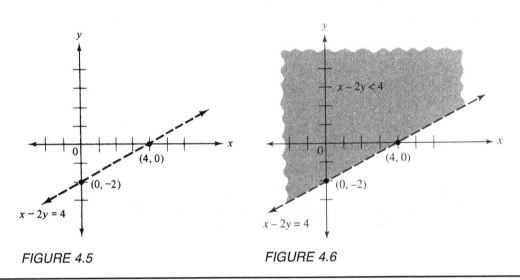

FIGURE 4.5                    FIGURE 4.6

## EXERCISES

*Sketch the graphs of the following inequalities.*

**33.** $x + y \geq 3$

**34.** $x + 2y \leq 6$

**35.** $x - y \geq 4$

**36.** $2x + y < 8$

**37.** $2x - y < 1$

**38.** $x + 3y < 6$

**39.** $x \geq 0$

**40.** $y \geq 0$

**41.** $3x - y \leq 3$

**42.** $4x - 3y > 12$

**43.** $x + y \geq 0$

**44.** $x - y < 0$

**45.** $-x + 4y \geq 0$

**46.** $4x + 5y \geq 75$

**47.** $2x + y \geq 23$

### Systems of Linear Inequalities

Two or more linear inequalities involving the same variables, considered as a unit, are said to form a **system of linear inequalities.** For example,

$$2x - 3y \leq 6 \tag{4.1}$$

$$5x + 2y \geq 10 \tag{4.2}$$

and

$$x \geq 0 \tag{4.3}$$

$$y \geq 0 \tag{4.4}$$

$$x + 3y \leq 90 \tag{4.5}$$

$$2x + y \leq 80 \tag{4.6}$$

are systems of 2-variable linear inequalities while

$$x + 2y + z \le 8 \tag{4.7}$$

$$2x - y - \frac{1}{2}z \le 6 \tag{4.8}$$

and

$$x \ge 0 \tag{4.9}$$

$$y \ge 0 \tag{4.10}$$

$$z \ge 0 \tag{4.11}$$

$$2x + 3y + z \le 11 \tag{4.12}$$

$$x + y + 3z \le 10 \tag{4.13}$$

$$2x + 2y + z \le 10 \tag{4.14}$$

are systems of three-variable linear inequalities.

The notion of solution of a system of inequalities is analogous to the notion of solution of a system of equations. For example, any ordered pair of numbers that satisfies all inequalities in the system

$$2x - 3y \le 6 \tag{4.1}$$

$$5x + 2y \ge 10 \tag{4.2}$$

is said to be a solution of this two-variable system. Thus $(2, 1)$ is a solution, since the substitution of 2 for $x$ and 1 for $y$ in (4.1) and (4.2) yields

$$1 \le 6$$

$$12 \ge 10$$

both of which are correct statements. The ordered pair $(1, 0)$ is not a solution since the substitution of 1 for $x$ and 0 for $y$ in (4.1) and (4.2) yields the statements

$$2 \le 6$$

$$5 \ge 10$$

not both of which are correct.

Any ordered triple of numbers that satisfies all inequalities in the system

$$x \ge 0 \tag{4.9}$$

$$y \ge 0 \tag{4.10}$$

$$z \ge 0 \tag{4.11}$$

$$2x + 3y + z \le 11 \tag{4.12}$$

$$x + y + 3z \le 10 \tag{4.13}$$

$$2x + 2y + z \le 10 \tag{4.14}$$

is said to be a solution of this three-variable system. Thus $(1, 2, 1)$ is a solution, since it satisfies inequalities (4.9) through (4.14), whereas $(2, 1, 3)$ is not a solution since it does not satisfy (4.13) (substitution yields $12 \le 10$). More generally, a **solution** of an $n$-variable system of linear inequalities is any ordered collection of $n$ numbers that satisfies all inequalities in the system.

# EXERCISES

**48.** Which of $(2, 3), (4, 1), (-1, 3), (\frac{1}{2}, 2), (6, -1)$, and $(3, 1)$ are solutions of the following system?

$$2x + 4y \leq 15 \qquad (4.15)$$

$$3x + y \leq 10 \qquad (4.16)$$

**49.** Which of $(8, 2), (8, 17), (20, 5), (9, 12), (5, 4), (12, 2)$, and $(14, 7)$ are solutions of the following system?

$$x \geq 0 \qquad (4.17)$$

$$y \geq 0 \qquad (4.18)$$

$$x + y \leq 25 \qquad (4.19)$$

$$-x + 4y \geq 0 \qquad (4.20)$$

$$x \geq 8 \qquad (4.21)$$

**50.** Which of $(2, 1, 3), (1, 4, 1), (4, 3, 2), (3, 1, 1), (4, 2, 2,), (1, 5, 1), (1, 2, 1)$, and $(0, 4, 3)$ are solutions of the following system?

$$x + 2y + 3z \leq 12 \qquad (4.22)$$

$$5x - y + z \leq 9 \qquad (4.23)$$

**51.** Which of $(1, 1, 2), (0, 1, 3), (2, 1, 1), (3, -1, 2), (4, 1, 0), (2, 1, 2), (1, 3, 1)$, and $(0, 2, 3)$ are solutions of the following system?

$$x \geq 0, \quad y \geq 0, \quad z \geq 0$$

$$2x + 3y + z \leq 11$$

$$x + y + 3z \leq 10$$

$$2x + 2y + z \leq 10$$

## Graphs of Systems of Two-Variable Linear Inequalities

To graph a system of two-variable inequalities we graph the individual inequalities on the same coordinate system and then pick out the region which is common to them—that is, the overlap.

**EXAMPLE 7**

Sketch the graph of the system

$$x + 2y \leq 8$$

$$3x + y \leq 9$$

We first graph $x + 2y \leq 8$. The boundary line $x + 2y = 8$ is shown in Figure 4.7. Since the test point $(0.0)$ satisfies $x + 2y \leq 8$, the graph of $x + 2y \leq 8$ consists of all points on and below $x + 2y = 8$ as shown in Figure 4.8.

Sketching the graph of $3x + y \leq 9$ on the same coordinate system yields Figure 4.9. The overlap consists of all points on and below both boundary lines, shown in Figure 4.10.

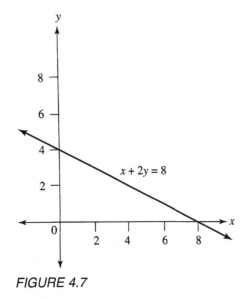

FIGURE 4.7

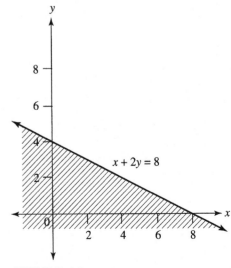

FIGURE 4.8

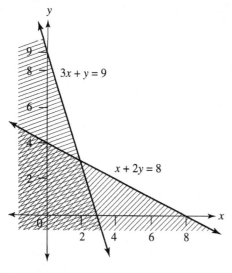

FIGURE 4.9

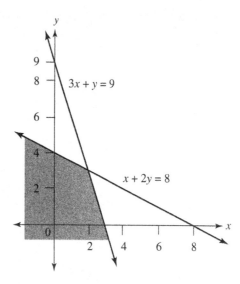

FIGURE 4.10

The nonnegativity conditions $x \geq 0$, $y \geq 0$ restrict us to the first quadrant. It is useful to make mental note of this because many problems of interest require such conditions. We encounter such problems in Chapter 4.

## EXAMPLE 8

Sketch the graph of the system

$$x \geq 0$$
$$y \geq 0$$
$$x + 2y \leq 8$$
$$3x + y \leq 9$$

All we have to do is restrict the graph shown in Figure 4.10 to the first quadrant. The result is shown in Figure 4.11.

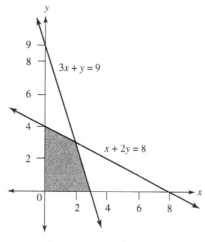

FIGURE 4.11

# EXAMPLE 9

Sketch the graph of the system

$$x \geq 0$$

$$y \geq 0$$

$$2x + y \geq 23$$

$$4x + 5y \geq 75$$

Since $x \geq 0$ and $y \geq 0$ restrict us to the first quadrant, it suffices to graph $2x + y \geq 23$ and $4x + 5y \geq 75$ restricted to the first quadrant. The graph of $2x + y \geq 23$ restricted to the first quadrant is shown in Figure 4.12(a), and the graphs of $2x + y \geq 23$ and $4x + 5y \geq 75$, restricted to the first quadrant, are shown in Figure 4.12(b). The graph of our system is the region in the first quadrant that is both shaded and lined in (the region in the first quadrant above and including both boundary lines) and is shown in Figure 4.13.

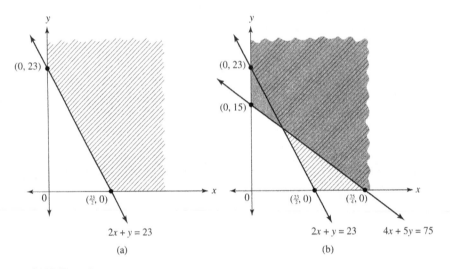

FIGURE 4.12

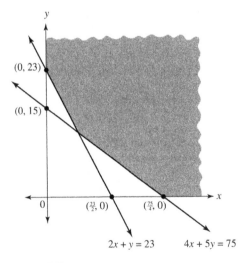

FIGURE 4.13

# EXERCISES

*Sketch the graphs of the following systems of inequalities.*

**52.** $x \geq 0, \quad y \geq 0$
$x + y \geq 3$

**53.** $x \geq 0, \quad y \geq 0$
$2x + y \geq 6$
$x \geq 3$

**54.** $x + y \leq 5$
$2x + 3y \leq 12$

**55.** $x \geq 0, \quad y \geq 0$
$x + y \leq 5$
$2x + 3y \leq 12$

**56.** $2x + y \geq 10$
$x + 2y \leq 12$

**57.** $2x + y \leq 10$
$x + 2y \geq 12$

**58.** $x + y \geq 5$
$2x + 3y \geq 12$

**59.** $x + y \leq 5$
$2x + 3y \geq 12$

**60.** $x + 4y \leq 10$
$3x + y \leq 8$

**61.** $x \geq 0, \quad y \geq 0$
$x + 2y \geq 10$
$3x + y \geq 15$

**62.** $x \geq 0, \quad y \geq 0$
$x + 2y \geq 4$
$x - 2y \geq 2$

**63.** $x \geq 0, \quad y \geq 0$
$2x + y \geq 4$
$x + y \geq 3$
$x + 2y \geq 4$

**64.** $x \geq 0, \quad y \geq 0$
$x + y \leq 25$
$-x + 4y \geq 0$
$x \geq 8$

**65.** $x \geq 0, \quad y \geq 0$
$5x + 4y \leq 120$
$x \geq 8, \quad x \leq 16,$
$y \geq 6$

**66.** $x \geq 0, \quad y \geq 0$
$4x + 3y \leq 320$
$5x + 2y \leq 330$

**67.** $x \geq 0, \quad y \geq 0$
$2x + 3y \leq 1100$
$5x + 3y \leq 1400$
$4x + y \leq 756$
$x \geq 25, \quad y \geq 40$

**68.** $x \geq 0, \quad y \geq 0$
$8x + 5y \leq 2210$
$3x + 2y \leq 860$
$x \geq 50, \quad y \geq 50$

# Interlude: Practice Problems 4.1

1. Sketch graphs of the following systems of inequalities. Label all corner points with exact coordinates.

   (a) $2y < 6x - 12$

   $x - y \leq 6$

   $x \leq 8$

   $y \leq 4$

   (b) $2x + y \geq 6$

   $-5x + y \leq 6$

   $x \leq y$

2. Weerahl Nuts has 75 pounds of cashews and 120 pounds of peanuts that will be mixed into one-pound packages of two different mixtures. The first mixture contains 4 ounces of cashews and 12 ounces of peanuts, while the second mixture contains equal amounts of cashews and peanuts. Set up a system of inequalities that models how many packages of each type of mix can possibly be assembled and sketch the corresponding feasible region. Declare what variables represent and label all corner points with exact coordinates.

3. Consider the feasible region shown to the right. Find a system of linear inequalities that generates this feasible region. Assume that gridlines each mark one unit.

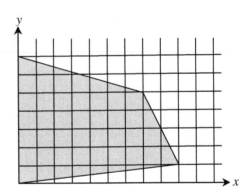

## 4.2 THE BIRTH AND REBIRTH OF LINEAR PROGRAMMING

The seed from which the discipline called linear programming first germinated was planted in the late 1930's when the Leningrad Plywood Trust approached the Mathematics and Mechanics Department of Leningrad University for help in solving a production scheduling problem of the following nature. The Plywood Trust had different machines for peeling logs for the manufacture of plywood. Various kinds of logs were handled and the productivity of each kind of machine (that is, the number of logs peeled per unit of time) depended on the wood being worked on. The problem was to determine how much work time each kind of machine should be assigned to each kind of log so that the number of peeled logs produced is maximized. A basic condition which had to be satisfied is that if logs of a given type of wood make up a specified percent of the input, then peeled logs of that type would also constitute that percent of the output.

The germination of this seed is due to Leonid Kantorovich who saw that it together with a wide variety of economic planning problems can be formulated in terms of what are today called linear program models. These problems involved the optimum distribution of worktime of machines, minimization of scrap in manufacturing processes, best utilization of raw materials, optimum distribution of arable land, optimal fulfillment of a construction plan with given construction materials, and the minimal cost plan for shipping freight from given sources to given destinations.

In 1939 Kantorovich published a report* on his discoveries that included a method sufficient for solving all the linear program models he had formulated for the aforenoted problems. The chaos of the Second World War and the postwar intellectual climate in the Soviet Union were highly unfavorable for the development and implementation of Kantorovich's linear programming methods in the Soviet economic scene. Independently of the Soviet scene, linear programming methods were developed in the United States and Western Europe in the late 1940's, and the 1950's and 60's saw the development of a wide variety of linear program models for problems arising in such areas as economic planning, accounting, banking, finance, industrial engineering, and marketing. The thaw in the Soviet Union's intellectual climate which followed Joseph Stalin's death in 1953 saw the development and implementation of Kantorovich's ideas in the economic life of the U.S.S.R.

In 1975 Kantorovich was a co-recipient of the Nobel Prize in economics for his development of linear programming methods and their application in economics.

## 4.3 A TALE OF TWO LINEAR PROGRAMS

The Austin Company, a producer of high quality electronic home entertainment equipment, has decided to enter the digital tape player market by introducing two models, DT-1 and DT-2. Their problem is to determine the number of units of each model to be produced to maximize profit.

The Company's operations research department was asked to study the situation and make recommendations. The OR department began their analysis by collecting data. They divided the manufacturing process into three phases; construction, assembly, and finishing. The data collected and their analysis led them to make the following assumptions.

(a) In the construction phase each DT-1 unit requires 2 hours of labor and each DT-2 unit requires 3 hours of labor. At most 1,100 hours of construction time are available per week.

(b) In the assembly phase each DT-1 unit requires 5 hours of labor and each DT-2 unit requires 3 hours of labor. At most 1,400 hours of assembly time are available per week.

(c) In the finishing phase each DT-1 unit requires 4 hours of labor and each DT-2 unit requires 1 hour of labor. At most 756 hours of finishing time are available per week.

---

*L. V. Kantorovich, *Mathematical Methods of Organizing and Planning Production,* Leningrad University, 1939. For an English translation, see *Management Science,* vol. 6, no. 4 (July 1960), pp. 363–422; or V. S. Nemchinov, ed., *The Use of Mathematics in Economics* (Cambridge, Mass.: MIT Press, 1964).

(d) After taking cost and revenue factors into consideration the anticipated profit for each DT-1 unit is $150 and the anticipated profit for each DT-2 unit is $120. In order for these unit profit values to hold the Company must produce at least 25 DT-1 and 40 DT-2 units per week.

(e) There is an unlimited market for the DT-1 and DT-2 models.

(f) All factors other than the ones considered in the analysis of the production of the DT-1 and DT-2 models are negligible.

Its next task was to translate these assumptions into mathematical form, being careful to include everything stated in the assumptions and not go beyond them. The OR department began by introducing variables for the quantities it sought to determine; let $x$ denote the number of DT-1 and $y$ the number of DT-2 units to be made weekly. There is a fair amount of data contained in the assumptions and it is useful to make it available at a glance in tabular form. This is done in Table 4.1.

The key to expressing profit and the conditions that have emerged in terms of $x$ and $y$ is the information stated on unit profits and unit construction, assembly and finishing times for DT-1 and DT-2.

**Table 4.1**

|  | No. of Units to Be Made | Profit Per Unit | Construction Time Per Unit (hrs) | Assembly Time Per Unit (hrs) | Finishing Time Per Unit (hrs) |
|---|---|---|---|---|---|
| DT-1 | $x$ | $150 | 2 | 5 | 4 |
| DT-2 | $y$ | $120 | 3 | 3 | 1 |

We have:

$$\text{profit} = \begin{bmatrix} \text{profit on} \\ \text{DT-1} \end{bmatrix} + \begin{bmatrix} \text{profit on} \\ \text{DT-2} \end{bmatrix}$$

$$= \begin{bmatrix} \text{profit on} \\ \text{one DT-1} \\ \text{unit} \end{bmatrix} \cdot \begin{bmatrix} \text{no. of} \\ \text{units} \\ \text{made} \end{bmatrix} + \begin{bmatrix} \text{profit on} \\ \text{one DT-2} \\ \text{unit} \end{bmatrix} \cdot \begin{bmatrix} \text{no. of} \\ \text{units} \\ \text{made} \end{bmatrix}$$

$$= 150x + 120y$$

The profit obtained by making $x$ DT-1 and $y$ DT-2 units per week is expressed by the linear function:

$$P(x, y) = 150x + 120y$$

As to the conditions that $x$ and $y$ must satisfy, since the number of units made must be non-negative, we have:

$$x \geq 0$$
$$y \geq 0$$

The construction time condition is that

$$(\text{total construction time used}) \leq 1100.$$

In terms of unit construction times, 2 hours are needed for one unit of DT-1 and 3 hours are needed for one unit of DT-2, $2x$ hours are needed for $x$ DT-1 units and $3y$ hours are needed for $y$ DT 2 units. The total construction time utilized is expressed by $2x + 3y$. Thus, the construction time utilized is expressed by $2x + 3y$. Thus, the construction time condition is:

$$2x + 3y \leq 1100$$

Similarly, the assembly and finishing time conditions are stated by the inequalities:

$$5x + 3y \leq 1400$$

$$4x + y \leq 756$$

The conditions that at least 25 DT-1 and 40 DT-2 units must be produced weekly are expressed by the inequalities:

$$x \geq 25$$

$$y \geq 40$$

We thus emerge with the following mathematical structure, called linear program model LP-1, as a translation of the assumptions made by the OR department of the Austin Company.

$$\text{Maximize } P(x, y) = 150x + 120y$$

subject to

$$x \geq 0, \quad y \geq 0$$

$$2x + 3y \leq 1100$$

$$5x + 3y \leq 1400$$

$$4x + y \leq 756$$

$$x \geq 25, \quad y \geq 40$$

Here $x$ represents the number of DT-1 and $y$ the number of DT-2 units to be made weekly.

More generally, a **linear program** is a mathematical problem with the following structure: there is specified a linear function of a number of variables that are required to satisfy linear conditions described by some mixture of linear inequalities and linear equations, called **constraints.** The problem is to find values for these variables which satisfy the constraints and yield the maximum, or minimum, value of the function, which is called an **objective function.** LP-1 is a 2-variable linear program, but the same kind of problems may involve 200, or 2000, or even 200,000 or more variables.

The Austin Company also hired the Marks Company, a consulting operations research firm, to independently study the digital tape player situation and make recommendations. The Marks OR group divided the manufacturing process into two phases: construction (which included assembly) and finishing. The data collected and their analysis led them to make the following assumptions:

(a) In the construction phase each DT-1 unit requires 8 hours of labor and each DT-2 unit requires 5 hours of labor. At most 2,210 hours of construction time are available per week.

(b) In the finishing phase each DT-1 unit requires 3 hours of labor and each DT-2 unit requires 2 hours of labor. At most 860 hours of finishing time are available per week.

(c) The anticipated profit for each DT-1 unit is \$140 and the anticipated profit for each DT-2 unit is \$150. In order for these unit profit values to hold the company must produce at least 50 DT-1 and 50 DT-2 units per week.

(d) There is an unlimited market for the DT-1 and DT-2 models.

(e) All factors other than the ones considered in the analysis of the production of the DT-1 and DT-2 models are negligible.

The same sort of analysis that leads to LP-1 from the assumptions made by the Austin Company's operation research department leads to the Marks OR group's linear program model LP-2:

$$\text{Maximize } P(x, y) = 140x + 150y$$

subject to

$$x \geq 0, \quad y \geq 0$$

$$8x + 5y \leq 2210$$

$$3x + 2y \leq 860$$

$$x \geq 50, \quad y \geq 50,$$

where $x$ represents the number of DT-1 and $y$ the number of DT-2 units to be made weekly.

We turn our attention to developing a systematic approach to solving such problems in the next section.

## EXERCISE

1. Set up LP-2 from the assumptions formulated by the Marks Company.

# Interlude: Practice Problems 4.3, Part 1

1. The Wisconsin Temperance Foundation is organizing a trip to Lambeau Field for a football game and plans to rent vans and busses for the event. Each bus has 40 regular seats and 1 handicapped seat; each van has 8 regular seats and 3 handicapped seats. It costs $350 to rent a van and $975 to rent a bus. At least 320 regular seats and 36 handicapped seats will be necessary for the participants. The organization wishes to determine how many of each type of vehicle to rent to minimize rental costs. You are to set up a linear program to model the problem.

   (a) Give the objective function for the linear program. Declare what variables represent, and indicate if you are to maximize or minimize the objective function.

   (b) Write all constraints for this linear program.

   (c) Sketch the feasible region for the system of constraints. Label all corner points with exact coordinates.

2.  A small furniture manufacturing business is planning to manufacture only rocking chairs and coffee tables this week. Each rocking chair requires five hours of material preparation, two hours of assembly, and three hours of finishing work. Each coffee table requires one hour each of material preparation, assembly, and finishing work. The employees that handle the three production phases are available 80 hours, 50 hours, and 70 hours, respectively, this week. Profit on a rocking chair is $170 and profit on a coffee table is $80. The business wishes to determine how many of each item to manufacture this week in order to maximize profit. You are to set up a linear program to model the problem.

    (a) Give the objective function for the linear program. Declare what variables represent, and indicate if you are to maximize or minimize the objective function.

    (b) Write all constraints for this linear program.

    (c) Sketch the feasible region for the system of constraints. Label all corner points with exact coordinates.

# Interlude: Practice Problems 4.3, Part 2

1. Goodfellas Financial is looking to offer up to $720,000 dollars in small business loans to local entrepreneurs. Goodfellas expects an annual return of 6.5% on the long-term loans they provide and an annual return of 5.25% on the short-term loans they provide. A minimum of $150,000 will be set aside for each type of loan, but Goodfellas will not allocate more than twice as much on long-term loans as compared to short-term loans. Goodfellas Financial needs to decide how much to allocate to the two types of loans so as to maximize expected annual returns. You are to set up a linear program to model the problem.

   (a) Give the objective function for the linear program. Declare what variables represent, and indicate if you are to maximize or minimize the objective function.

   (b) Write all constraints for this linear program.

   (c) Sketch the feasible region for the system of constraints. Label all corner points with exact coordinates.

2.  A dietician working for NDSU athletics is preparing a supplemental food paste to be used in each meal provided to athletes at the dining center. Each serving of this paste needs to have a minimum of 400 mg of calcium, 10 mg of iron, and 40 mg of vitamin C. She will be preparing the paste from containers of premixed sludge and goop. Each ounce of sludge contains 30 mg of calcium, 1 mg of iron, 2 mg of vitamin C, and 2 mg of cholesterol. Each ounce of goop contains 25 mg of calcium, 0.5 mg of iron, 5 mg of vitamin C, and 5 mg of cholesterol. The dietician needs to determine how many ounces each of sludge and goop to combine for each meal in order to meet the nutrient requirements while minimizing cholesterol. You are to set up a linear program to model the problem.

    (a) Give the objective function for the linear program. Declare what variables represent, and indicate if you are to maximize or minimize the objective function.

    (b) Write all constraints for this linear program.

    (c) Sketch the feasible region for the system of constraints. Label all corner points with exact coordinates.

# Interlude: Practice Problems 4.3, Part 3

1. The Metal Shop manufactures four different varieties of metal shelving units that require the use of three different machines in the manufacturing process, as summarized in the table below.

| Machine \ Shelf | Light Home | Heavy Home | Light Industrial | Heavy Industrial |
|---|---|---|---|---|
| Fabricating | 2 hours | 2.5 hours | 2 hours | 2 hours |
| Treatment | 1 hour | 1.5 hours | 1 hour | 2 hours |
| Packaging | 0.5 hour | 0.5 hour | 1 hour | 2 hours |

The fabricating, treatment, and packaging machines are available up to 4000, 1800, and 1000 hours per week, respectively. The profit from each Light Home, Heavy Home, Light Industrial, and Heavy Industrial shelving units is $50, $85, $120, and $160, respectively. The Metal Shop wants to determine how many of each type of shelving unit to produce weekly to maximize profit. You are to set up a linear program to model the problem.

(a) Give the objective function for the linear program. Declare what variables represent, and indicate if you are to maximize or minimize the objective function.

(b) Write all constraints for this linear program.

2. A food processing company has two processing plants, Plant I and Plant II, and two warehouses, Warehouse A and Warehouse B. The company can produce at most 450 cases of canned spinach per week at Plant I and most 350 cases of canned spinach per week at Plant II. It costs $0.75 per case to ship the product to Warehouse A regardless of the plant, but it costs $0.98 per case to ship from Plant I to Warehouse B and $1.14 per case to ship from Plant II to Warehouse B. Each week Warehouse A needs at least 300 cases of the spinach delivered and Warehouse B needs at least 500 cases delivered to meet retail demands. The food processing company needs to determine how to organize production and shipping of cases of canned spinach from each plant to each warehouse to meet shipping demands at minimum shipping cost. You are to set up a linear program to model the problem.

   (a) Give the objective function for the linear program. Declare what variables represent, and indicate if you are to maximize or minimize the objective function.

   (b) Write all constraints for this linear program.

## 4.4   THE CORNER POINT SOLUTION METHOD

We next address the problem of solving LP-1 and LP-2. To do this we develop a simple method for solving linear programs called the **corner point method.** It has an appealing geometric flavor and is effective for 2-variable linear programs.

As a working vehicle consider the linear program:

$$\text{Maximize } F(x, y) = 5x + 8y$$

subject to

$$x \geq 0, \quad y \geq 0$$
$$x + 2y \leq 8$$
$$3x + y \leq 9.$$

The points which satisfy the constraints of a linear program, called **feasible points,** are the points that the objective function to be optimized may be applied to.

Our problem here is to determine that feasible point (or those feasible points, if there is more than one) which yields the maximum value for the objective function $F(x, y) = 5x + 8y$.

Our first step in solving this problem is to obtain a geometric representation of the feasible points by graphing the constraints. This is done in Example 8 of Section 4.1.

The intersection points of at least two boundary lines which come out of the equality conditions of the constraints of a 2-variable linear program are called **corner points.** The corner points of our linear program are $(0, 0), (0, 4), (3, 0)$ and $(2, 3)$—obtained by solving the system of equations $x + 2y = 8$ and $3x + y = 9$; they are shown in Figure 4.14.

The significance of corner points is made clear by the following argument. As a starting point consider any feasible point $P(a, b)$ in the region of feasible points of our linear program; see Figure 4.14. If we move up the vertical line $L$ passing through $P(a, b)$ to $Q(a, c)$, (see Figure 4.15), our objective function $F(x, y) = 5x + 8y$ increases in value from $5a + 8b$ to $5a + 8c$ since $c$ is larger than $b$. By taking feasible points higher and higher on $L$ we increase $F(x, y) = 5x + 8y$. Since we must remain within the set of feasible points in taking points higher and higher on $L$, we can go as far as $R(a, d)$ on the boundary line $x + 2y = 8$ (Figure 4.16). From there we can move in one of two directions on the boundary line until we come to a corner point (see Figure 4.17).

The question that this raises is, how does a linear function behave as we take points in one direction or the other along a boundary line? We shall not do so here, but one can prove that one of two things happens: (a) The linear function increases in value as we take points in one direction along the boundary line and decreases as we take points in the other direction, or (b) the linear function has the same value at all the points on the boundary line. In either case we are led to a corner point. This argument suggests the following theorem.

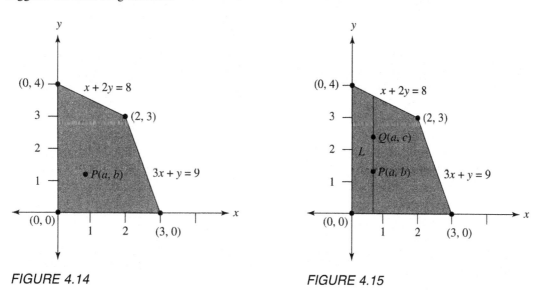

FIGURE 4.14                                    FIGURE 4.15

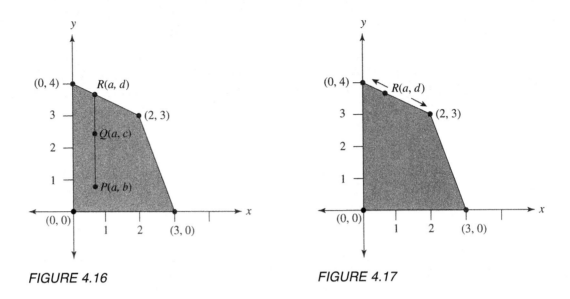

FIGURE 4.16                    FIGURE 4.17

---

**Corner Point Theorem:** If a 2-variable linear program has an optimal value (maximum or minimum value, depending on the nature of the linear program), then a solution yielding this optimal value can be found from among the corner points of the linear program.

---

The corner point theorem's hypothesis presupposes that the linear program under consideration has a solution. It does not say that a solution cannot occur at a feasible point which is not a corner point; this does happen with some linear programs. We are, however, assured of a solution at a corner point, assuming that the linear program has a solution to begin with.

Implementation of the corner point method to solve a 2-variable linear program involves the following sequence of steps:

---

**Corner Point Method Steps**

1. Graph the feasible points of the linear program.

2. Locate its corner points on the graph.

3. Determine the coordinates of all corner points. For a corner point which is not on either of the coordinate axes this is done by solving the system of equations which describe a pair of boundary lines which intersect at the corner point.

4. Compute the value of the objective function at each corner point.

5. From these values pick out the largest or smallest value, depending on the nature of the linear program, and the solution(s) which yields it.

---

To illustrate these procedures we return to our vehicle:

$$\text{Maximize } F(x, y) = 5x + 8y$$

subject to

$$x \geq 0, \quad y \geq 0$$
$$x + 2y \leq 8$$
$$3x + \ y \leq 9$$

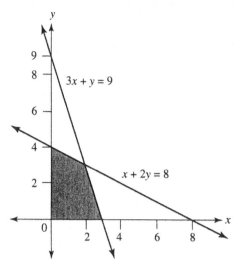

FIGURE 4.18

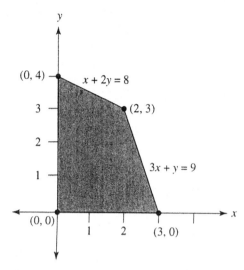

FIGURE 4.19

The graph of its feasible points is reproduced as Figure 4.18. The corner points, displayed in Figure 4.19, are $(0, 0)$, $(3, 0)$, $(2, 3)$ and $(0, 4)$; as was previously noted, $(2, 3)$ is obtained by solving the system of boundary line equations $3x + y = 9$ and $x + 2y = 8$.

The computation of the value of the objective function $F(x, y) = 5x + 8y$ at the corner points yields the results summarized in Table 4.2, from which we see that 34 is the maximum value and $(2, 3)$ is the solution.

**Table 4.2**

| Corner Point | $F(x, y) = 5x + 8y$ |
| --- | --- |
| $(0, 0)$ | 0 |
| $(3, 0)$ | 15 |
| $(2, 3)$ | 34 |
| $(0, 4)$ | 32 |

## EXAMPLE 1

Solve LP-1, the linear program model obtained by the operations research department of the Austin Company.

$$\text{Maximize } P(x, y) = 150x + 120y$$

subject to

$$x \geq 0, \quad y \geq 0$$
$$2x + 3y \leq 1100$$
$$5x + 3y \leq 1400$$
$$4x + y \leq 756$$
$$x \geq 25, \quad y \geq 40$$

Our first step is to sketch the graph of the feasible points. For this see Exercise 67 of Section 4.1 .

Locate the corner points on the graph and solve the appropriate systems of equations to determine their coordinates. There are five corner points, shown in Figure 4.20: $(25, 40)$, $(25, 350)$, $(100, 300)$—obtained by solving $2x + 3y = 1100$ and $5x + 3y = 1400$, $(124, 260)$—obtained by solving $4x + y = 756$ and $5x + 3y = 1400$, and $(179, 40)$.

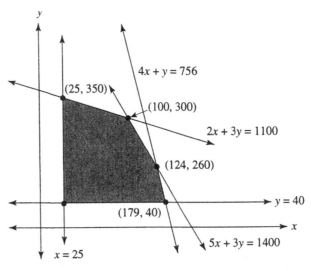

FIGURE 4.20

The computation of $P(x, y) = 150x + 120y$ at the corner points, summarized in Table 4.3, yields the solution $(100, 300)$ with maximum value 51,000.

**Table 4.3**

| Corner Point | $P(x, y) = 150x + 120y$ |
|:---:|:---:|
| $(25, 40)$ | 8,550 |
| $(25, 350)$ | 45,750 |
| $(100, 300)$ | 51,000 |
| $(124, 260)$ | 49,800 |
| $(179, 40)$ | 31,650 |

## EXAMPLE 2

Solve LP-2, the linear program model obtained by the Marks Company for the Austin Company:

$$\text{Maximize } P(x, y) = 140x + 150y$$

subject to

$$x \geq 0, \quad y \geq 0$$
$$8x + 5y \leq 2210$$
$$3x + 2y \leq 860$$
$$x \geq 50, \quad y \geq 50$$

Our first step is to sketch the graph of the feasible points. For this see Exercise 68 of Section 4.1.

Locate the corner points on the graph and solve the appropriate systems of equations to determine coordinates. There are four corner points, shown in Figure 4.21: $(50, 50)$, $(245, 50)$, $(120, 250)$, and $(50, 355)$.

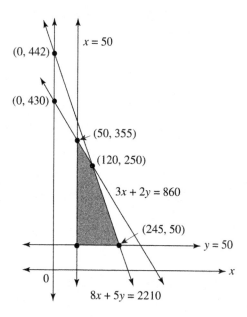

**FIGURE 4.21**

The computation of $P(x, y) = 140x + 150y$ at the corner points, summarized in Table 4.4, yields the solution $(50, 355)$ with maximum value $60,250$.

**Table 4.4**

| Corner Point | $P(x, y) = 140x + 150y$ |
| --- | --- |
| $(50, 50)$ | 14,500 |
| $(245, 50)$ | 41,800 |
| $(120, 250)$ | 54,300 |
| $(50, 355)$ | 60,250 |

Model LP-1 has solution $(100, 300)$ with maximum value $51,000$ whereas model LP-2 has solution $(50, 355)$ with maximum value $60,250$. In terms of the Austin Company's situation, to implement LP-1's conclusion the production schedule would have to be set to manufacture 100 DT-1 and 300 DT-2 units per week with an anticipated weekly profit of $51,000; to implement LP-2's conclusion the production schedule would have to be set to manufacture 50 DT-1 and 355 DT-2 units per week with an anticipated weekly profit of $60,250.

Which solution should be implemented? It may seem foolish even to ask, but is it? We explore this issue in the next section.

# EXERCISES

*Solve, if possible, the following linear programs.*

**1.** Max. $F(x, y) = 5x + 4y$

subject to

$x \geq 0, \quad y \geq 0$
$2x + 3y \leq 15$
$\phantom{2}x + \phantom{3}y \leq 6$

**2.** Min. $G(x, y) = 10x + 12y$

subject to

$x \geq 0, \quad y \geq 0$
$3x + 5y \leq 12$
$\phantom{3}x + \phantom{5}y \geq 3$

**3.** Min. $C(x, y) = 1.50x + 1.10y$

subject to

$x \geq 0, \quad y \geq 0$
$2x + \phantom{2}y \geq 4$
$\phantom{2}x + \phantom{2}y \geq 3$
$\phantom{2}x + 2y \geq 4$

**4.** Min. $C(x, y) = 8x + 12y$

subject to

$x \geq 0, \quad y \geq 0$
$\phantom{2}x + y \geq 250{,}000$
$2x + y \leq 400{,}000$

**5.** Max. $I(x, y) = 0.10x + 0.08y$

subject to

$x \geq 0, \quad y \geq 0$
$\phantom{-}x + \phantom{3}y \leq 60$
$-x + 3y \geq 0$
$\phantom{-}x \phantom{+ 3y} \geq 15$

**6.** Max. $F(x, y) = 3x + 2y$

subject to

$x \geq 0, \quad y \geq 0$
$x + 2y \geq 4$
$x - 2y \geq 2$

**7.** Max. $I(x, y) = 110x + 70y$

subject to

$x \geq 0, \quad y \geq 0$
$\phantom{2}x \phantom{+ 4y} \leq 25$
$2x + 4y \leq 120$
$4x + 2y \leq 110$

**8.** Max. $C(x, y) = 39{,}600x + 54{,}000y$

subject to

$x \geq 0, \quad y \geq 0$
$2x + 3y \leq 50$
$\phantom{2}x + \phantom{3}y \geq 30$

**9.** Max. $P(x, y) = 180x + 120y$

subject to

$x \geq 0, \quad y \geq 0$
$4x + 3y \leq 320$
$5x + 2y \leq 330$

**10.** Max. $P(x, y) = 190x + 110y$

subject to

$x \geq 0, \quad y \geq 0$
$\phantom{3.2}5x + 2y \leq 330$
$3.25x + 2y \leq 225$
$\phantom{3.2}4x + 3y \leq 320$

**11.** Min. $C = -5x + 12{,}795$

subject to

$x \geq 0, \quad y \geq 0$
$x + y \leq 100$
$x + y \geq 15$
$x \leq 75, \quad y \leq 90$

**12.** Min. $G(s, t) = 90s + 80t$

subject to

$s \geq 0, \quad t \geq 0$
$\phantom{3}s + 2t \geq 10$
$3s + \phantom{2}t \geq 15$

# Interlude: Practice Problems 4.4

1. Consider the objective function

$$H(x, y) = 10x + 18y$$

subject to the constraints

$$x + 2y \geqslant 1$$
$$x + 2y \leq 10$$
$$x + y \leqslant 8$$
$$y \geqslant 2x$$
$$x \geq 0, \quad y \geq 0$$

(a) Sketch the feasible region and label all corner points with exact coordinates.

(b) Find the maximum and minimum values of $H(x, y)$ subject to the given constraints. If you believe one of those values does not exist, explain why.

2.  Maximize and minimize

$$z = 2x + 9y$$

on the feasible region shown to the
right. Assume that gridlines each
mark one unit.

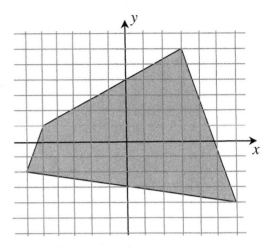

3.  Give an example of a linear program (with two variables) such that the objective function has a minimum value but not a maximum value. Besides showing your objective function and all constraints, sketch a graph of the feasible region.

# 4.5 THE SCOPE OF LINEAR PROGRAMMING APPLICATIONS

Linear programming has turned out to have a wide spectrum of applications. To obtain some sense of this spectrum we look at six case studies. The case studies are all realistic, but are presented in miniature for the sake of manageability. Actual real-life situations that emerge have the same structure and tone, but are more complex in that more factors are generally considered and more variables are required.

We view all of these situations through the eyes of others in much the same way that we see events through the eyes of a reporter or observer by reading his account of them in a newspaper, journal or book. Just as the reporter has selected what he believes are important features surrounding the events and has omitted those he considers unessential, we too are looking at features considered crucial to the situations we examine as seen by someone who has made such a selection. This selection reflects assumptions that have been made. To maintain a proper perspective on this it is important to keep in mind that other analysts, as other reporters, might see things in a different light and, accordingly, make other assumptions.

## Case 1   Production Planning

The Austin Company's problem of determining the number of DT-1 and DT-2 digital tape players to be made per week so as to maximize profit, considered in Section 4.2, is a production scheduling problem which we expressed in linear program terms under the assumptions introduced. The background that led to the Austin Company's linear programs is illustrative of situations with the following general features: A firm makes a number of products or models of a product and utilizes a number of resources in their manufacture, such as raw materials, labor, capital, different machines, storage facilities. It is assumed that for each product made a fixed amount of each resource is required to make a unit of that product. Within the production time frame a fixed amount of each resource is available and cannot be exceeded. It is also assumed that for a range of possible output levels there is a fixed profit per unit of each product which does not depend on the number of units produced. Under these conditions the problem of determining output levels of the products produced so as to maximize total profit can be formulated in terms of a linear program.

As a further illustration of a production planning situation we turn to the Ramuné Company's problem.

The Ramuné Company makes stereo systems. Two new models, RA5 and RA9, are to be mass produced. Both models pass through assembly and finishing plants of the company. In the assembly plant an RA5 unit is worked on for 1 hour; an RA9 unit is worked on for 3 hours. In the finishing plant an RA5 unit is worked on for 2 hours; an RA9 unit is worked on for 1 hour. At most, 90 hours of assembly time and 80 hours of finishing time are available per week. The anticipated profit on an RA5 unit is $10 and the anticipated profit on an RA9 unit is $15.

The problem is to determine, with respect to the given assumptions, how many RA5 and RA9 units should be made per week so as to maximize profit.

We begin by introducing variables to stand for the quantities we wish to determine. Let $x$ denote the number of RA5 units to be made and let $y$ denote the number of RA9 units to be made. To make needed information available at a glance, we express the basic data in tabular form, as shown in Table 4.5.

**Table 4.5**

|  | No. of Units to Be Made | Profit Per Unit | Assembly Time Per Unit (hours) | Finishing Time Per Unit (hours) |
|---|---|---|---|---|
| Model RA5 | $x$ | $10 | 1 | 2 |
| Model RA9 | $y$ | $15 | 3 | 1 |

Since profit is to be maximized, we must express profit in terms of $x$ and $y$. It is useful to note that

$$\text{profit} = \begin{bmatrix} \text{profit on} \\ \text{model RA5} \end{bmatrix} + \begin{bmatrix} \text{profit on} \\ \text{model RA9} \end{bmatrix}$$

$$= \begin{bmatrix} \text{profit on} \\ \text{one RA5} \\ \text{unit} \end{bmatrix} \bullet \begin{bmatrix} \text{no. of} \\ \text{units} \\ \text{made} \end{bmatrix} + \begin{bmatrix} \text{profit on} \\ \text{one RA9} \\ \text{unit} \end{bmatrix} \bullet \begin{bmatrix} \text{no. of} \\ \text{units} \\ \text{made} \end{bmatrix}$$

$$= 10x + 15y$$

The profit obtained by making $x$ RA5 units and $y$ RA9 units is expressed by the linear function

$$P(x, y) = 10x + 15y$$

Our next task is to describe the conditions that $x$ and $y$ must satisfy. Since the number of units made is nonnegative, we have

$$x \geq 0$$

$$y \geq 0$$

To express assembly plant operation time in terms of $x$ and $y$, we note that

$$\begin{bmatrix} \text{assembly} \\ \text{plant} \\ \text{time} \end{bmatrix} = \begin{bmatrix} \text{assembly} \\ \text{time on} \\ \text{RA5} \end{bmatrix} + \begin{bmatrix} \text{assembly} \\ \text{time on} \\ \text{RA9} \end{bmatrix}$$

$$= \begin{bmatrix} \text{assembly} \\ \text{time for} \\ \text{1 RA5 unit} \end{bmatrix} \bullet \begin{bmatrix} \text{no. of} \\ \text{RA5 units} \\ \text{made} \end{bmatrix} + \begin{bmatrix} \text{assembly} \\ \text{time for} \\ \text{1 RA9 unit} \end{bmatrix} \bullet \begin{bmatrix} \text{no. of} \\ \text{RA9 units} \\ \text{made} \end{bmatrix}$$

$$= 1 \bullet x + 3 \bullet y$$

Since at most 90 hours of assembly time per week is available, we have

$$x + 3y \leq 90$$

Similarly, the condition that at most 80 hours of finishing time are available per week is expressed in terms of $x$ and $y$ by the inequality

$$2x + y \leq 80$$

In summary, the assumptions made lead to the linear program

$$\text{Maximize } P(x, y) = 10x + 15y$$

subject to

$$x \geq 0$$

$$y \geq 0$$

$$x + 3y \leq 90$$

$$2x + y \leq 80,$$

where $x$ and $y$ denote the number of RA-5 and RA-9 units to be made per week, respectively.

To solve the Ramunė Company's linear program we first sketch the graph of its feasible points, shown in Figure 4.22.

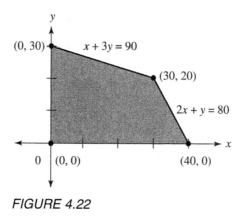

*FIGURE 4.22*

From Table 4.6 we see that (30, 20) is the solution of the Ramuné Company's linear program and 600 is its maximum value.

**Table 4.6**

| Corner Point | $P(x, y) = 10x + 15y$ |
|:---:|:---:|
| (0, 0) | 0 |
| (0, 30) | 450 |
| (40, 0) | 400 |
| (30, 20) | 600 |

Implementation of this solution by the Ramuné Company calls for making 30 RA5 and 20 RA9 units per week. Whether this solution should be implemented or not is, of course, another issue. If a careful review of this linear program's assumptions confirms the assessment that they are realistic, then implementation of its solution makes good sense.

## Case 2   Diet Problems

A sack of animal feed is to be put together from linseed oil meal and hay. It is required that each sack of feed contain at least 2 pounds of protein, 3 pounds of fat, and 8 pounds of carbohydrate. It is estimated that each unit (a unit is 30 pounds) of linseed oil meal contains 1 pound of protein, 1 pound of fat, 2 pounds of carbohydrate, and that each unit of hay contains 1/2 pound of protein, 1 pound of fat, and 4 pounds of carbohydrate. Linseed oil meal costs $1.50 per unit and hay costs $1.10 per unit.

The problem is to determine how many units of linseed oil meal and hay should be used to make up a sack of animal feed that satisfies the nutritional requirements at minimal cost.

To translate this problem and the assumptions which underlie it into a linear program, let $x$ and $y$ denote the number of units of linseed oil meal and hay, respectively, to be used in making up a sack of animal feed. The basic data are summarized in Table 4.7.

**Table 4.7**

| | No. Units Used | Cost Per Unit | Protein Per Unit (lbs) | Fat Per Unit (lbs) | Carbohydrate Per Unit (lbs) |
|:---|:---:|:---:|:---:|:---:|:---:|
| Linseed oil meal | $x$ | $1.50 | 1 | 1 | 2 |
| Hay | $y$ | $1.20 | 1/2 | 1 | 4 |

Since the total cost equals the cost of linseed oil meal, $1.50x$, plus the cost of hay, $1.10y$, the cost function to be minimized is

$$C(x, y) = 1.50x + 1.10y.$$

The number of pounds of protein in the mixture equals the amount contributed by the linseed oil meal, $x$ pounds, plus the amount contributed by the hay, $(1/2)y$ pounds. The sack of cattle feed must contain at least 2 pounds of protein, which translates to

$$x + (1/2)y \geq 2,$$

or equivalently

$$2x + y \geq 4.$$

Similarly, the fat and carbohydrate requirements are expressed by the constraints

$$x + y \geq 3$$

and

$$2x + 4y \geq 8,$$

or equivalently

$$x + 2y \geq 4.$$

We thus obtain the linear program model

$$\text{Minimize } C(x, y) = 1.50x + 1.10y$$

subject to

$$x \geq 0, \quad y \geq 0$$
$$2x + \phantom{2}y \geq 4$$
$$x + \phantom{2}y \geq 3$$
$$x + 2y \geq 4,$$

which has solution $(1, 2)$ and minimum value 3.7 (see Section 4.4, Exercise 3).

To implement this valid conclusion of the model we would use 1 unit of linseed oil meal (30 pounds) and 2 units of hay (60 pounds). Based on the assumed costs of the ingredients, the anticipated cost of a sack of animal feed is $3.70.

The problem considered illustrates "diet problems" with the following general features: A diet, or food substance, is to be put together from a number of available foods. It is required that the diet be balanced in the sense that it must contain minimal amounts of stated nutrients—proteins, fats, carbohydrates, minerals, vitamins, etcetera. It is assumed that each food unit contains a known fixed amount of each nutritional unit and that the unit prices of the food items are known and fixed within the time period considered. The problem is to determine the minimal cost diet which satisfies the prescribed nutritional requirements.

## Case 3 Environmental Protection

The Saxon Company must produce at least 250 thousand tons of paper annually. From the current operating system 10 pounds of chemical residue is deposited into a neighboring water system for each ton of paper produced. The resulting pollution has become a problem of serious concern, and to remain eligible for state tax benefits the Saxon Company must restrict the chemical residue emitted into the state's water system to not exceed 200 tons per year. Two filtration systems, Delta and Beta, have emerged for consideration. It is estimated that the installation of the Delta system would reduce emissions to 2 pounds for each ton of paper produced, and installation of the Beta system would reduce emissions to 1 pound for each ton of paper produced. Capital and operating costs for the Delta and Beta systems have been estimated at $8 and $12, respectively, per ton of paper produced.

The problem is to determine how many tons of paper should be produced subject to the Delta system and how many should be produced subject to the Beta system so that the emissions standard is met at minimal cost.

Let $x$ and $y$ denote the number of tons of paper to be produced annually subject to the Delta and Beta systems, respectively. The cost function to be minimized is

$$C(x, y) = 8x + 12y.$$

The condition that the Saxon Company must produce at least 250 thousand tons of paper annually is expressed by

$$x + y \geq 250,000.$$

The total amount of chemical residue produced annually is the number of pounds produced through use of the Delta system, $2x$ pounds, plus the amount produced through use of the Beta system, $y$ pounds. Since this cannot exceed 200 tons, we have

$$2x + y \leq 400,000.$$

We thus emerge with the linear program model

$$\text{Minimize } C(x, y) = 8x + 12y$$

subject to

$$x \geq 0, \quad y \geq 0$$
$$x + y \geq 250,000$$
$$2x + y \leq 400,000,$$

which has solution (150000, 100000) and minimum value 2,400,000 (see Section 4.4, Exercise 4).

To implement this result the Saxon Company would have to produce 150,000 tons of paper annually subject to the Delta system and 100,000 tons of paper annually subject to the Beta system. The anticipated cost would be $2.4 million.

## Case 4 Bank Portfolio Management

The Charles National Bank has assets in the form of loans and negotiable securities which, it is assumed, bring returns of 10 and 8 percent, respectively, in a certain time period. The bank has a total of $60 million to allocate between loans and securities. To meet unanticipated deposit withdrawals the bank maintains a securities balance greater than or equal to 25 percent of total assets. Lending is the bank's most important activity and to satisfy its clients it requires that at least $15 million be available for loans.

The bank wishes to determine, under these conditions, how funds should be allocated to maximize total investment income.

Let $x$ and $y$ denote the amount, in millions of dollars, to be allocated for loans and securities, respectively. The income function to be maximized is

$$I(x, y) = 0.10x + 0.08y.$$

The following constraints emerge:

$x + y \leq 60$: $60 million is available for investment in loans and securities.

$y \geq (1/4)(x + y)$, or equivalently, $-x + 3y \geq 0$: A securities balance greater than or equal to 25% of total assets must be maintained. Note, total assets is defined as the sum of the amounts invested in loans and securities, which is $x + y$.

$x \geq 15$: at least $15 million must be available for loans.

We thus obtain the linear program model

$$\text{Maximize } I(x, y) = 0.10x + 0.08y$$

subject to

$$x \geq 0, \quad y \geq 0$$
$$x + y \leq 60$$
$$-x + 3y \geq 0$$
$$x \geq 15,$$

which has solution (45, 15) and maximum value 5.7 (see Section 4.4, Exercise 5).

To implement this result the Charles Bank would have to allocate $45 million to loans and $15 million to securities. The anticipated interest on investment is $5.7 million.

## Of Interest

A. Broaddus, "Linear programming: A New Approach to Bank Portfolio Management," *Federal Reserve Bank of Richmond: Monthly Review,* vol. 58, No. 11 (Nov. 1972), pp. 3–11. This article provides an introductory nontechnical discussion of linear programming for bank portfolio management.

K. J. Cohen and F. S. Hammer, "Linear Programming and Optimal Bank Asset Management Decisions," *Journal of Finance,* vol. 22 (May 1967), pp. 147–165. This paper describes a linear program model that had been used for several years by Bankers Trust Company in New York to assist in reaching portfolio decisions.

## Case 5   Transportation

Heavy duty transformers made by the Thomas Company are to be sent from their plants in Dobsville and Watertown to distribution centers in New York, Chicago and Detroit. There are 100 transformers in Dobsville and 150 transformers in Watertown. The distribution centers in New York, Chicago and Detroit are to receive 75, 90 and 85 transformers, respectively. It costs $50, $52 and $54 to ship a transformer from Dobsville to New York, Chicago and Detroit, respectively. It costs $51, $48 and $50 to ship a transformer from Watertown to New York, Chicago and Detroit, respectively.

The problem is to determine how many transformers should be sent from each plant to each distribution center so that total cost is minimized.

This situation has the special feature that the total number of transformers to be sent from the plants (250) is equal to total number to be received by the destinations. This allows us to analyze the problem in terms of two variables as opposed to six variables (one linking each source to each destination) which would be needed if this equilibrium condition were not satisfied.

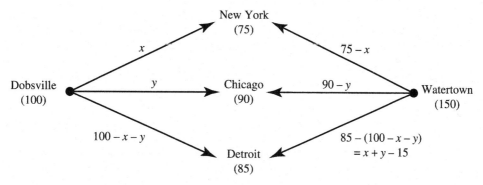

*FIGURE 4.23*

Let $x$ and $y$ denote the number of transformers to be sent from Dobsville to New York and Chicago, respectively. Send what remains at Dobsville, $100 - x - y$ transformers, to Detroit. From Watertown we send to New York, Chicago and Detroit the difference between what they should receive and what they have been sent from Dobsville. Thus, from Watertown we send $75 - x$ transformers to New York, $90 - y$ transformers to Chicago, and $85 - (100 - x - y) = x + y - 15$ transformers to Detroit. In summary, we have the shipping schedule shown in Figure 4.23.

The cost of shipping the transformers from Dobsville to New York is $50x$ dollars, the cost of shipping one transformer, $50, times the number being sent, $x$. Similarly, the cost of shipping the transformers from Dobsville to Chicago and Detroit is $52y$ and $54(100 - x - y)$ dollars, respectively. The cost of shipping the transformers from Watertown to New York, Chicago and Detroit is $51(75 - x)$, $48(90 - y)$ and $50(x + y - 15)$ dollars, respectively.

The total cost function $C(x, y)$, obtained by adding up the costs of shipping the transformers from the sources to the destinations is

$$C(x, y) = -5x + 12{,}795.$$

Since we have taken into account the number of transformers at each source and the number to be received by each destination, the only condition that remains to be stated is that the direction of the flow is from the sources to the destinations; no back-flow. Doing so yields the linear program model

$$\text{Minimize } C(x, y) = -5x + 12{,}795$$

subject to

$$x \geq 0, \quad y \geq 0$$
$$100 - x - y \geq 0$$
$$75 - x \geq 0$$
$$90 - y \geq 0$$
$$x + y - 15 \geq 0,$$

or equivalently,

$$\text{Minimize } C(x, y) = -5x + 12{,}795$$

subject to

$$x \geq 0, \quad y \geq 0$$
$$x + y \leq 100$$
$$x + y \geq 15$$
$$x \leq 75, \quad y \leq 90,$$

which has solutions $(75, 0)$ and $(75, 25)$ and minimum value 12,420. Moreover, all points of the form $(75, y)$, where $y$ is an integer value between 0 and 25, are also solutions of this linear program which could be implemented as a shipping schedule (see Section 4.4, Exercise 11).

Implementation of the solution $(75, 0)$, for example, requires that $75, 0$ and 25 transformers be sent from Dobsville to New York, Chicago, and Detroit, respectively, and that $0, 90$ and 60 transformers be sent from Watertown to New York, Chicago, and Detroit, respectively. The anticipated total cost is $12,420.

More generally, a transportation problem has the following features. Given amounts of a commodity are available at a number of sources of supply, such as warehouses. Specified amounts are required by various destinations, such as retail outlets. The total amount required by the destinations may or may not be equal to the total amount available at the sources. Estimates (assumptions) are available on the cost of sending one unit of the commodity from each source to each destination. The problem is to determine the least cost shipping schedule. When the problem involves very few sources and destinations, as in Case 5, it can be solved by inspection. When the number of sources and destinations is large, inspection will not do, and linear programming provides us with a systematic approach to such problems. We discuss this further in the context of the Vytis Publishing Company's shipping problem in Section 4.6.

## Case 6 An Assignment Problem

The Rasa Publishing Company has two positions to fill, editor of the mathematics list (job 1) and editor of the social science list (job 2), and is considering three candidates, Albert Roberts (candidate 1), Rita O'Brien (candidate 2), and Martin Thorp (candidate 3). After considering resumes, letters of recommendation, and conducting interviews the editorial board of the company has assigned a numerical rating to each person's qualifications for each position as stated in Table 4.8. These ratings serve as a quantitative measure of each candidate's potential for each position as seen by the editorial board. The editorial board wishes to assign candidates to positions in such a way that total potential is maximized.

**Table 4.8**

| | Position | |
| --- | --- | --- |
| Candidate | Math Editor (job 1) | Soc. Sci. Editor (job 2) |
| Roberts (candidate 1) | 8 | 8 |
| O'Brien (candidate 2) | 7 | 9 |
| Thorp (candidate 3) | 9 | 8 |

To relate the candidates to the jobs we introduce $X_{ij}$ to relate candidate $i$ to job $j$. $X_{ij}$ can assume one of two values, 0 if candidate $i$ is not assigned job $j$, and 1 if candidate $i$ is assigned job $j$. In summary, we emerge with Table 4.9.

The function

$$P = 8X_{11} + 8X_{12} + 7X_{21} + 9X_{22} + 9X_{31} + 8X_{32},$$

**Table 4.8**

| | Position | |
| --- | --- | --- |
| Candidate | Math Editor (job 1) | Soc. Sci. Editor (job 2) |
| Roberts (candidate 1) | $X_{11}$ | $X_{12}$ |
| O'Brien (candidate 2) | $X_{21}$ | $X_{22}$ |
| Thorp (candidate 3) | $X_{31}$ | $X_{32}$ |

obtained by multiplying the variable that relates a candidate to a job by the candidate's potential for the job and adding, is the potential function to be maximized subject to two conditions:

1. Each candidate is assigned to at most one job.

The variables $X_{11}$ and $X_{12}$ (row 1 of Table 4.9) relate candidate 1 to the available jobs 1 and 2. The constraint

$$X_{11} + X_{12} \leq 1$$

expresses the requirement that candidate 1 be assigned to at most one job since it makes it impossible for candidate 1 to be assigned job 1 ($X_{11} = 1$) and job 2 ($X_{12} = 1$). This constraint comes from row 1 of Table 4.9, and in general the condition that each candidate be assigned at most one job is expressed by requiring that the sum of the variables in each row of Table 4.9 be less than or equal to one. Rows 2 and 3 yield the same condition for candidates 2 and 3:

$$X_{21} + X_{22} \leq 1$$
$$X_{31} + X_{32} \leq 1$$

2. Each job is filled by at most one person.

The variables $X_{11}, X_{21}$, and $X_{31}$ in column 1 of Table 4.9 relate candidates 1, 2, and 3 to job 1. The constraint

$$X_{11} + X_{21} + X_{31} \leq 1,$$

obtained by requiring that the sum of the variables in the first column 1 of Table 4.9 be less than or equal to one, expresses the requirement that job 1 be filled by at most one candidate since it makes it impossible for any two or all three candidates to be assigned to job 1 ($X_{11} = 1, X_{21} = 1, X_{31} = 1$). Column 2 yields the same condition for job 2:

$$X_{12} + X_{22} + X_{32} \leq 1$$

We thus obtain the following linear program: Find nonnegative integers (zeros and ones) that

$$\text{Maximize } P = 8X_{11} + 8X_{12} + 7X_{21} + 9X_{22} + 9X_{31} + 8X_{32}$$

subject to

$$X_{11} + X_{12} \leq 1$$
$$X_{21} + X_{22} \leq 1$$
$$X_{31} + X_{32} \leq 1$$
$$X_{11} + X_{21} + X_{31} \leq 1$$
$$X_{12} + X_{22} + X_{32} \leq 1.$$

By inspection we can see from Table 4.8 that the potential function $P$ is maximized when candidate 3 (Thorp) is assigned to job 1 (Math Editor) and candidate 2 (O'Brien) is assigned to job 2 (Social Science Editor). That is, $X_{11} = 0, X_{12} = 0, X_{21} = 0, X_{22} = 1, X_{31} = 1, X_{32} = 0$ maximizes potential. When the number of candidates and jobs is large such problems cannot be handled by inspection, but can be handled by linear programming methods.

Assignment problems may involve the most efficient assignment of people to jobs, machines to tasks, project leaders to projects, police cars to city sectors, departments to store locations, sales people to territories, and so on. The objective might involve maximizing effectiveness in some sense or minimizing cost or travel time.

## EXERCISES

*The situations presented in the following exercises reflect assumptions made by an individual or group. Set up linear programs for the problems that arise, if possible solve them, and state how the solutions obtained would be implemented. What concerns would you want to have satisfactorily addressed before implementing a solution obtained?*

1. The Algis Company makes radios. Two new portable stereo models, K15, and K31 are to be introduced. Both models pass through assembly and finishing plants of the company. In the assembly plant a K15 unit is worked on for 2 hours; a K31 unit is worked on for 3/2 hours. In the finishing plant a K15 unit is worked on for 5/2 hours; a K31 unit is worked on for 1 hour. At most 160 hours of assembly time and 165 hours of finishing time are available per week. The anticipated profit is $180 per K15 unit and $120 per K31 unit.

    The problem is to determine, under such conditions, how many K15 and K31 units should be made weekly so as to maximize profit.

2. A fruit juice is to be made from orange juice concentrate and apricot juice concentrate. Particular attention is being paid to the vitamin A, C, and D content of the fruit juice. Each container of fruit juice is to contain at least 120 units of vitamin A, 150 units of vitamin C, and 55 units of vitamin D. One ounce of orange juice concentrate contains 2 units of vitamin A, 3 units of vitamin C, and 1 unit of vitamin D. One ounce of apricot juice concentrate contains 3 units of vitamin A, 2 units of vitamin C, and 1 unit of vitamin D. Orange juice concentrate costs 3¢ per ounce and apricot juice concentrate costs 2¢ per ounce.

    The problem is to determine how many ounces of each concentrate should be used to make a least-cost container of juice that satisfies the vitamin requirements.

3. Cans of meat are to be mass produced and made available for distribution in emergency situations (such as earthquakes and floods). Each can of meat is to be a mixture of pork and beef and must contain at least 12 ounces of protein and 9 ounces of fat. It is estimated that a pound of the beef to be used contains 5 ounces of protein and 3 ounces of fat while a pound of pork contains 3 ounces of protein and 3 ounces of fat.

    If beef costs 50¢ a pound and pork costs 45¢ a pound, how many pounds of each should be used to make up a can of meat so that the nutritional requirements are met and the cost is minimal? What is the minimal cost?

4. The Saturn Company makes refrigerators and air conditioners. Two plants, I and II, are used. The assembly work is done in plant I, and it is estimated that 5 labor-hours of work are required to produce a refrigerator and 2 labor-hours of work are required to produce an air-conditioner. The finishing work is done in plant II, and it is estimated that 3 labor-hours of work are needed to finish a refrigerator and 2 labor-hours are needed to finish an air-conditioner. Plant I has 220 labor-hours per week available and plant II has 180 labor-hours per week available. A market survey indicates that there is an unlimited market for these products.

    If the Saturn Company makes a profit of $50 on each refrigerator and $30 on each air-conditioner, how many of each should be produced weekly so as to maximize profit?

5. The Andrius Bank has assets in the form of loans and negotiable securities. For a certain time period it is assumed that loans and securities bring returns of 9 and 6 percent, respectively. The bank has a total of $25 million, provided by demand deposit accounts and time deposit accounts, to allocate between loans and securities. To meet unanticipated deposit withdrawals the bank always maintains a securities balance equal to or greater than 20 percent of total assets. Since lending is the bank's most important activity, it imposes certain restrictions on its loan balance to satisfy its principal clients. Specifically, it requires that at least $8 million be available for loans.

   Under the given conditions, how should the bank allocate funds between loans and securities so that total investment income is maximized?

6. The Jay Toy Store plans to invest up to $2200 in buying and stocking two popular children's toys. The first toy costs $4 per unit and occupies 5 cubic feet of storage space; the second toy costs $6 per unit and occupies 3 cubic feet of storage space. The store has 1400 cubic feet of storage space available. The owner expects to make a profit of $1.50 on each unit of the first toy he buys and stocks and a profit of $2.00 on each unit of the second toy.

   How many units of each should be bought and stocked so that profit is maximized?

7. At Ecap University discussion has centered on determining the number of openings, called slots, to be made available in the forthcoming year at the associate and full-professor ranks. Each person promoted to associate professor is to receive a merit increment of $500, and each person promoted to full professor is to receive a merit increment of $1000. At most $15,000 is available for merit increments. A long-standing guideline is that the number of full-professor slots is not to exceed one fourth the number of associate professor slots. The university senate has recommended that at least 3 slots at the full-professor rank be established and that not more than 22 slots at the associate professor rank be established.

   Of interest to the faculty council is the question of how many slots should be established at each rank so as to maximize the total number of promotions. Administration has raised the question of how many slots at each rank should be established so as to minimize the total cost of increments.

8. The Brooks and Darius mines of Lexington Mines, Inc., produce high-grade and medium-grade silver ore. The Brooks mine yields 1 ton of high-grade ore and 4 tons of medium-grade ore per hour. The Darius mine yields 2 tons of high-grade ore and 3 tons of medium-grade ore per hour. To meet its commitments the company needs at least 40 tons of high-grade and 100 tons of medium-grade ore per day. It costs $500 per hour to operate the Brooks mine and $700 per hour to operate the Darius mine.

   Lexington Mines, Inc., would like to determine how many hours per day each mine should be operated if their ore requirements are to be met at minimal cost.

9. The Petrovski Steel Company produces 2 million tons of steel annually. In the current operation of the blast and open-heart furnaces 50 pounds of particulate matter is emitted into the atmosphere for every ton of steel produced. The resulting air pollution has become a problem of serious concern and efforts are being directed at curbing the emissions. On the basis of studies that have been conducted, it is estimated that installation of the F14 filter system would reduce emissions to 20 pounds of particulate matter per ton of steel produced, and installation of the F24 filter system would reduce emissions to 18 pounds of particulate matter per ton of steel produced. Capital and operating costs for the F14 and F24 filter systems are estimated at $1.2 and $1.8, respectively, per ton of steel produced. It is desired that particulate emissions be reduced by 62,400,000 pounds or better per annum. At the same time cost is an important factor if the company is to remain competitive.

   The problem is to determine how many tons of steel should be produced annually subject to the F14 system and how many tons should be produced subject to the F24 system so that the desired reduction in particulate emissions is achieved at minimal total cost.

10. The Rasa Publishing Company has warehouses in Dallas and Kansas City. Copies of its newly published best seller are to be sent to dealers in New York, San Francisco, and Atlanta. The Dallas warehouse has 15,000 copies and the Kansas City warehouse has 20,000 copies. Orders for 10,000 copies, 12,000 copies, and 13,000 copies have been placed by New York, San Francisco, and Atlanta book dealers, respectively. It costs 5¢, 7¢, and 4¢ per copy to ship the books from Dallas to New York, San Francisco, and Atlanta, respectively; it costs 4¢, 5¢, and 5¢ per copy to ship the books from Kansas City to these respective destinations.

    How should the books be shipped so as to minimize the total shipping cost?

11. The Linnus Company is planning to put into production three minicomputer models, C15, C24, and C51. Each unit must pass through assembly, finishing, and inspection plants of the company. In the assembly plant, a C15 unit is worked on for 2 hours, a C24 unit is worked on for 3 hours, and a C51 unit is worked on for 1 hour. In the finishing plant, it takes 1 hour to finish a C15 unit, 1 hour to finish a C24 unit, and 3 hours to finish a C51 unit. In the inspection plant, it takes 2 hours to inspect a C15 unit, 1 hour to inspect a C24 unit, and 1 hour to inspect a C51 unit. At most 16, 17, and 12 hours of assembly time, finishing time, and inspection time, respectively, are available per day. The anticipated profit is $24, $30, and $26 per unit of C15, C24, and C51, respectively.

The problem is to determine, under these conditions, how many units of each model should be made daily so as to maximize profit.

12. Soybeans are to be shipped from New York and New Orleans to Dakar, Marseille, and Odessa. 8000 tons are in New York and 11,000 tons are in New Orleans. 6000 tons are to be sent to Dakar, 7000 tons to Marseille, and 6000 tons to Odessa. The cost (in dollars) of shipping one ton of soybeans from each distribution point to each destination is given in Table 4.10.

**Table 4.10**

|             | Dakar | Marseille | Odessa |
|-------------|-------|-----------|--------|
| New York    | 10    | 11        | 12     |
| New Orleans | 10.5  | 11.2      | 12.3   |

The problem is to determine how many tons should be shipped from each distribution center to each destination so that the total shipping cost is minimized.

13. Up to $50,000 of a client's money is to be invested by the Onute Investment Company in low-risk bonds with a 10 percent yield, and high-risk speculative stock with an anticipated 18 percent yield. The client has insisted that the amount invested in stock be no greater than one fourth the amount invested in bonds. The managers of the investment company feel that at most $41,000 should be invested in bonds and that at least $5000 should be invested in stock.

The problem is to determine the investment plan that satisfies these conditions and maximizes return on investment.

14. The Selby and Turchin plants of the Asta Paper Company produce three grades of paper. The Selby plant produces 5 tons of grade A paper, 3 tons of grade B paper, and 7 tons of grade C paper per hour. The Turchin plant produces 3 tons of grade A paper, 5 tons of grade B paper, and 6 tons of grade C paper per hour. To satisfy existing contracts, the company must produce at least 95 tons of grade A paper, 88 tons of grade B paper, and 160 tons of grade C paper per day. It costs $200 per hour to operate the Selby plant and $150 per hour to operate the Turchin plant.

The Asta Paper Company would like to determine the number of hours per day that each plant should be kept in operation if their paper needs are to be met at minimum cost.

15. A railroad company has freight cars of two types. Model AA1 has 8000 cubic feet of refrigerated space and 10,000 cubic feet of nonrefrigerated space; model AA2 has 6000 cubic feet of refrigerated space and 15,000 cubic feet of nonrefrigerated space. Audre Food Distributors deal in meat products, which require refrigeration, and a variety of nonrefrigerated foods. 60,000 cubic feet of meat products and 120,000 cubic feet of nonrefrigerated foods are to be shipped from Chicago to Louistown. A model AA1 car rents for $0.8 per mile, and a model AA2 car rents for $1.25 per mile.

The problem is to determine the number of freight cars of each type that should be rented so that the total cost of shipping the foods is minimized.

16. The Stillwell Company produces lubricating oil for machine tools (BV3 oil) and steel mills (BV7 oil). The production of a unit of BV3 oil requires 1 hour of refining and 0.5 hours of blending; the production of a unit of BV7 oil requires 1.2 hours of refining and 1 hour of blending. The refinery can be kept in operation at most 12 hours per day, and the blending plant can be kept in operation at most 8 hours per day. BV3 oil brings a profit of $15 per unit, and BV7 oil brings a profit of $20 per unit.

The problem is to determine the number of units of each type of oil that should be produced daily to maximize profit.

17. The Clinton Company produces a portable X-ray unit that it sells to two types of outlets, hospitals and medical supply houses. The profit margin varies between the two types of outlets owing to differences in order sizes, selling costs, and credit policies. It is estimated that the profit per unit is $100 and $120, respectively, for the hospitals and medical supply houses. Sales promotion is carried out by personal sales force calls and media advertising. The company has eight salespersons on its marketing staff, representing 12,000 hours of available customer contact time during the next year. $30,000 has been allocated for media advertising during the next year. An examination of past data indicates that a unit sale to a hospital requires about a 1/2-hour sales call and $2 worth of advertising; a unit sale to a medical supply house requires a 1-hour sales call and $1 worth of advertising. The company would like to achieve sales of at least 5000 units in each customer segment.

The problem is to determine what sales volume (in product units) it should seek to develop in each segment in order to maximize total profit.

# Interlude: Practice Problems 4.5, Part 1

1. Tots a' Truckin' manufactures a standard and a deluxe model of its toy dump truck. In the manufacturing process each standard model requires two hours of metal work and two hours of finishing work, whereas each deluxe model requires two hours of metal work and four hours of finishing. The company has two people available to do the metal work and three people available to do the finishing work, each of whom can work a maximum of 40 hours per week. Each standard model dump truck generates a profit of $6 and each deluxe model generates a profit of $8. How many of each type should the company manufacture each week to maximize profit?

   (a) Write an objective function and all constraints for this linear program. Declare what all variables represent.

   (b) Sketch the appropriate feasible region, label all corner points with exact coordinates, and solve the linear program. State your conclusion in a complete sentence.

2.  Cobb County operates solid waste incinerators in Hicksburg and Booniesville. Each incinerator cogenerates a certain amount of electricity. The Hicksburg site can burn 10 tons of trash per hour while cogenerating 6 kilowatts of electricity. Meanwhile, the Boooniesville site can burn 5 tons of trash per hour while cogenerating 4 kilowatts of electricity. The county needs to be able to burn at least 70 tons of trash per day and cogenerate at least 48 kilowatts of electricity. It costs $80 per hour to operate the Hicksburg incinerator and $50 per hour to operate the Booniesville incinerator. The county wants to know how many hours per day to operate each incinerator to minimize operating costs of the incinerators.

    (a)  Write an objective function and all constraints for this linear program. Declare what all variables represent.

    (b)  Sketch the appropriate feasible region, label all corner points with exact coordinates, and solve the linear program. State your conclusion in a complete sentence.

# Interlude: Practice Problems 4.5, Part 2

1. Juice De Fruit makes two different types of juice. A can of Citrus Crush contains 10 ounces of pineapple juice, 3 ounces of orange juice, and one ounce of apricot juice. A can of Perky Punch contains 10 ounces of pineapple juice, 2 ounces of orange juice, and 2 ounces of apricot juice. A can of Citrus Crush earns a profit of 40¢ while a can of Perky Punch earns a profit of 60¢. Each week the company has 9000 ounces of pineapple juice, 2400 ounces of orange juice, and 1400 ounces of apricot juice available. The company would like to determine how many cans of each type of juice to produce weekly in order to maximize expected profit. You are to determine how to meet that goal.

   (a) Write an objective function and all constraints for this linear program. Declare what all variables represent.

   (b) Sketch the appropriate feasible region, label all corner points with exact coordinates, and solve the linear program. State your conclusion in a complete sentence.

2. Frank jogs, plays handball, and swims at the athletic club. Jogging burns 18 calories per minute, handball burns 12 calories per minute, and swimming burns 10 calories per minute. He always swims for at least 30 minutes and plays handball at least twice as long as he jogs. Frank is wondering about two things.

   (a) If he only has 90 minutes to exercise, how long should he participate in each activity in order to maximize the number of calories burned? Write an objective function and all constraints for this linear program. Declare what all variables represent.

   (b) How long should he participate in each activity to burn at least 900 calories in the least amount of time? Write an objective function and all constraints for this linear program. Declare what all variables represent.

# Interlude: Practice Problems 4.5, Part 3

1. Prodigal Financial has $30 million to invest on behalf of a state employees union. The money will be divided among Treasury notes, bonds, and stocks. The union stipulates that at least $3 million be invested in each of the three types of investments, and that at least $15 million be allocated to some combination of Treasury notes and bonds, with the amount in bonds not to exceed twice the amount invested in Treasury notes. The expected annual return for the three types of investment are 7% for Treasury notes, 8% for bonds, and 9% for stocks. How should the $30 million be invested in order to maximize expected annual return?

   (a) Write an objective function and all constraints for this linear program. Declare what all variables represent.

   (b) Sketch the appropriate feasible region, label all corner points with exact coordinates, and solve the linear program. State your conclusion in a complete sentence.

2. Creedy's Cruisers is supplying riverboats under a charter agreement with a travel agency. Creedy's has two types of vessels. The Raging Red model has 60 deluxe cabins and 160 standard cabins, while the Moshing Missouri model has 80 deluxe cabins and 120 standard cabins. The travel agency needs to have a minimum of 360 deluxe and 680 standard cabins available for a planned cruise. It will cost $44,000 to operate a Raging Red and $54,000 for operate a Moshing Missouri for the duration of the cruise. How many of each type of riverboat should Creedy's provide to the travel agency to keep operating costs minimized?

   (a) Write an objective function and all constraints for this linear program. Declare what all variables represent.

   (b) Sketch the appropriate feasible region, label all corner points with exact coordinates, and solve the linear program. State your conclusion in a complete sentence.

# 4.6   PROBLEMS REQUIRING SOLUTIONS IN INTEGERS

**EXAMPLE 1 ◆ How Many Coats and Dresses Should be Made?**

The Hoffman Clothing Manufacturers, Inc., has available 120 square yards of cotton and 100 square yards of wool for the manufacture of coats and dresses. Two square yards of cotton and 4 square yards of wool are used in making a coat while 4 square yards of cotton are used in making a dress. Cotton costs $5 per square yard and wool cost $20 per square yard. Four hours of labor are needed to make a coat and 2 hours of labor are needed to make a dress. The cost of labor is $25 per hour. At most 110 hours of labor are available for the manufacture of the coats and dresses.

If a coat sells for $300 and a dress sells for $140, how many of each should be made if net income is to be maximized?

Let $x$ and $y$ denote the number of coats and dresses to be made, respectively. The data given are summarized in Table 4.11.

**Table 4.11**

|  | Selling Price | Cotton Used Per Item | Wool Used Per Item | Labor-Hours Per Item | Number Made |
|---|---|---|---|---|---|
| Coat | $300 | 2 | 4 | 4 | $x$ |
| Dress | $140 | 4 | 0 | 2 | $y$ |
|  |  | $5 per sq yd; 120 sq yd available | $20 per sq yd; 100 sq yd available | $25 per labor-hour; 110 labor-hours available |  |

Since net income is to be maximized we turn our attention to expressing net income in terms of $x$ and $y$.

$$\text{Net income} = \text{Amt from sales} - \text{Production cost}$$

$$I(x, y) = \underbrace{300x + 140y}_{\text{sales}} - \underbrace{[5(2)x + 20(4)x]}_{\substack{\text{cost of coat} \\ \text{material}}} - \underbrace{5(4)y}_{\substack{\text{cost of} \\ \text{dress} \\ \text{material}}}$$

$$\underbrace{-[25(4)x + 25(2)y]}_{\text{labor cost}}$$

By multiplying and collecting terms we obtain:

$$I(x, y) = 110x + 70y$$

The constraint $4x \leq 100$ or, equivalently, $x \leq 25$, expresses the condition that the amount of wool used cannot exceed 100 square yards; $2x + 4y \leq 120$ expresses the condition that the amount of cotton used cannot exceed 120 square yards; $4x + 2y \leq 110$ expresses the condition that the number of labor-hours employed cannot exceed 110.

We thus emerge with the following linear program:

$$\text{Maximize } I(x, y) = 110x + 70y$$

subject to

$$x \geq 0, \quad y \geq 0$$
$$x \leq 25$$
$$2x + 4y \leq 120$$
$$4x + 2y \leq 110$$

From Section 4.4, Exercise 7 we see that (50/3, 65/3) yields the maximum value 3350.

What is actually required is a feasible point expressed in integers which maximizes $I(x, y)$. As one would expect, such linear programs are called **integer programs.** Sometimes, as in the case of the Austin Company's linear programs, it turns out that solutions in integers are obtained when a linear program solution method is applied. Sometimes not, as we have just seen.

We know that a solution of a linear program can be found among its corner points. Thus if an integer solution is desired and some of the corner points are not integers, the idea of modifying the given linear program by appending to it new constraints with the property that no integer feasible points are lost and the resulting corner points involve only integers is naturally suggested. Upon implementation of this idea, linear programming methods can then be applied to the modified linear program to obtain a solution in integers which will also be a solution of the given integer program. We illustrate the implementation of this idea in the case of two-variable integer programs by returning to the Hoffman Clothing Manufacturer's integer program.

Find integer values for $x$ and $y$ which

$$\text{Maximize } I(x, y) = 110x + 70y$$

subject to

$$x \geq 0, \quad y \geq 0$$
$$x \leq 25$$
$$2x + 4y \leq 120$$
$$4x + 2y \leq 110.$$

The feasible points of this problem viewed as a linear program are shown in Figure 4.24. Figure 4.25 shows in detail the region surrounding $(50/3, 65/3)$.

From Figure 4.25 we see that if the new constraints to be appended are to

1.   eliminate $(\frac{50}{3}, \frac{65}{3})$ as a corner point,
2.   not eliminate any integer feasible points,
3.   yield corner points which are integers,

then we need only introduce the additional constraint $x + y \leq 38$, based on the boundary line $(x + y = 38)$ passing through $(16, 22)$ and $(17, 21)$.

Thus the modified linear program is the following:

$$\text{Maximize } I(x, y) = 110x + 70y$$

subject to

$$x \geq 0, \quad y \geq 0$$
$$x \leq 25$$
$$2x + 4y \leq 120$$
$$4x + 2y \leq 110$$
$$x + \ \ y \leq 38.$$

The feasible points of this modified program are shown in Figure 4.26. All of its corner points, $(0, 0)$, $(0, 30)$, $(16, 22)$, $(17, 21)$, $(25, 5)$, and $(25, 0)$, are expressed in terms of integers. From Table 4.12 we

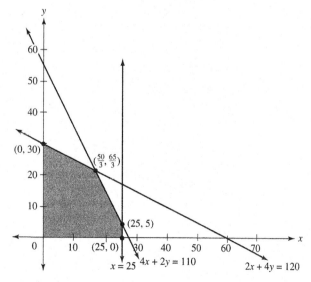

*FIGURE 4.24*

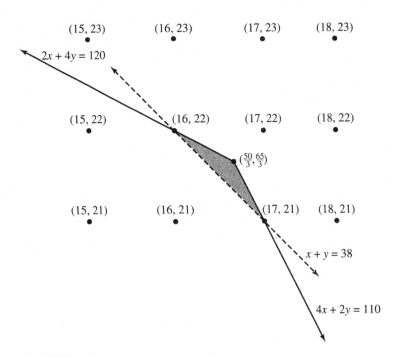

*FIGURE 4.25*

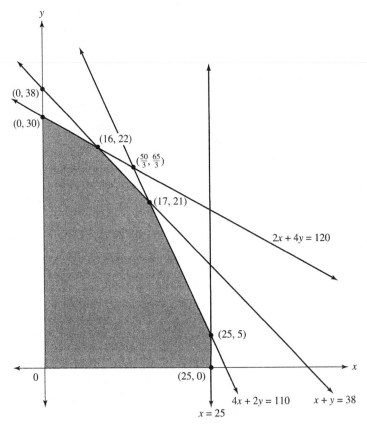

*FIGURE 4.26*

**Table 4.12**

| Corner Point | $I(x, y) = 110x + 70x$ |
|:---:|:---:|
| $(0, 0)$ | 0 |
| $(0, 30)$ | 2100 |
| $(16, 22)$ | 3300 |
| $(17, 21)$ | 3340 |
| $(25, 5)$ | 3100 |
| $(25, 0)$ | 2750 |

see that $(17, 21)$ is the solution and 3340 is the maximum value. Since this linear program differs from the original only in the loss non-integer feasible points (those in the triangular region with vertices $(16, 22)$, $(17, 21)$, $(50/3, 65/3)$, above the line $x + y = 38$; see Figure 4.26), its solution $(17, 21)$ is also a solution of the Hoffman clothing manufacturer's integer program.

When the number of variables is greater than two, graphical techniques like the one employed to determine the new constraints for obtaining a solution in integers cannot be used and other methods must be sought. Algebraic techniques for the determination of the required new constraints which are computationally effective in many situations have been developed. However, such techniques are beyond the scope of this book.

The following situation illustrates problems which lead to integer programs which can be solved by linear programming methods. Due to certain structural features integer solutions are guaranteed.

## EXAMPLE 2 ◆ The Vytis Company's Shipping Problem

The Vytis Publishing Company has warehouses W1 in Williamstown and W2 in Jamesville. Copies of a newly published book which is in great demand are to be sent via air freight to distributors D1 in Chicago, D2 in New York, and D3 in Detroit. The Williamstown warehouse has 20,000 copies and the Jamesville warehouse has 15,000 copies. The Chicago distributor needs 8000 copies while the New York and Detroit distributors have each requested 10,000 copies. These distributors are able to accept larger supplies if it is economically advantageous to ship larger numbers. Table 4.13 is a shipping cost schedule which specifies the cost of shipping a book from each warehouse to each distributor.

How should the shipments be made so as to minimize the total cost?

First, let us observe that we cannot analyze this problem in terms of two variables as we have other shipment problems since the total number of items needed $(8000 + 10,000 + 10,000 = 28,000)$ is not equal to the number available $(20,000 + 15,000 = 35,000)$. Our approach is to introduce a variable to stand for the number of books to be sent from each warehouse to each distributor. Let $X_{11}$ denote the number of books to be sent from W1 to D1, $X_{12}$ denote the number of books to be sent from W1 to D2, and so on. In general, $X_{ij}$ is used to denote the number of items shipped from source $i$ to destination $j$,

**Table 4.13**

| From Warehouse | To Distributor | | |
|:---:|:---:|:---:|:---:|
| | **D1** | **D2** | **D3** |
| W1 | 5¢ | 8¢ | 7¢ |
| W2 | 4¢ | 9¢ | 6¢ |

**Table 4.14**

| Shipped From | Received At | | |
| --- | --- | --- | --- |
| | D1 | D2 | D3 |
| W1 | $X_{11}$ | $X_{12}$ | $X_{13}$ |
| W2 | $X_{21}$ | $X_{22}$ | $X_{23}$ |

and the number of variables is equal to the product of the number of sources multiplied by the number of destinations. Thus for the problem at hand, six variables are needed (see Table 4.14).

The shipping cost function to be minimized is

$$C = 5X_{11} + 8X_{12} + 7X_{13} + 4X_{21} + 9X_{22} + 6X_{23}.$$

In addition to the nonnegativity of the variables, there are two conditions which must be satisfied.

First, the number of items that can be shipped from a source cannot exceed the number available at the source.

Thus we have

$$X_{11} + X_{12} + X_{13} \leq 20{,}000$$

$$X_{21} + X_{22} + X_{23} \leq 15{,}000.$$

Second, at least the number of items required at a destination shall be sent to the destination.

Thus we have

$$X_{11} + X_{21} \geq 8000$$

$$X_{12} + X_{22} \geq 10{,}000$$

$$X_{13} + X_{23} \geq 10{,}000.$$

The rows of Table 4.14 yield the first two constraints whereas the columns yield the last three.

The integer program that emerges is the following:

Find nonnegative integer values for $X_{ij}$ which

$$\text{minimize } C = 5X_{11} + 8X_{12} + 7X_{13} + 4X_{21} + 9X_{22} + 6X_{23}$$

subject to

$$X_{11} + X_{12} + X_{13} \leq 20{,}000$$

$$X_{21} + X_{22} + X_{23} \leq 15{,}000$$

$$X_{11} + X_{21} \geq 8000$$

$$X_{12} + X_{22} \geq 10{,}000$$

$$X_{13} + X_{23} \geq 10{,}000.$$

# EXERCISES

1. Martins Plant Foods, Inc., plans to introduce Martins Miracle, hailed as a "blossom booster" for outdoors and indoors flowering plants. Martins Miracle is to be a mixture of bonemeal and processed vegetable matter. Each can of Martins Miracle is to contain at least 12 units of nitrogen and 9 units of phosphorus. Each pound of bonemeal contains 3 units of nitrogen and 1 unit of phosphorus; each pound of vegetable matter contains 2 units of nitrogen and 2 units of phosphorus. Bonemeal costs $1 per pound and vegetable matter costs $1.50 per pound.

   The problem is to determine the number of pounds of bonemeal and vegetable matter that should be used to make up a can of Martins Miracle so that the nutrient requirements are met at minimal cost.

   (a) Set up and solve the linear program model for this situation. What is the minimal cost in terms of this model and what mix would achieve it?
   (b) Management has been advised to obtain the best solution in integers to its component mixture problem. What additional constraint(s) must be added to the linear program obtained from (a) to guarantee a solution in integers?
   (c) Determine the solution to the integer program model that emerges. What does it mean to Martins Plant Foods, Inc.?

*For the integer programs given in Exercises 2–4, determine the additional constraints that must be imposed to guarantee a solution in integers.*

2. Find nonnegative integers which minimize $F(x, y) = 3x + 2y$ subject to

$$3x + y \geq 6$$
$$x + y \geq 5.$$

3. Find nonnegative integers which maximize $P(x, y) = 16x + 6y$ subject to

$$3x + y \leq 6$$
$$x + y \leq 5.$$

4. Find nonnegative integers which minimize $C(x, y) = 4x + 5y$ subject to

$$3x + y \geq 6$$
$$x + 3y \geq 6.$$

5. The Hudson Furniture Manufacturing Company, Inc., has available 300 square yards of pine veneer and 228 square yards of walnut veneer for the manufacture of model D-1 desks and model B-14 bookcases. 8 square yards of pine veneer and 6 square yards of walnut veneer are used in making a desk; 10 square yards of pine veneer and 7 square yards of walnut veneer are used in making a bookcase. Pine veneer costs $2 per square yard and walnut veneer costs $4 per square yard. 5 labor-hours are needed to manufacture a desk and 4 labor-hours are needed to manufacture a bookcase. The cost of labor is $4 per labor-hour. At most 150 labor-hours are available for the manufacture of the desks and bookcases.

A desk sells at $100 and a bookcase sells at $108.

   The problem is to determine the number of each that should be made and sold to maximize net income.

   (a) Determine the additional constraint(s) that must be imposed to guarantee a solution in integers.
   (b) Solve the integer program that emerges.
   (c) What advice would you give to the Hudson Company?

6. Formulate the following problem as an integer program but do not solve.

   Cartons of citrus fruits are to be shipped from Lakeland and Orlando to distributors in Portland, New York, and Baltimore. Lakeland has available 10,000 cartons and Orlando has available 18,000 cartons. The Portland distributor needs 1000 cartons, the New York distributor needs 14,000 cartons, and the Baltimore distributor needs 9000 cartons. The shipping cost Table 4.15 specifies the cost of shipping a carton from each source to each destination. How should the shipments be made so as to minimize cost?

**Table 4.15**

|          | Portland | New York | Baltimore |
|----------|----------|----------|-----------|
| Lakeland | 7¢       | 6¢       | 5¢        |
| Orlando  | 5¢       | 5¢       | 6¢        |

7. Formulate the following problem as a linear program but do not solve.

   Bauxite ore is to be shipped from the Turin and Johnston mines of the Alexander Aluminum Company to refineries in Baltimore, Cincinnati, and Pittsburgh. The Turin mine has 15,000 tons and the Johnston mine has 20,000 tons. The Baltimore refinery requires at least 8000 tons, the Cincinnati refinery at least 10,000 tons, and the Pittsburgh refinery at least 12,000 tons. Table 4.16 specifies the cost (in dollars) of shipping 1 ton of bauxite from each source to each destination.

**Table 4.16**

|          | Baltimore | Cincinnati | Pittsburgh |
|----------|-----------|------------|------------|
| Turin    | 2         | 3          | 2          |
| Johnston | 3         | 4          | 2          |

The problem is to determine how the shipments should be made so as to satisfy the requirements of the refineries at minimum total cost.

Structurally speaking, this problem is the same as problem 6. In this case the shipping units (tons) admit noninteger solutions. This would not be the case had we been shipping cartons, automobiles, appliances, machinery, etc. which require integer values.

## Problems Leading to 0–1 Integer Programs

A number of problems lead to integer program models which, more specifically, require that the integers be 0's and 1's. The assignment situation considered in Case 6 of Section 4.5 is one such problem. A value of 1 for an assignment variable means that the person or item or task is to be included in the assignment; a value of 0 means that it is not to be included.

As another example of a 0–1 integer program situation consider the following selection problem.

---

**EXAMPLE 3  ◆  The Marsden Company's Selection Problem**

---

The Marsden Company has seven items $I_1, I_2, I_3, I_4, I_5, I_6, I_7$, that are to be shipped from New York to Detroit as quickly as possible. A company airplane with a freight capacity of 1500 pounds is available for this purpose. Table 4.17 gives the weight and value of each of the seven items. Since the sum of the weights of these items is 2340 pounds and the freight capacity of the plane is 1500 pounds, not all of the items can be taken.

**Table 4.17**

|       | Weight (lb) | Value ($) |
|-------|-------------|-----------|
| $I_1$ | 300         | 600       |
| $I_2$ | 350         | 610       |
| $I_3$ | 400         | 650       |
| $I_4$ | 250         | 400       |
| $I_5$ | 260         | 405       |
| $I_7$ | 280         | 410       |
| $I_7$ | 500         | 660       |

How should the plane be loaded so that the value of its contents is maximized and its freight capacity is not exceeded?

Corresponding to items $I_1, \ldots, I_7$, we introduce variables $X_1, \ldots, X_7$, respectively. $X_1$ can take on one of two values, 0 if item $I_1$ is not taken and 1 if item $I_1$ is taken; $X_2$ can take on one of two values, 0 if item $I_2$ is not taken and 1 if item $I_2$ is taken; and so on.

The function

$$V = 600X_1 + 610X_2 + 650X_3 + 400X_4 + 405X_5 + 410X_6 + 660X_7$$

is formed by multiplying $X_1$ by the value of item $I_1$ ($600), multiplying $X_2$ by the value of item $I_2$ ($610), and so on, then adding. $V$ is the value function to be maximized. Whenever $X_1, \ldots, X_7$ are given values (0's and 1's), $V$ becomes a sum of item values. For example, when $X_1 = X_2 = X_3 = 1$ and $X_4 = X_5 = X_6 = X_7 = 0$, then

$$V = 600 + 610 + 650 = 1860$$

the sum of the values of $I_1, I_2$, and $I_3$.

The problem of loading the plane so that the value of its contents is maximized and its freight capacity is not exceeded is expressed by the following integer program.

Find nonnegative integers which maximize

$$V = 600X_1 + 610X_2 + 650X_3 + 400X_4 + 405X_5 + 410X_6 + 660X_7$$

subject to

$$X_1 \leq 1$$

$$X_2 \leq 1$$

$$X_3 \leq 1$$

$$X_4 \leq 1$$

$$X_5 \leq 1$$

$$X_6 \leq 1$$

$$X_7 \leq 1$$

$$300X_1 + 350X_2 + 400X_3 + 250X_4 + 260X_5 + 280X_6 + 500X_7 \leq 1500.$$

The nonnegativity statement and the first seven constraints express the requirement that the integer values of the variables be 0's or 1's; the last one expresses the condition that the freight capacity of the plane (1500 pounds) not be exceeded.

---

Example 3 illustrates problems with the following general structure. A container of some sort (truck, car, plane) is to be loaded with items of various values and weights. For the items involved there is a limitation on the weight that can be loaded in the container but not on the volume. The problem is to load the container in such a way that its weight limit is not exceeded and the value of the items loaded is the largest possible.

# EXERCISES

*Formulate the following problems as integer programs.*

**8.** The Vroman Institute has two jobs to fill, physiologist (job 1) and biochemist (job 2), and is considering three candidates, Ann (candidate 1), Gena (candidate 2), and Marty (candidate 3), for the jobs. Each candidate's qualifications for each of the jobs have been assigned a numerical rating (see Table 4.18).

**Table 4.18**

| | Job | |
|---|---|---|
| Candidate | Physiologist (job 1) | Biochemist (job 2) |
| Ann (candidate 1) | 3 | $\frac{5}{2}$ |
| Gena (candidate 2) | 2 | $\frac{5}{2}$ |
| Marty (candidate 3) | 2 | $\frac{3}{2}$ |

These ratings are interpreted as a measure of a candidate's potential for a particular job. The Institute's problem is to fill the jobs in such a way that potential is maximized.

**9.** A legal advisory group is to make recommendations on two positions, State Supreme Court Judge and Civil Court Judge, and is considering three candidates, M. Jones, R. Johnson, and A. Marks. Table 4.19 describes potential rat-

ings that have been assigned by the advisory group as a quantitative measure of each person's qualifications for each position.

**Table 4.19**

| | Job | |
|---|---|---|
| Candidate | Supreme Court Judge | Civil Court Judge |
| Jones | 9 | 8 |
| Johnson | 8 | 9 |
| Marks | 10 | 8 |

The advisory group wishes to make its recommendations on how these positions should be filled on the basis of maximization of potential.

**10.** The Onute Land Development Company has identified five sites in Albuquerque Dallas, Miami, Phoenix, and San Diego for the construction of condominiums. The anticipated cost of construction on these sites (in millions of dollars) and the expected profit to be realized from each development (in millions of dollars) are described in Table 4.20. The company can commit at most $32 million to these developments, which is insufficient to undertake them all. The problem is to determine the selection of sites that yields the largest total expected profit, but for which the total cost does not exceed the amount available.

**Table 4.20**

| Sites | Cost | Profit |
|---|---|---|
| $(S_1)$ Albuquerque | 10 | 0.19 |
| $(S_2)$ Dallas | 12 | 0.23 |
| $(S_3)$ Miami | 11 | 0.20 |
| $(S_4)$ Phoenix | 15 | 0.30 |
| $(S_5)$ San Diego | 9 | 0.16 |

11. Ecap University has two jobs to fill, Dean of the Graduate School (job 1) and Dean of the School of Arts and Science (job 2), and is considering three candidates, J. Frank (candidate 1), M. Smith (candidate 2), and T. James (candidate 3). The search committee of the university, which is to make recommendations to the president, has assigned a numerical rating to each candidate's qualifications for each job (see Table 4.21). These ratings are interpreted as a quantitative measure of each candidate's potential for each job. The search committee wishes to make its recommendations in such a way that potential is maximized.

**Table 4.21**

| | Job | |
|---|---|---|
| Candidate | Dean, Graduate School (job 1) | Dean, Arts and Science School (job 2) |
| Frank (candidate 1) | 9 | 8 |
| Smith (candidate 2) | 7 | 7 |
| James (candidate 3) | 6 | 8 |

12. The Birute Investment Company has $50,000 available for investment. Four stocks, $S_1$ (oil), $S_2$ (computers), $S_3$ (airlines), and $S_4$ (steel), are being considered. Table 4.22 specifies the current price per share of each stock and the expected net profit per share of stock. At most 200 shares of $S_1$ stock, 300 shares of $S_2$ stock, 400 shares of $S_3$ stock, and 300 shares of $S_4$ stock are to be purchased.

**Table 4.22**

| Stock | Price | Profit |
|---|---|---|
| $S_1$ (oil) | $75 | $6.00 |
| $S_2$ (computers) | $80 | $5.60 |
| $S_3$ (airlines) | $50 | $4.00 |
| $S_4$ (steel) | $60 | $4.50 |

The problem is to determine the number of shares of each stock that should be purchased so as to maximize the total return.

# Chapter 4  Progress Check, Part 1

1. On St. Patrick's Day, the school cafeteria serves a traditional meal of corned beef and cabbage. In an effort to satisfy health concerns, the cafeteria wants each plate served to have sufficient food so as to provide at least 275 units of Vitamin C, but no more than 190 units of Vitamin A, and the minimal amount of sodium. Each slice of corned beef contains 30 units of Vitamin C, 50 units of Vitamin A, and 60 units of sodium. Each cup of cabbage contains 50 units of Vitamin C, 20 units of Vitamin A, and 30 units of sodium. You are to determine how much of each food should be served on each plate in order to meet the stated health goals.

   (a) Write an objective function and all constraints for this linear program. Declare what all variables represent.

   (b) Sketch the appropriate feasible region, label all corner points with exact coordinates, and solve the linear program. State your conclusion in a complete sentence.

2. A discount store will stock two styles of coffee tables that are received from the manufacturer with slight defects. The store has storage space for as many as 80 tables and is willing to allocate up to 110 labor hours to repair the defects. Each table of Style A requires one hour of labor to repair, while each table of Style B requires two hours of labor to repair. The Style A tables are sold for $48 each, while the Style B tables are sold for $36 each. You are to determine how many of each style of table should be stocked in order to maximize gross sales.

(a) Write an objective function and all constraints for this linear program. Declare what all variables represent.

(b) Sketch the appropriate feasible region, label all corner points with exact coordinates, and solve the linear program. State your conclusion in a complete sentence.

# Chapter 4 Progress Check, Part 2

1. Tamara is raising a northern strain largemouth bass in hopes that one day it will reach a size of 16 pounds. In order to accomplish this, she is producing a special fish food mixture from two brands of raw fish food, Brand A and Brand B. She wants to make sure that the mixture contains at least 700 units of vitamin $B_1$, at least 800 units of protein, and at least 500 units of fat, but also wants to minimize the total cost of the mixture. Brand A fish food costs $3 per pound and each pound contains 3 units of vitamin $B_1$, 5 units of protein, and 8 units of fat. Brand B fish food costs $6 per pound and each pound contains 2 units of vitamin $B_1$, 8 units of protein, and 6 units of fat. Tamara needs your help with this linear programming problem.

   (a) Write an objective function and all constraints for this linear program. Declare what all variables represent.

   (b) Sketch the appropriate feasible region, label all corner points with exact coordinates, and solve the linear program. State your conclusion in a complete sentence.

2. Megafast Food Corporation is planning on opening several different fast food restaurants near college campuses in the upper Midwest. A Quickwich restaurant requires an initial investment of $600,000, requires 15 employees, and is expected to generate an annual profit of $40,000. A Meximart restaurant requires an initial investment of 400,000, requires 9 employees, and is expected to generate an annual profit of $30,000. A Frozone restaurant requires an initial investment of $300,000, requires 5 employees, and is expected to generate an annual profit of $25,000. The corporation will spend up to $48,000,000 in initial investment capital, will hire up to 1000 new employees, and wants to open no more than 70 restaurants. The corporation wishes to determine how many of each type of restaurant to open in order to maximize annual profits.

(a) Give the objective function for the linear program. Declare what variables represent, and indicate if you are to maximize or minimize the objective function.

(b) Write all constraints for this linear program.

# ANSWERS TO SELECTED EXERCISES

## Section 4.1

1. $(0, 2), (3, -2), (-4, 0)$
3. $(2, 3), (6, 2)$
5. $(3, -2), (-4, 0), (2, 3), (6, 2)$
7. $(3, -1, 4), (\frac{1}{2}, \frac{1}{3}, -3), (-2, 3, 1)$
9. $(2, 3, 1), (1, 3, -1), (4, 2, 1), (\frac{1}{2}, \frac{1}{3}, -3)$
11. $(2, 3, 1), (3, -1, 4), (\frac{1}{2}, \frac{1}{3}, -3), (-2, 3, 1)$

## Section 4.1

21. $x > -\frac{2}{3}$
23. $x \le 2$
25. $y < 2$
27. $x < -15$
29. $x > \frac{5}{2}$
31. $y \ge 2$

## Section 4.1

33.

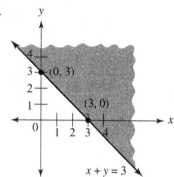

35.

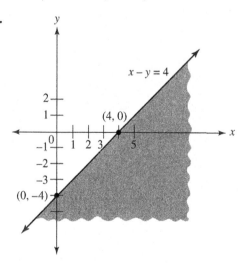

37.

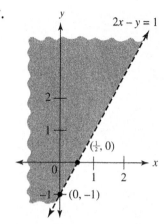

39.

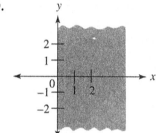

41.

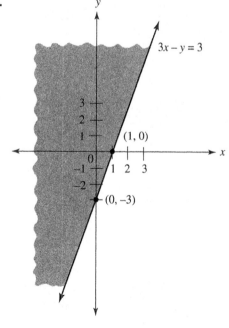

**43.**

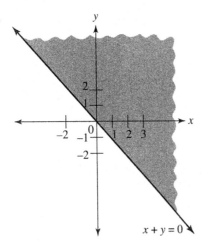

**45.**

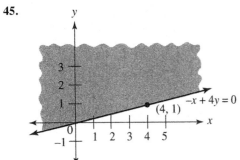

**47.**

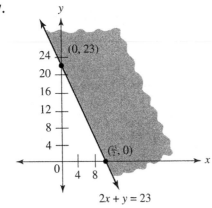

## Section 4.1

**49.** $(8, 2), (8, 17), (20, 5), (9, 12), (14, 7)$

**51.** $(1, 1, 2), (0, 1, 3), (2, 1, 1), (4, 1, 0), (2, 1, 2)$

**53.**

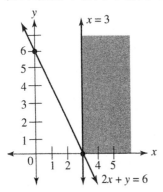

**55.**

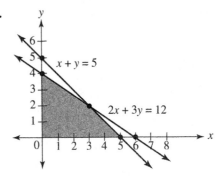

**57.**

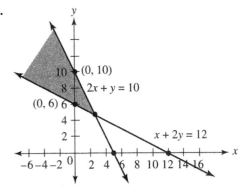

**59.**

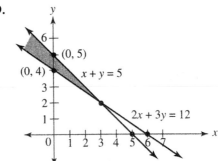

**61.**

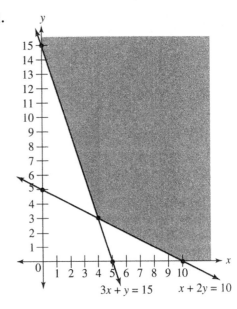

**63.**

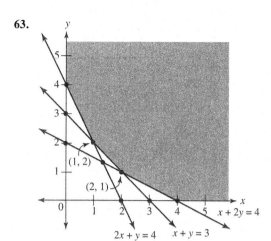

**65.**

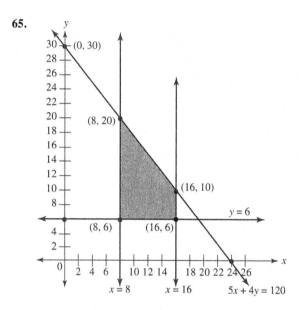

**67.**

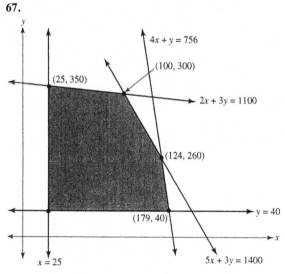

## Section 4.3

**1.** The basic data are summarized in Table 1.

**Table 1**

| Product | No. Made | Profit Per Unit | Construction Time Per Unit (hours) | Finishing Time Per Unit (hours) |
|---------|----------|-----------------|-------------------------------------|----------------------------------|
| DT-1 | $x$ | 140 | 8 | 3 |
| DT-2 | $y$ | 150 | 5 | 2 |
| | | | $\leq 2210$ | $\leq 860$ |

The assumptions underlying these data and the conditions that at least 50 DT-1 units and 50 DT-2 units must be made per week lead to LP-2.

## Section 4.4

**1.** $(6, 0)$; 30

**3.** $(1, 2)$; 3.7

**5.** $(45, 15)$; 5.7

**7.** $(\frac{50}{3}, \frac{65}{3})$; 3350

**9.** $(50, 40)$; 13,800

**11.** $(75, 0)$ and, more generally, $(75, y)$, where $y$ is an integer between 0 and 25, inclusive; 12,420.

## Section 4.5

**1.**
$$\text{Maximize } P = 180x + 120y$$

subject to

$$x \geq 0, \quad y \geq 0$$
$$4x + 3y \leq 320$$
$$5x + 2y \leq 330$$

$x$ and $y$ are the number of $K15$ and $K31$ units, respectively, to be made per week.

Sol. $(50, 40)$; max. value 13,800 (see Exercise 9 of Section 3.4 (p. 86).

**3.**
$$\text{Minimize } C = 45x + 50y$$

subject to

$$x \geq 0, \quad y \geq 0$$
$$3x + 5y \geq 12$$
$$x + y \geq 3$$

$x$ and $y$ are the number of pounds of pork and beef, respectively, to be used in putting together a can of meat.

Sol. $(\frac{3}{2}, \frac{3}{2})$; minimum value 142.5.

**5.**                    Maximize $I = 0.09x + 0.06y$

subject to

$$x \geq 0, \quad y \geq 0$$

$$x + y \leq 25$$

$$-x + 4y \geq 0$$

$$x \geq 8$$

$x$ and $y$ are the amounts (in millions of dollars) allocated for loans and securities, respectively.
Sol. (20, 5); maximum value 2.1.

**7.** Two linear programs emerge, one for the faculty council and the other for the administration. The constraints are the same for both linear programs since the underlying conditions are the same.

$$\text{Max. } F = x + y$$

subject to

$$x \geq 0, \quad y \geq 0$$

$$x \leq 22, \quad y \geq 3$$

$$x + 2y \leq 30$$

$$-x + 4y \leq 0$$

Sol. (22, 4); maximum value 26.

$$\text{Min. } C = 500x + 1000y$$

subject to

(same constraints)

$x$ and $y$ denote the number of associate and full-professor slots, respectively, to be established.
Sol. (12, 3); minimum value 9000.

**9.**                    $\text{Min. } C = 1.2x + 1.8y$

subject to

$$x \geq 0, \quad y \geq 0$$

$$x + y = 2,000,000$$

$$30x + 32y \geq 62,400,000$$

$x$ and $y$ denote the number of tons of steel produced subject to the F14 and F24 filter systems, respectively.
Sol. (800,000; 1,200,000); minimum value 3,120,000.

**11.**                    $\text{Max. } P = 24x + 30y + 26z$

subject to

$$x \geq 0, \quad y \geq 0, \quad z \geq 0$$

$$2x + 3y + z \leq 16$$

$$x + y + 3z \leq 17$$

$$2x + y + z \leq 12$$

$x, y,$ and $z$ denote the number of C15, C24, and C51 minicomputer units, respectively, to be made.

**13.**                    $\text{Max. } I = 0.1x + 0.18y$

subject to

$$x \geq 0, \quad y \geq 0$$

$$x \leq 41,000; \quad y \geq 5,000$$

$$x + y \leq 50,000$$

$$-x + 4y \leq 0$$

$x$ and $y$ denote the amounts to be invested in bonds and stock, respectively.
Sol. (40,000; 10,000); Max. value 5800.

**15.**                    $\text{Min. } C = 0.8x + 1.25y$

subject to

$$x \geq 0, \quad y \geq 0$$

$$4x + 3y \geq 30$$

$$2x + 3y \geq 24$$

$x$ and $y$ denote the number of AA1 and AA2 freight cars, respectively, to be rented.
Sol. (12, 0); Min. value 9.6.

**17.**                    $\text{Max. } P = 10x + 12y$

subject to

$$x \geq 0, \quad y \geq 0$$

$$x + 2y \leq 24,000$$

$$2x + y \leq 30,000$$

$$x \geq 5000, \quad y \geq 5000$$

$x$ and $y$ denote the target sales volume (in product units) for hospitals and medical supply houses, respectively.

In all of these situations we are translating assumptions made into, a linear program, omitting nothing that was explicitly stated and adding nothing of our own. Our concern is that the assumptions made are realistic and that the factors that were left out because they were viewed as negligible may be realistically treated as such.

## Section 4.6

**1. (a)**                    $\text{Min. } C = x + 1.5y$

subject to

$$x \geq 0, \quad y \geq 0$$

$$x + 2y \geq 9$$

$$3x + 2y \geq 12$$

$x$ and $y$ denote the number of pounds of bonemeal and processed vegetable matter, respectively, to be used to make up a can of Martins Miracle.
Sol. (3/2, 15/4); min. value 7.125.

**(b)** $x + y \geq 6$

**(c)** $(3, 3)$

**3.** $2x + y \leq 5$; sol. $(1, 3)$, max value 34.

**5. (a)** $x + y \leq 33$

**(b)** Max. $I = 40x + 44y$

subject to

$$x \geq 0, \quad y \geq 0$$

$$4x + 5y \leq 150$$

$$6x + 7y \leq 228$$

$$5x + 4y \leq 150$$

$$x + y \leq 33$$

$x$ and $y$ denote the number of desks and bookcases, respectively, to be made. Sol. $(15, 18)$; max. value 1392.

**7.** The variables defined in Table 2 denote the number of tons of Bauxite to be shipped from the sources to the destinations.

**Table 2**

| | Destination | | |
|---|---|---|---|
| Source | Baltimore $(D_1)$ | Cincinnati $(D_2)$ | Pittsburgh $(D_3)$ |
| Turin $(S_1)$ | $X_{11}$ | $X_{12}$ | $X_{13}$ |
| Johnston $(S_2)$ | $X_{21}$ | $X_{22}$ | $X_{23}$ |

The following linear program emerges:

Minimize $C = 2X_{11} + 3X_{12} + 2X_{13} + 3X_{21} + 4X_{22} + 2X_{23}$

subject to

$$X_{11} \geq 0, \quad X_{12} \geq 0, \text{etc.}$$

$$\left.\begin{array}{l} X_{11} + X_{21} \geq 8000 \\ X_{12} + X_{22} \geq 10{,}000 \\ X_{13} + X_{23} \geq 12{,}000 \end{array}\right\} \begin{array}{l}\text{At least the amts} \\ \text{required will be} \\ \text{received.}\end{array}$$

$$\left.\begin{array}{l} X_{11} + X_{12} + X_{13} \leq 15{,}000 \\ X_{21} + X_{22} + X_{23} \leq 20{,}000 \end{array}\right\} \begin{array}{l}\text{The capacity of the} \\ \text{sources cannot be} \\ \text{exceeded.}\end{array}$$

**9.** To relate candidates 1 (Jones), 2 (Johnson), and 3 (Marks) to jobs 1 (Supreme Court Judge) and 2 (Civil Court Judge) we introduce variables $X_{11}, X_{21}$, and more generally $X_{ij}$, to relate candidate $i$ to job $j$, as summarized in Table 3. $X_{ij}$ can assume one of two values, 0 if candidate $i$ is not assigned

job $j$, 1 if candidate $i$ is assigned job $j$.

**Table 3**

| | Job | |
|---|---|---|
| Candidate | Supreme Court Judge (job 1) | Civil Court Judge (job 2) |
| Jones (candidate 1) | $X_{11}$ | $X_{12}$ |
| Johnson (candidate 2) | $X_{21}$ | $X_{22}$ |
| Marks (candidate 3 | $X_{31}$ | $X_{32}$ |

The problem of filling the positions so that the total potential rating is maximized in such a way that each candidate is assigned to at most one job and each job is filled by at most one person is expressed by the following $0-1$ integer program:

Max. $P = 9X_{11} + 8X_{21} + 10X_{31} + 8X_{12} + 9X_{22} + 8X_{32}$

subject to

$$\left.\begin{array}{l} X_{11} + X_{12} \leq 1 \\ X_{21} + X_{22} \leq 1 \\ X_{31} + X_{32} \leq 1 \end{array}\right\} \begin{array}{l}\text{Each candidate is} \\ \text{assigned to at most} \\ \text{one job}\end{array}$$

$$\left.\begin{array}{l} X_{11} + X_{21} + X_{31} \leq 1 \\ X_{12} + X_{22} + X_{32} \leq 1 \end{array}\right\} \begin{array}{l}\text{Each job is filled} \\ \text{by at most one} \\ \text{candidate}\end{array}$$

**11.** Max. $P = 9X_{11} + 10X_{21} + 8X_{31} + 6X_{12} + 10X_{22} + 3X_{32}$

subject to

$$X_{11} \geq 0, \quad X_{21} \geq 0, \text{etc.}$$

$$X_{11} + X_{12} \leq 1$$

$$X_{21} + X_{22} \leq 1$$

$$X_{31} + X_{32} \leq 1$$

$$X_{11} + X_{21} + X_{31} \leq 1$$

$$X_{12} + X_{22} + X_{32} \leq 1$$

# PART II

# Probability and Statistics

**CHAPTER**

# 5

# Introduction to Probability Theory

In this chapter we will discuss sets (and the related concepts of relations and functions), an important idea used throughout the language of mathematics. We will use the language of sets at several points in this book, such as when we look carefully at the properties of whole numbers, when we study probability, and when we examine the fundamental concepts of geometry. However, be assured that the language of sets has much broader use than this. The language of sets lives at the heart of all discussions in higher mathematics and has found application in such diverse disciplines as computer science and the social sciences.

## 5.1 AN INTRODUCTION TO SETS

**Problem**   Place eight tennis balls in three baskets of different sizes so that there is an odd number of balls in each basket.

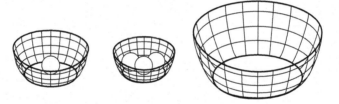

**Overview**   The preceding is a trick problem involving subsets.

**Goals**   Use the Communication Standard: ● organize and consolidate mathematical thinking through communication; ● communicate mathematical thinking coherently and clearly to peers, teachers, and others; ● analyze and evaluate the mathematical thinking and strategies of others; ● use the language of mathematics to express mathematical ideas precisely.

Although this book gives little emphasis to the formal theory of sets, some familiarity with the notation and language of sets is useful and important. During the latter part of the 19th century, while working with mathematical entities called *infinite series,* Georg Cantor found it helpful to borrow a word from common usage to describe a mathematical idea. The word he borrowed was **set**.

### Describing Sets

Our purposes will be served if we intuitively describe a set as a collection of objects possessing a property that enables us to determine whether a given object is in the collection. We sometimes say that a set is a *well-defined* collection—meaning that, given an object and a set, we are able to determine whether or not the object is in the set. The individual objects in a set are called **elements** of the set. They are said **to belong to** or **to be members of** or **to be in** the set. The relationship between objects in a set and the set itself is expressed in the form *is an element of or is a member of*.

To denote this relationship we use the following notation:

$x \in A$ means $x$ is an **element** of set $A$.

$x \notin A$ means $x$ is **not an element** of set $A$.

Often it is possible to specify a set by listing its members within braces. This method of describing a set is called the **tabulation** method (sometimes called the *roster method*). For example, the set of counting numbers less than 10 can be written as $\{1, 2, 3, 4, 5, 6, 7, 8, 9\}$.

A set remains the same, regardless of the order in which the elements are tabulated. Thus $\{1, 2, 3\}$ is the same set as $\{2, 1, 3\}$, $\{3, 2, 1\}$, $\{3, 1, 2\}$, $\{1, 3, 2\}$, or $\{2, 3, 1\}$. In fact, two sets are said to be **equal** if they contain exactly the same elements. $A = B$, **if and only if $A$ and $B$ have exactly the same elements**. If there is at least one element in either $A$ or $B$ that is not in the other, then $A \neq B$.

## EXAMPLE 1

Glenda, Mark, and Martia are the only counselors in the admissions office. They constitute the set $A = \{$Glenda, Mark, Martia$\}$. Mark $\in A$; Linda $\notin A$. Can you identify other elements of set $A$?

Sometimes sets have so many elements that it is tedious, difficult, or even impossible to tabulate them. Sets of this nature may be indicated by a descriptive statement or a rule. The following sets are well specified without a tabulation of members: the counting numbers less than 10, the even numbers less than 1000, the past presidents of the United States, and the football teams in Pennsylvania.

## Then and Now

### Georg Cantor
### 1845–1918

An exciting feature that distinguishes mathematics from other disciplines is that mathematics deals with ideas of the infinite. Concepts of infinity plagued many mathematicians until the seminal work of Georg Cantor. We struggle with the notion of infinity because things that are infinite are beyond our realm of normal physical experience; they are therefore abstractions that we can deal with only in our minds. Yet Cantor showed that it makes sense to talk about the number of elements in any set, finite or infinite, and, surprisingly, showed that an infinite sequence of higher infinities can also be described.

Cantor was born in St. Petersburg, Russia, to Norwegian parents. The family moved to Frankfurt, Germany, when Georg was 11 years old. Resisting his father's encouragement to study engineering, Georg focused on mathematics, philosophy, and physics in his studies at the Universities of Zurich, Gottingen, and Berlin. He obtained a position at the University of Halle, in Wittenberg, Germany. Although his goal was an appointment to the University of Berlin, he spent most of his life in Wittenberg.

For not attaining his goal to be appointed to the University of Berlin, Cantor blamed a Berlin mathematician named Leopold Kronecker, who did not accept any of Cantor's work on infinite sets. Kronecker openly attacked Cantor's findings, contributing to Cantor's mental collapse in 1884. This collapse was the first of many that occurred throughout the rest of his life. He died in an institution in Wittenberg in 1918, before his results were widely accepted.

In the last decade, a new chapter has been added to Cantor's work. The study of chaos theory and its geometric progeny, fractals, has renewed interest in some of Cantor's most unusual work.

The difficulty of tabulating sets can be minimized by using **set-builder notation**, which encloses within braces a letter or symbol representing an element of the set followed by a qualifying description of the element. For example, let A represent the set of counting numbers less than 10; then

$$A = \{n \mid n \text{ is a counting number less than } 10\}$$

This notation is read "the set of all elements $n$ such that $n$ is a counting number less than 10." Notice that the vertical line is read "such that."

## EXAMPLE 2

Use set-builder notation to denote the set of current United States senators.

**Solution**

$\{x \mid x \text{ is a U.S. senator}\}$. The set is read "the set of all $x$ such that $x$ is a U.S. senator."

Frequently, three dots (called an **ellipsis**) are used to indicate the omission of terms. The set of even counting numbers less than 100 may be written as $\{2, 4, 6, \ldots, 98\}$. This notation saves time in tabulating elements of large sets, but it can be ambiguous unless the set has been specified completely by another description. For example, $\{2, 4, \ldots, 16\}$ could be $\{2, 4, 8, 16\}$ or it could be $\{2, 4, 6, 8, 10, 12, 14, 16\}$.

An ellipsis is also used to indicate that a sequence of elements continues indefinitely. For example, consider the set of **natural** (or **counting**) **numbers**.

$$\{1, 2, 3, 4, 5, \ldots\}$$

The set of natural numbers is an example of an **infinite set**, described informally as one that contains an unlimited number of elements. In contrast, a **finite set** contains zero elements or a number of elements that can be specified by a natural number.

**DEFINITION** | **Empty Set**

A set that contains no elements is called the *empty* or *null set* and is denoted by either $\varnothing$ or $\{ \}$.

The relationship between two sets such as $A = \{1, 3, 5, 7\}$ and $B = \{1, 2, 3, 4, 5, 6, 7, 8\}$ is described by the term *subset*.

**DEFINITION** | **Subset**

Set $A$ is said to be a *subset* of set $B$, denoted by $A \subseteq B$, if and only if each element of $A$ is an element of $B$.

## EXAMPLE 3

If $A = \{x, y\}$ and $B = \{w, x, y, z\}$, then $A \subseteq B$ because each element of $A$ is an element of $B$.

## EXAMPLE 4

If $P = \{1, 4, 7\}$ and $Q = \{4, 7, 1\}$, then $P \subseteq Q$ because each element of $P$ is an element of $Q$. Moreover, $Q \subseteq P$, and $P \subseteq P$.

To show that $A$ is not a subset of $B$ (denoted by $A \not\subset B$), we must find at least one element in $A$ that is not in $B$. Let's use this idea to examine whether $\varnothing$ is a subset of some set $B$. Using our problem-solving techniques, let's list all possibilities:

1.  Either $\varnothing$ is a subset of $B$
2.  Or $\varnothing$ is not a subset of $B$

In Possibility 2, if $\varnothing$ is not a subset of $B$, then there must be an element of $\varnothing$ not in $B$. But $\varnothing$ has no elements. Consequently, Possibility 2 cannot be true. The only alternative is for Possibility 1 to be true.

---

**Subset $\varnothing$**

For any set $B$, $\varnothing \subseteq B$.

---

Because all dogs are animals, the set of dogs is a subset of the set of animals. Moreover, we call the set of dogs a "proper" subset of the set of animals because some animals are not dogs.

**DEFINITION**

---

**Proper Subset**

Set $A$ is said to be a *proper subset* of set $B$, denoted by $A \subset B$, if and only if each element of $A$ is an element of $B$ and at least one element of $B$ is not an element of $A$; that is, $A \subset B$ if $A \subseteq B$ and $A \neq B$.

---

## Operations on Sets

If a discussion is limited to a fixed set of objects and if all elements to be discussed are contained in this set, then this overall set is called the **universal set**, or simply the **universe**. A very useful device for visualizing and discussing sets is the **Venn diagram**. Circles or other closed curves are used in Venn diagrams (named after the English logician, Robert Venn) to represent sets. The universe can be represented as the region bounded by a rectangle, and the set under consideration as the region bounded by a circle (or some other closed region) within the rectangle. In Figure 5.1, $x \in A$ means that $x$ is some point in the circular region $A$. Also in Figure 5.1, set $B$ is a proper subset of set $A$.

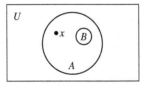

FIGURE 5.1

The region outside set $A$ and inside the universe represents the complement of $A$, denoted by $\overline{A}$ (or sometimes by $A'$ or $\sim A$) and read "$A$ bar" or "bar $A$."

**DEFINITION**

---

**Complement of a Set**

The *complement* of set $A$, denoted by $\overline{A}$, is the set of elements in the universe that are not in set $A$. If $A$ is a subset of the universe $U$, then

$$\overline{A} = \{x \mid x \in U \text{ and } x \notin A\}.$$

---

## EXAMPLE 5

If $U = \{a, b, c, d\}$ and $A = \{b, c\}$, find $\overline{A}$.

**Solution**

$\overline{A} = \{a, d\}$.   **Elements in $U$ but not in $A$**

## EXAMPLE 6

If the universe is all college students and if $A$ is the set of college students who have made all A's, then all college students who have made at least one grade lower than an A is the complement of $A$ (that is, $\overline{A}$) and is represented by the shaded region of Figure 5.2.

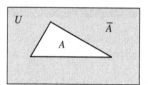

*FIGURE 5.2*

---

### Complement of $A$ Relative to $B$

The complement of $A$ relative to $B$ is the set of elements in $B$ that are not in $A$. This may be written as $B - A$ or in set-builder notation as

$$\{x \mid x \in B \quad \text{and} \quad x \notin A\}$$

---

$B - A$ is sometimes read "the set difference of $B$ and $A$." $B - A$, represented by the shaded region in Figure 5.3, is sometimes called a relative complement.

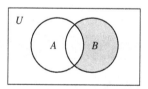

*FIGURE 5.3*

## EXAMPLE 7

If $A = \{x, y, z, w\}$ and $B = \{u, v, x, y\}$, then

$$B - A = \{u, v\}$$
$$A - B = \{z, w\}$$
$$B - B = \varnothing$$

Special notations are used for discussing the relationships among members of two or more sets.

DEFINITION

> **Intersection of Sets**
>
> The *intersection* of any two sets $A$ and $B$, denoted by $A \cap B$, is the set of all elements common to both $A$ and $B$. $A \cap B = \{x \mid x \in A \text{ and } x \in B\}$.

DEFINITION

> **Union of Sets**
>
> The *union* of any two sets $A$ and $B$, denoted by $A \cup B$, is the set of all elements in set $A$ or in set $B$ or in both $A$ and $B$. $A \cup B = \{x \mid x \in A \text{ or } x \in B\}$.

## EXAMPLE 8

Let $A$ represent a committee consisting of {Rose, Dave, Sue, John, Jack}, and let $B$ represent a second committee consisting of {Sue, Edward, Cecil, John}. The intersection of these two sets, $A \cap B$, is {Sue, John}.

## EXAMPLE 9

If $A = \{1, 2, 3\}$ and $B = \{6, 7\}$, find $A \cup B$.

**Solution**

$A \cup B = \{1, 2, 3, 6, 7\}$   **Elements in either $A$ or $B$**

Recall that $A \cup B$ is the set of all elements that belong to $A$ or to $B$ or to both $A$ and $B$. Any elements common to both sets are listed only once in the union. Thus, for

$$A = \{a, b, c, d, e\} \quad \text{and} \quad B = \{c, d, e, f, g\} \text{ we find that}$$

$$A \cup B = \{a, b, c, d, e, f, g\}$$

## EXAMPLE 10

If $A = \{1, 2\}$, then $A \cup A = \{1, 2\}$.

The shaded regions in Figure 5.4 compare intersection and union under different situations for sets $A$ and $B$. Notice in (a) that $A$ and $B$ overlap or have elements in common. In (b), $A$ is a proper subset of $B$. In (c), $A$ and $B$ have no elements in common, so $A \cap B = \varnothing$; here $A$ and $B$ are said to be **disjoint**.

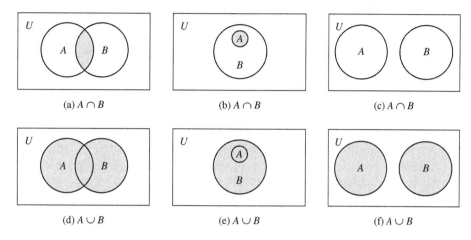

*FIGURE 5.4*

**DEFINITION**   Disjoint Sets

Two sets $A$ and $B$ are said to be *disjoint* if and only if $A \cap B = \varnothing$—that is, if the two sets have no elements in common.

## EXAMPLE 11

If $U = \{x \mid x$ is a counting number less than $10\}$, $A = \{2, 4, 6\}$, $B = \{1, 2, 3, 4, 5\}$, and $C = \{3, 5, 7\}$, find $A \cap B$, $A \cap C$, $A \cup C$, and $\overline{B}$.

**Solution**

$$A \cap B = \{2, 4\} \qquad \textbf{Elements in common}$$

$$A \cap C = \varnothing \qquad \textbf{No elements in common}$$

$$A \cup C = \{2, 3, 4, 5, 6, 7\} \qquad \textbf{Elements in either } \boldsymbol{A} \textbf{ or } \boldsymbol{C} \textbf{ or both}$$

$$\overline{B} = \{6, 7, 8, 9\} \qquad \textbf{Elements in the universe, not in } \boldsymbol{B}$$

*Practice Problem*

Find the intersection and union of $A = \{1, 2, 3, \ldots, 100\}$ and $B = \{60, 61, \ldots, 1000\}$.

**Answer**

$A \cap B = \{60, 61, \ldots, 100\}$, $A \cup B = \{1, 2, 3, \ldots, 1000\}$

## Properties of Set Operations

The following properties of the intersection and union of sets can be illustrated by using Venn diagrams.

---

**Properties of Intersection and Union**

For all sets $A$, $B$, and $C$:

1. *Commutative properties:* $A \cup B = B \cup A$; $A \cap B = B \cap A$
2. *Associative properties:* $A \cup (B \cup C) = (A \cup B) \cup C$; $A \cap (B \cap C) = (A \cap B) \cap C$
3. *Identity properties:* $A \cup \varnothing = A$; $A \cap U = A$

---

Notice that the *commutative* properties of intersection and union indicate that order is not important in performing these operations. The *associative* properties of intersection and union indicate that grouping is not important in performing these operations. The *identity* property of union indicates that there is a special set (the null set) with the property that its union with a set is always that same set. The universal set serves the "identity" role for intersection.

Figure 5.5(a) represents the formation of $(A \cap B) \cap C$. Similarly, Figure 5.5(b) represents the formation of $A \cap (B \cap C)$. The double-shaded region representing $(A \cap B) \cap C$ is the same as the region representing $A \cap (B \cap C)$; this illustrates the associative property, $(A \cap B) \cap C = A \cap (B \cap C)$.

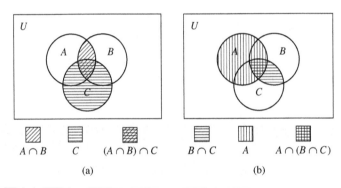

FIGURE 5.5

---

## EXAMPLE 12

Let $A = \{x \mid x$ is a state whose name begins with A$\}$ and $B = \{x \mid x$ is a state east of the Mississippi River$\}$. $A \cap B = \{$Alabama$\}$, and $B \cap A = \{$Alabama$\}$; therefore, $A \cap B = B \cap A$, illustrating the commutative property of intersection.

---

The commutative property of union is demonstrated in the next example.

## EXAMPLE 13

Consider $A = \{1, 3\}$ and $B = \{3, 5, 7\}$. Then $A \cup B = \{1, 3, 5, 7\}$ and $B \cup A = \{1, 3, 5, 7\}$; thus, $A \cup B = B \cup A$.

---

## EXAMPLE 14

If $A = \{1, 2, 3, 4\}$ and $B = \varnothing$, then

$$A \cup B = \{1, 2, 3, 4\} \cup \varnothing = \{1, 2, 3, 4\} = A$$

---

**EXAMPLE 15**

In Figure 5.6, (a) and (b) represent the formation of $(A \cup B) \cup C$, and (c) and (d) represent $A \cup (B \cup C)$. Notice in (b) and (d) that $(A \cup B) \cup C$ and $A \cup (B \cup C)$ are represented by the same shaded region. This demonstrates that the associative property holds for union.

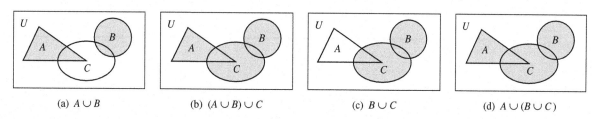

(a) $A \cup B$        (b) $(A \cup B) \cup C$        (c) $B \cup C$        (d) $A \cup (B \cup C)$

*FIGURE 5.6*

**Practice Problem**

Shade the portion of the diagram in Figure 5.7 that represents $A \cap (B \cap C)$.

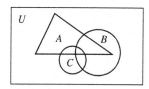

*FIGURE 5.7*

**Answer**

The answer is given in two steps in Figure 5.8(a) and (b).

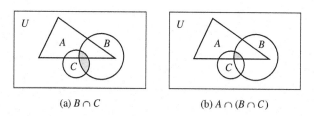

(a) $B \cap C$        (b) $A \cap (B \cap C)$

*FIGURE 5.8*

**EXAMPLE 16**

The introductory problem asks that we place eight tennis balls in three baskets of different sizes so that there is an odd number of balls in each basket.

**Solution**

The problem did not state that the baskets must be disjoint. In fact, the hint suggested using subsets. So place one basket in another. One solution is given in Figure 5.9.

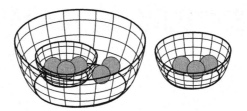

FIGURE 5.9

## Just for Fun

Use problem-solving techniques to discover the number of subsets in a set with *n* elements. (*Hint:* Use the strategy of trying simple versions of the problem; then look for a pattern.)

## EXERCISES

*In this exercise set we shall use the counting numbers* {1, 2, 3, . . .}, *the odd counting numbers* {1, 3, 5, 7, . . .}, *and the even counting numbers* {2, 4, 6, . . .}.

1. Let *A* be the set of all counting numbers less than 16. Which of the following statements are true, and which are false? Justify your answer.

   (a)  $11 \in A$        (b)  $81 \in A$
   (c)  $\{5\} \in A$       (d)  $\{1, 2, 3, \ldots, 15\} \in A$
   (e)  $14 \in A$       (f)  $0 \in A$

2. List within braces the members of the following sets.

   (a)  The counting numbers less than or equal to 16.

   (b)  The set of even counting numbers.

   (c)  The set of women presidents of the United States.

3. Express the sets in Exercise 2 in set-builder notation.

4. For each set in the left column, choose the sets from the right column that are subsets of it.

   (a)  $\{a, b, c, d\}$     (i)  { }
   (b)  $\{o, p, k\}$      (ii)  $\{1, 4, 8, 9\}$
   (c)  Set of letters in the    (iii)  $\{o, k\}$
        word *book*        (iv)  $\{12\}$
   (d)  $\{2, 4, 6, 8, 10, 12\}$   (v)  $\{b\}$

5. Classify each statement as true or as false.

   (a)  $\{z, r\} \subseteq \{x, y, z, r\}$
   (b)  {Brenda, Sharon, Glenda} $\subseteq$ {Brenda}
   (c)  $\{7, 2, 6\} \subseteq \{2, 6, 7\}$
   (d)  $6 \subseteq \{4, 5, 6, 7\}$
   (e)  $\{p, q, r\} \subseteq \{p, q, r, s\}$

6. Which of the following sets are well defined?

   (a)  The set of great baseball players

   (b)  The set of beautiful horses

   (c)  The set of students in this class

   (d)  The set of counting numbers smaller than a million

7. Insert the appropriate symbol $\{\in, \notin, \subset, \text{or } \subseteq\}$ to make the following statements true:

   (a)  3 _____ $\{1, 2, 3\}$
   (b)  $\{2\}$ _____ $\{1, 2, 3\}$
   (c)  $\{1, 2, 3\}$ _____ $\{1, 2, 3\}$
   (d)  0 _____ $\{1, 2, 3\}$

8. Form the union and the intersection of the following pairs of sets:

   (a)  $R = \{5, 10, 15\}, T = \{15, 20\}$
   (b)  $M = \{1, 2, 3\}, N = \{101, 102, 103, 104\}$
   (c)  $A = \{0, 10, 100, 1000\}, B = \{10, 100\}$
   (d)  $G = \{$odd counting numbers less than 100$\}$
        $H = \{$even counting numbers between 1 and 31$\}$
   (e)  $A = \{x, y, z, t\}, B = \{x, y, r, s\}$

9. Using sets *A* and *B*, describe in words the elements in regions (a), (b), (c), and (d).

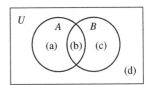

10. If $U = \{a, b, c, d, e, f, g, h\}, A = \{b, d, f, h\}$, and $B = \{a, b, e, f, g, h\}$, find the following:

    (a)  $\overline{A}$           (b)  $\overline{B}$
    (c)  $A \cap B$      (d)  $\overline{A \cap B}$
    (e)  $\overline{A} \cap \overline{B}$      (f)  $\overline{A \cup B}$

**11.** Let $U$ be the set of students at Myschool; let $A$ be the set of female students, $B$ the set of male students, $C$ the set of students who ride the bus, and $D$ the set of members of athletic teams. Denote each of the following symbolically:

**(a)** The set of female athletes

**(b)** The set of male athletes who ride the bus

**(c)** The set of all males who neither ride the bus nor are athletes

**(d)** The set of females who either ride the bus or are not athletes

**12.** Shade the portion of the diagram that illustrates each of the following sets:

**(a)** $A \cap B$

**(b)** $\overline{A} \cap C$

**(c)** $\overline{A} \cap \overline{B}$

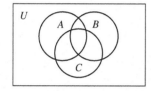

**13.** Shade the portion of the diagram that will illustrate each of the following sets:

**(a)** $A \cup B$

**(b)** $\overline{A} \cap \overline{C}$

**(c)** $A \cap (B \cap C)$

**(d)** $A \cup (B \cup C)$

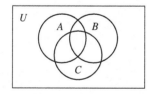

**14. (a)** Is there any distinction between $\{\varnothing\}$ and $\varnothing$?

**(b)** Is $\varnothing \in \{\ \}$?

**(c)** Is $0 \in \{\ \}$?

**(d)** Is $\varnothing \subseteq \{0\}$?

**(e)** If $A$ is any set, is $A \subset \varnothing$?

**(f)** What is the distinction between 3 and $\{3\}$?

**15.** Draw a Venn diagram that illustrates each situation.

**(a)** In Ourtown, no elementary teacher teaches in the high school, but some high school teachers teach in the college. Let $U$ be all teachers in Ourtown.

**(b)** At Ourtown High, all mathematics teachers have a chalkboard. Some have an overhead projector, and some have a video cassette recorder, but no mathematics teacher has all three. Let $U$ be all teachers at Ourtown High.

**16.** In the accompanying figure, the sets of elements of closed regions are indicated by (a), (b), . . . , (h). Express the following in terms of these sets:

**(a)** $\overline{A}$

**(b)** $A \cap B$

**(c)** $A \cap B \cap C$

**(d)** $\overline{A} \cup \overline{B}$

**(e)** $A \cup B \cup C$

**(f)** $A \cap \overline{B}$

**(g)** $\overline{A \cup B}$

**(h)** $\overline{A \cup B \cup C}$

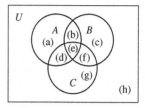

**17. (a)** List all the subsets of $\{a, b\}$.

**(b)** List all the subsets of $\{a, b, c\}$.

**(c)** List all proper subsets of $\{a, b, d\}$.

**18.** Let $A = \{3, 4, 5, 6\}$ and $B = \{5, 6, 7, 8\}$. Find

**(a)** $B - A$

**(b)** $A - B$

**19.** In a Venn diagram, a single set partitions the universe into two distinct regions. Two sets partition the universe into a maximum of four regions.

**(a)** What is the maximum number of regions into which three sets will partition the universe?

**(b)** Four sets?

**(c)** Five sets?

**(d)** Use problem-solving techniques to conjecture an answer for $n$ sets.

**20.** Using $A$, $B$, and $C$, describe the elements of the closed regions denoted by (a), (b), (c), (d), (e), (f), (g), and (h).

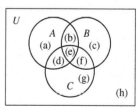

**21.** Use Venn diagrams to illustrate De Morgan's laws.

**(a)** $\overline{A \cap B} = \overline{A} \cup \overline{B}$

**(b)** $\overline{A \cup B} = \overline{A} \cap \overline{B}$

**22.** Demonstrate with Venn diagrams the following property, called the *distributive property of union over intersection*.

$$A \cup (B \cap C) = (A \cup B) \cap (A \cup C)$$

**23.** Demonstrate with Venn diagrams the following property, called the *distributive property of intersection over union*.

$$A \cap (B \cup C) = (A \cap B) \cup (A \cap C)$$

**24. (a)** Illustrate the associative property of union where $U = \{a, b, c, d, e, f, g\}$, $A = \{a, b\}$, $B = \{b, c, d, e\}$, and $C = \{d, e, f\}$.
**(b)** Use the sets listed in (a) to illustrate the associative property of intersection.

**25. (a)** If $x \in A \cap B$, is $x \in A \cup B$? Explain your answer.
**(b)** If $x \in A \cup B$, is $x \in A \cap B$? Explain your answer.

**26.** How many subsets does $\{a, b, c, d\}$ have? Which are proper subsets?

*Set theory can be applied to analyze the power of voting coalitions. A winning coalition consists of any set of voters who can carry a proposal. Answer the questions about voting coalitions in Exercises 27 and 28.*

**27.** A committee has five members: *A*, *B*, *C*, *D*, and *E*. For a proposal to be passed, it must have at least three votes. List all the possible winning coalitions.

*Example:* $\{A, B, D\}$ is a winning coalition; $\{B, D\}$ is not a winning coalition.

**28.** A town council consists of 5 members whose votes are weighted according to the number of citizens in their districts. Member *A* has 6 votes; member *B*, 5 votes; *C*, 4 votes; *D* and *E*, 1 vote apiece. Nine or more votes are required to carry an issue.

**(a)** List all winning coalitions.

*Example:* $\{A, B\}$ and $\{A, C, D\}$ are winning coalitions.

**(b)** Review the list in (a), and pick all winning coalitions having the property that if any voter were removed from this coalition, the coalition would fail to win.

*Example:* $\{A, B\}$ has this property, but if *D* were removed from $\{A, C, D\}$, the resulting coalition $\{A, C\}$ would still win.

**(c)** Do members *D* and *E* appear in any coalition that you listed in (b)?

**(d)** What does this say about the power of *D* and *E*? (*Note:* Similar arguments have been used in the courts to force modification of certain voting schemes.)

**29.** Describe in symbols the shaded portion of each of the following Venn diagrams.

(a)  (b)  (c)

**30. A.** Consider the sequence $1, 8, 16, 24, 32, \ldots$. Now consider the sequence of partial sums

$$1 = 1$$
$$1 + 8 = 9$$
$$1 + 8 + 16 = 25$$

**(a)** Write the next three partial sums.
**(b)** Associate with these partial sums the number of dots around closed figures in a square such as

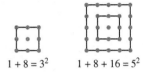

Draw the geometric figure for $1 + 8 + 16 + 24$, and find an expression for this sum.
**(c)** Draw a $9 \times 9$ square and find the corresponding sum; an $11 \times 11$ square.
**(d)** What is the sum of the first *n* terms of the sequence $1, 8, 16, 24, 32, \ldots$?

**B.** We can represent the partial sums of the sequence $1, 2, 3, 4, \ldots$ by using blocks.

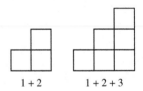

**(a)** Draw a similar diagram to represent $1+2+3+4$.
**(b)** You can use this idea to find the sum of a given number of counting numbers. For example, find the sum of the first 4 counting numbers. To do this, place together, as shown, two diagrams representing $1 + 2 + 3 + 4 = \frac{4 \cdot 5}{2}$.

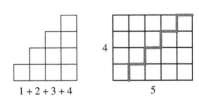

Use this reasoning to find the sum of the first 5 counting numbers.
**(c)** Verbally describe the pattern you have discovered.
**(d)** Find the sum of the first 100 counting numbers.

# Interlude: Practice Problems 5.1

In each of the following, we use $n(A)$ to denote the number of elements in set $A$.

1. For two sets $A$ and $B$, $n(B) = 47$, $n(A \cup B) = 386$, and $n(A \cap B) = 12$. Find $n(A)$.

2. If a universal set contains 200 elements, $n(A) = 112$, $n(A \cup B) = 130$, and $n(A \cap B) = 8$, then find $n(\overline{A} \cap \overline{B})$ and $n(\overline{B})$.

3. The following information gives the number of elements in some subsets of $A$, $B$, and $C$. Represent this information in a Venn diagram and find $n(C)$.

$$n(A) = 12, n(B) = 12, n(A \cap B) = 5, n(A \cap C) = 5, n(B \cap C) = 5,$$

$$n(A \cap B \cap C) = 2, n(A \cup B \cup C) = 25$$

4. Recently, 100 students on campus that were taking some sort of mathematics class were surveyed to determine what sort of reading materials were used in the class they were taking. These are the results:

   - 40 students use a textbook
   - 30 students use an online reference source
   - 25 students use instructor notes
   - 15 students use both a textbook and an online reference source
   - 12 students use both a textbook and instructor notes
   - 10 students use both an online reference source and instructor notes
   - 4 students use all of the three types of reading materials

   (a) How many students use none of the three types of reading materials?

   (b) How many students use at least one of the three types of reading materials?

# 5.2   THE LANGUAGE OF PROBABILITY

**Problem**   A card is dealt from a shuffled deck of 52 cards. What is the probability that the card is an ace of hearts?

**Overview**   As this problem suggests, we shall be studying the language of uncertainty in this chapter. One significant characteristic of our increasingly complex society is that we must deal with questions for which there is no known answer but instead one or more probable (or improbable) answers.

**Goals**   Use the Data Analysis and Probability Standard: ● formulate questions that can be addressed with data and collect, organize, and display relevant data to answer them; ● select and use appropriate statistical methods to analyze data; ● develop and evaluate inferences and predictions that are based on data; ● understand and apply basic concepts of probability.
Illustrate the Representation Standard.
Illustrate the Problem Solving Standard.

## Outcomes

In this section, we discuss procedures for assigning probabilities, rules that govern these probabilities, and conclusions that can legitimately be deduced from a probability once it is assigned. Because probability is a language of uncertainty, any discussion of probability presupposes a process of observation or measurement in which the outcomes are not certain. Such a process is called an **experiment**. A result of an experiment is called an **outcome**. A list of all possible outcomes in an experiment is called a **sample space**.

### EXAMPLE 1

*Experiment:* A coin is tossed.
*Possible outcomes:* Heads (H) or tails (T). (See Figure 5.10.)
*Sample space:* {H, T}.

*FIGURE 5.10*

### EXAMPLE 2

*Experiment:* A die is tossed.
*Possible outcomes:* The top side of the die shows 1, 2, 3, 4, 5, or 6 dots. (See Figure 5.11.)
*Sample space:* {1, 2, 3, 4, 5, 6}.

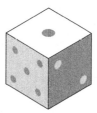

*FIGURE 5.11*

Assuming that the coin is fair, the outcomes (getting heads or getting tails) in the first experiment are said to be **equally likely**, because one outcome has the same chance of occurring as the other. Likewise, the outcomes in the second experiment (getting a 1, 2, 3, 4, 5, or 6 in the roll of a die) are equally likely.

## EXAMPLE 3

Suppose that we have a spinner whose needle is as likely to stop at one place as another. Suppose further that the spinner is divided into three sections labeled R, B and W. (See Figure 5.12.) Let an experiment consist of spinning the needle and observing the label of the region where it stops. A sample space for this experiment would be the list of labels:

$$\{R, W, B\}$$

Note that since the portion of the spinner labeled $W$ is much larger than the portions of the circle labeled $R$ or $B$, the outcomes of this experiment are not equally likely.

*FIGURE 5.12*

## Tree Diagrams

Tree diagrams often serve to help list all possible outcomes of a multistep experiment. The procedure for drawing a tree diagram is illustrated in the following example.

## EXAMPLE 4

Joe must guess at both questions on a quiz. The first question is a True/False question and the second is a multiple choice question with possible answers a through d. What are the outcomes of this experiment?

**Solution**

From a single point, draw a line to each of the possible choices on the True/False question. From each of these answers draw four lines to each possible answer on the multiple choice question. Do you see that the tree diagram in Figure 5.13 shows eight possible outcomes?

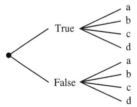

*FIGURE 5.13*

## Choosing a Sample Space

A first step in analyzing an experiment is the selection of a sample space. Different sample spaces can result from the same experiment, depending on how the observer chooses to record the outcomes.

**EXAMPLE 5**

A fair coin is flipped twice. Find three different sample spaces.

**Solution**

The sample spaces for this experiment can best be understood with the aid of a diagram (Figure 5.14) in which we record possible results of each flip.

*Sample space A:* One complete listing of the outcomes is

$$S = \{(H, H), (H, T), (T, H), (T, T)\}$$

The letter listed first in each pair indicates the result of the first flip, and the letter listed second gives the result of the second flip. Each outcome of this sample space is equally likely.

*Sample space B:* An alternative way to list the outcomes is to ignore the order in which the heads and tails occur and to record only how many of each appear. Thus, another possible sample space is

$$S = \{(2H), (1H \text{ and } 1T), (2T)\}$$

These outcomes are not equally likely. Because (1H and 1T) can occur in two ways (as HT and as TH), it is more likely to occur than the other two outcomes.

*Sample space C:* Another way to tabulate the same outcomes is to list the number of heads that occur: $\{0, 1, 2\}$. Again, the outcomes of this sample space are not equally likely. Zero heads can occur in only one way (as tails on one coin and tails on the other coin). Likewise, two heads can occur in only one way. But a result of heads on one coin and tails on the other can occur in two ways (as heads on the first coin and tails on the second, or as tails on the first flip and heads on the second).

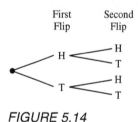

First          Second
Flip            Flip

FIGURE 5.14

Although the three sample spaces in the previous example are different, they share certain properties. Each set of outcomes classifies completely what can happen if the experiment is performed. Further, the members of each set of outcomes are distinct; that is, they do not overlap. In other words, each possible result of the experiment is represented by exactly one member of the set. This discussion suggests the following definition of a sample space.

**DEFINITION**

### Sample Space

A *sample space* (denoted by $S$) is a list of the possible outcomes of an experiment, constructed in such a way that it has the following characteristics:

1. The categories do not overlap.
2. No result is classified more than once.
3. The list is complete (exhausts all the possibilities).

## Probability

To each outcome of a sample space we can assign a number, called its **probability**, that measures how likely that outcome is to occur. The probability of an outcome is a number between 0 and 1 inclusive. If the probability of an outcome is near 1, that outcome will likely occur when the experiment is performed. If the probability of an outcome is near 0, it is unlikely that that outcome will occur when the experiment is performed. A probability of zero indicates there is no outcome, and a probability of 1 indicates the outcome is the entire sample space.

---

Properties of Probability

A **probability assignment on a sample space** must satisfy two properties:

1.  If $A$ is a possible outcome, then its probability $P(A)$ is between 0 and 1 inclusive.
2.  The sum of the probabilities of all outcomes in the sample space equals 1.

---

For some sample spaces we are able to assign probabilities to outcomes based on our understanding of the properties of the sample space. Such assignments are called **theoretical probabilities**.

### EXAMPLE 6

In the experiment of tossing a fair coin the outcomes of a head and a tail are equally likely to occur. Hence we will assign probabilities $P(H) = \frac{1}{2}$ and $P(T) = \frac{1}{2}$. Note that these assignments satisfy the properties of probability:

(a)  $0 \leq P(T) \leq 1$   and   $0 \leq P(H) \leq 1$

(b)  $P(S) = P(H) + P(T) = \frac{1}{2} + \frac{1}{2} = 1$

---

### EXAMPLE 7

Consider the experiment of spinning the spinner in Figure 5.3. A sample space for this experiment would be $\{R, B, W\}$. If we believe that the spinner is twice as likely to stop on region $W$ as on regions $R$ or $B$, then we would assign a probability of $\frac{1}{2}$ to outcome $W$ and probabilities of $\frac{1}{4}$ to outcomes $R$ and $B$. Again,

(a)  $0 \leq P(R) \leq 1, 0 \leq P(W) \leq 1, 0 \leq P(B) \leq 1$

(b)  $P(R) + P(W) + P(B) = 1$

---

### Uniform Sample Space

There is one whole class of sample spaces, called *uniform sample spaces,* whose probability assignments are particularly easy to determine.

**DEFINITION**

Uniform Sample Space

If each possible outcome of the sample space is equally likely to occur, the sample space is called a *uniform sample space*.

Suppose that a uniform sample space consists of $m$ possible outcomes $\{A_1, A_2, \ldots, A_m\}$. Because each of the outcomes is equally likely, it seems reasonable to assign to each outcome $A_i$ the same probability, denoted by $P$. Because the sum of the probabilities of the $m$ individual outcomes must be 1, it follows that

$$P(A_1) + P(A_2) + P(A_3) + \cdots + P(A_m) = 1$$

$$\underbrace{P + P + P + \cdots + P}_{m \text{ times}} = 1$$

$$mP = 1 \quad \text{or} \quad P = \frac{1}{m}$$

Thus, each of the $m$ outcomes has probability $1/m$.

---

### Equal Probabilities

In a uniform sample space with $m$ possible outcomes, each outcome has probability $1/m$. This is sometimes written as

$$P(A) = \frac{1}{n(S)} = \frac{1}{m}$$

where $A$ represents one outcome and $n(S)$ represents the number of possible outcomes in the sample space.

---

## EXAMPLE 8

Eight identical balls numbered 1 to 8 are placed in a box. (See Figure 5.15.) Find a sample space and probability assignment describing the experiment of randomly drawing one of them from the box.

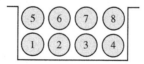

*FIGURE 5.15*

**Solution**

A suitable sample space is $\{1, 2, 3, 4, 5, 6, 7, 8\}$, each number representing one of the 8 balls. Because each ball is equally likely to be drawn, we assign a probability of $\frac{1}{8}$ to each outcome:

$$P(1) = \frac{1}{8}, \quad P(2) = \frac{1}{8}, \ldots, P(8) = \frac{1}{8}$$

---

## EXAMPLE 9

A card is drawn from a shuffled deck (52 cards with four suits: clubs, diamonds, hearts, and spades; each suit with 13 cards: 2–10, jack, queen, king, and ace). What is the probability of drawing the ace of hearts? (See Figure 5.16.)

*FIGURE 5.7*

**Solution**

A uniform sample space for this experiment would consist of a listing of all 52 cards. Hence each outcome would be assigned probability 1/52. In particular, the probability of drawing the ace of hearts is 1/52.

## Empirical Probability

An alternative way to assign probabilities to outcomes in an experiment involves performing the experiment many times and looking at the empirical data. Consider the following example.

## EXAMPLE 10

A fair die is rolled 10,000 times. Table 5.1 itemizes the number of times a 1 has occurred at various stages of the process.

**Table 5.1**

| Number of Rolls ($N$) | Number of 1's Occurring ($m$) | Relative Frequency ($m/N$) |
|---|---|---|
| 10 | 4 | .4 |
| 100 | 20 | .2 |
| 1000 | 175 | .175 |
| 3000 | 499 | .166333 . . . |
| 5000 | 840 | .168 |
| 7000 | 1150 | .164285714 . . . |
| 10,000 | 1657 | .1657 |

Notice that, as $N$ (number of rolls) becomes larger, the relative frequency ($m/N$) stabilizes in the neighborhood of $0.166 \approx 1/6$. Thus, we are willing to assign the probability

$$P(1) = \frac{1}{6}$$

Compare this thinking with our previous method of assigning probabilities. Assume that the die is constructed so that the outcomes are equally likely. Because the sample space $S = \{1, 2, 3, 4, 5, 6\}$ is a uniform sample space, the probability of obtaining a 1 in a sample space of 6 outcomes is 1/6, the same answer we obtained by looking at the empirical data.

In the previous example, we assigned probability to an outcome by assigning the fraction of times that the outcome occurred when the experiment was performed a large number of times. Similarly, suppose that a thumbtack lands with its point up 1000 times out of 10,000 trials. The relative frequency is $\frac{1000}{10,000} = \frac{1}{10}$. If we repeat the experiment 10,000 more times and find that the ratio is still approximately $\frac{1}{10}$, we will agree to assign this number as a measure of our degree of belief that it will land point up on the next toss. These examples suggest the following definition.

**DEFINITION**

### Empirical Probability

If an experiment is performed $N$ times, where $N$ is a very large number, the probability of an outcome $A$ should be approximately equal to the following ratio:

$$P(A) = \frac{\text{number of times } A \text{ occurs}}{N}$$

## EXAMPLE 11

A loaded die (one for which the outcomes are not equally likely) is thrown a number of times with results as shown in Table 5.2.

(a)  How many possible outcomes are there?

(b)  What are the possible outcomes?

(c)  Using the frequency table (Table 5.2), assign a probability to each of the outcomes.

**Table 5.2**

| Outcome | 1 | 2 | 3 | 4 | 5 | 6 |
|---|---|---|---|---|---|---|
| **Frequency** | 967 | 843 | 931 | 1504 | 1576 | 1179 |

**Solution**

There are 6 possible outcomes to this experiment: The die may show a 1, 2, 3, 4, 5, or 6. By adding the frequencies of the 6 outcomes, we see that the experiment was performed 7000 times. Thus the relative frequency of the outcome of 1 is 967/7000 or approximately 0.14. Similarly, the relative frequency of a 2 is 843/7000 or approximately 0.12. Continuing in this way, we make the following probability assignments:

$$P(1) = .14 \qquad P(2) = .12 \qquad P(3) = .13$$
$$P(4) = .21 \qquad P(5) = .23 \qquad P(6) = .17$$

# Then and Now

**Blaise Pascal**
**1623–1662**

Blaise Pascal was a frail youth whose mother died when he was 3 and whose upbringing fell to his father, Etienne, a French lawyer and mathematician. Blaise and his two sisters were educated at home by their father, and Blaise never attended a school or university.

His attraction to mathematics was evident by the time he was 12. At age 16, he had already published an original treatise on conic sections, a profound 1-page geometric result known today as Pascal's Theorem. Such a promising first paper showed indication of a brilliant career in geometry, but Pascal quickly turned to other mathematical interests. At age 18, he embarked on the unusual enterprise of designing, building, and selling mechanical calculating machines. Pascal's friend, the Chevalier de Mèrè, posed a gambling problem, whereupon Pascal—with Pierre de Fermat—originated the modern mathematical theory of probability. In this work, Pascal made use of the array that follows. Though this array was known by Jiǎ Xiàn in China in the 11th century, it is known as Pascal's Triangle because it was Pascal who developed its properties and showed its connections with other facets of mathematics.

Pascal's name first appears these days in our mathematics curriculum byway of Pascal's Triangle, usually in grades 7 and 8. Students are asked to write down the numbers in the next row by looking for the pattern in the first few rows. Then they are asked to find other patterns within Pascal's Triangle, or to find the sum of each row, and so on. Then, throughout the first 2 or 3 courses of algebra, Pascal's work reappears in various contexts.

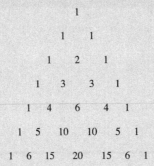

We owe to Pascal appreciation for his innovation in the science of computation (IBM has one of his mechanical calculators in its museum). A popular and widely used computer language is named PASCAL. A great deal of attention is now being paid to the subject of artificial intelligence, and Pascal pondered the question of whether machines could think three centuries ago.

# Just for Fun

Try to place the pennies in four equal columns subject to the following rules:

1.  Each penny can be moved only once.

2.  Each penny must be jumped over exactly two
    other pennies or stacks.

3.  If an empty space occurs, ignore it.

Good luck.

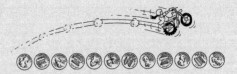

# EXERCISES

1. A box contains a yellow, a green, a black, and a blue marble.

   (a) How many outcomes are possible if you pick one marble from the box?

   (b) List the possible outcomes.

   (c) What is the probability of picking a blue marble?

   (d) What is the probability of picking a green marble?

   (e) Are the probabilities the same for each color of marble in the box?

2. Each letter of the word HOPEFUL is placed on a card, and the cards are placed in a basket. Sam draws a card without looking.

   (a) List the possible outcomes.

   (b) Are the outcomes equally likely?

   (c) What is the probability of drawing a *P?*

   (d) An *E?*

3. Yute has four belts of different colors (one of which is brown) in his drawer. He reaches into the drawer and picks a belt without looking. What is the probability that Yute has picked the brown belt?

4. The probability that Marsha will pick a blue button out of a jar is 1 of 2 or $\frac{1}{2}$. There are 16 buttons in the jar. How many of the buttons are blue?

5. Mark tossed a letter cube (with sides A, B, C, D, E, and F) and recorded the number of times (frequency) each letter occurred.

   | Letter | Frequency |
   |--------|-----------|
   | A | 8 |
   | B | 10 |
   | C | 9 |
   | D | 11 |
   | E | 12 |
   | F | 10 |

   (a) How many possible outcomes are there?

   (b) What are the possible outcomes?

   (c) From the frequency table, assign a probability of tossing a D.

   (d) Assign a probability to the outcome *D* based on the observation that each outcome is equally likely.

   (e) Why is there a difference?

   (f) Would the accuracy have improved if Mark had tossed the letter cube 100 times? 1000 times?

6. Suppose that you toss a coin 40 times and get 19 heads.

   (a) On the basis of this experiment, assign a probability of getting heads in one toss of a coin.

   (b) Using the fact that heads and tails are equally likely, compute the probability of getting heads in one toss.

   (c) Why do the two answers differ?

7. Give the sample space for each of the four spinners.

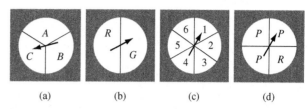

   (a)       (b)       (c)       (d)

8. Assign a probability to each outcome in the sample spaces of Exercise 7.

9. What is the probability of getting heads when a two-headed coin is tossed?

10. In Exercise 9, what is the probability of getting tails?

11. (a) A balanced coin is tossed 10,000 times. Approximately how many heads can one expect to occur?

    (b) A balanced coin is tossed 8 times and only 2 heads occur. Should we doubt that the coin is fair?

*In Exercises 12 through 15, tabulate a sample space and make probability assignments on the basis of the information given.*

12. Ten identically sized cards numbered 1 through 10 are placed in a box, and a single card is drawn from the box.

13. A die is found in a back alley and then rolled 1000 times with the following results.

    | Observed | Frequency |
    |----------|-----------|
    | 1 | 181 |
    | 2 | 152 |
    | 3 | 144 |
    | 4 | 138 |
    | 5 | 156 |
    | 6 | 229 |

14. The U-POT-EM manufacturing company makes clay pots. Of the last 1200 crates of pots shipped, 440 contained a broken pot when received. Muriel's Flower Shop receives a crate and prepares to open it.

15. Of the last 850 fish caught at White Lake in the traps of the Game and Wildlife Department, 200 have been bass, 278 have been bream, 122 have been carp, and the rest have been catfish.

16. Which of the following could not be a probability? Why?

    (a) $-\frac{1}{2}$

    (b) $\frac{17}{16}$

    (c) .001

    (d) 0

    (e) 1.03

    (f) .01

    (g) $\frac{5}{4}$

    (h) 1

**17. (a)** Use the tree diagram below to tabulate a uniform sample space for the experiment of tossing a coin three times. Express each outcome as an ordered triple, such as HHH or HTT.

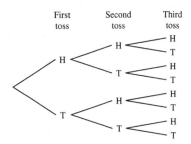

| First toss | Second toss | Third toss |

**(b)** What probability should be assigned to each outcome in this sample space?

*In Exercises 18 through 20 you may find a tree diagram to be helpful.*

**18.** A box contains 4 balls numbered 1 to 4. Record a sample space for the following experiments.

**(a)** A ball is drawn and the number is recorded. Then the ball is returned, and a second ball is drawn and recorded.

**(b)** A ball is drawn and recorded. Without replacing the first ball, the experimenter draws and records a second ball.

**19.** List the elements in a sample space for the simultaneous tossing of a coin and drawing of a card from a set of 6 cards numbered 1 through 6.

**20.** A box contains 3 red balls and 4 black balls. Let $R$ represent a red ball and $B$ a black ball. Tabulate a sample space if

**(a)** one ball is drawn at a time.

**(b)** two balls are drawn at a time.

**(c)** three balls are drawn at a time.

**(d)** Are these sample spaces uniform?

**21.** Three coins are tossed, and the number of heads is recorded. Which of the following sets is a sample space for this experiment? Why do the other sets fail to qualify as sample spaces?

**(a)** $\{1, 2, 3\}$

**(b)** $\{0, 1, 2\}$

**(c)** $\{0, 1, 2, 3, 4\}$

**(d)** $\{0, 1, 2, 3\}$

**22.** Three coins are tossed, and the number of heads is recorded. Which of the following sets are sample spaces for this experiment? If a set fails to qualify as a sample space, give the reason.

**(a)** $\{0, 2, \text{an odd number}\}$

**(b)** $\{x \mid x \text{ is a whole number and } x < 4\}$

**23.** If you flipped a fair coin 15 times and got 15 heads, what would be the probability of getting a head on the 16th toss?

## Laboratory Activities

**24.** Carefully label two pennies as Penny 1 and Penny 2. Flip them and classify each outcome by recording the results for each coin. More precisely, if Penny 1 shows a head and Penny 2 shows a tail, classify the outcome as HT. Similarly, if Penny 1 shows a tail and Penny 2 shows a head, classify the outcome as TH. All outcomes can be classified as HH, HT, TH, or TT. Perform this experiment 200 times and assign probabilities by using the empirical data. Do you have evidence that this is a uniform sample space?

**25.** Flip 2 pennies 200 times and classify each toss as HH, H&T, or TT depending on what shows. When both a head and tail occur on the same toss, make no effort to determine which coin shows a head and which coin shows a tail. Assign probabilities to the outcomes HH, H&T, and TT on the basis of the empirical data. Does this appear to be a uniform sample space? Compare to the results of Exercise 24.

**26.** Toss a paper cup into the air 100 times. After each toss record whether the cup lands on its bottom, upside down on its top, or on its side. Assign probabilities to these outcomes by using empirical probabilities.

**27.** Roll a pair of dice 300 times. How many times do you get a pair of 1's? What probability would you assign to this outcome?

# Interlude: Practice Problems 5.2, Part 1

1. Suppose that a pair of fair 6-sided dice are identical except that one is red and the other is green. An experiment consists of rolling the dice at the same time. The sum of the numbers shown on the upturned faces is then recorded.

   (a) Give the sample space for this experiment.

   (b) Assign to each outcome in the sample space an appropriate theoretical probability. Summarize your assignment in a table.

   | Outcome | |
   |---|---|
   | Probability | |

   (c) Is the sample space a uniform sample space? Explain.

2.  A bag contains five balls that are identical except that each ball is uniquely numbered 1, 2, 3, 4, or 5. An experiment consists of drawing a ball at random from the bag, replacing the ball, and then drawing a ball from the bag again. The sum of the numbers shown on each ball is then recorded.

(a)  Give the sample space for this experiment.

(b)  Assign to each outcome in the sample space an appropriate theoretical probability. Summarize your assignment in a table.

| **Outcome** | |
|---|---|
| **Probability** | |

(c)  Is the sample space a uniform sample space? Explain.

# Interlude: Practice Problems 5.2, Part 2

1. A bag contains four balls that are identical except that each ball is uniquely numbered 1, 2, 3, or 4. An experiment consists of drawing a pair of balls (both at the same time) at random from the bag and adding the numbers shown on each ball. This sum is then recorded.

   (a) Give the sample space for this experiment.

   (b) Assign to each outcome in the sample space an appropriate theoretical probability. Summarize your assignment in a table.

   | Outcome | |
   | --- | --- |
   | **Probability** | |

2. Two distinct fair coins are flipped in tandem several times. The results of these flips are recorded in the following table.

   | Outcome | HH | HT | TH | HH |
   | --- | --- | --- | --- | --- |
   | **Frequency** | 119 | 101 | 160 | 120 |

   Assign to each outcome in the sample space an appropriate empirical probability. Summarize your assignment in a table.

   | Outcome | |
   | --- | --- |
   | **Probability** | |

3.  Two tetrahedral dice (four-sided), each with equal sides numbered with one to four dots, are identical except that one is green and the other is blue. The two are tossed, and the total number of dots on the bottom faces is recorded.

(a)  Give the sample space for this experiment.

(b)  Assign to each outcome in the sample space an appropriate theoretical probability. Summarize your assignment in a table.

| Outcome | |
|---|---|
| **Probability** | |

4.  An experiment consists of tossing a coin twice and recording the sequence of upturned faces. The table below lists several different assignments of probabilities for the sample space. Determine which of the assignments are valid probability assignments. For each assignment that you determine to be invalid, explain what makes the assignment invalid.

Sample Space

| | | HH | HT | TH | TT |
|---|---|---|---|---|---|
| | I | 0 | $\frac{1}{4}$ | $\frac{1}{4}$ | $\frac{1}{2}$ |
| | II | $\frac{3}{8}$ | $\frac{3}{8}$ | $\frac{1}{2}$ | $\frac{1}{2}$ |
| Assignment | III | $\frac{1}{8}$ | $\frac{1}{4}$ | $\frac{3}{8}$ | $\frac{1}{2}$ |
| | IV | $\frac{1}{4}$ | $\frac{3}{8}$ | $\frac{1}{2}$ | $\frac{1}{8}$ |
| | V | $\frac{2}{9}$ | $\frac{2}{9}$ | $\frac{2}{9}$ | $\frac{2}{9}$ |
| | VI | 1 | 0 | 0 | 0 |

# 5.3  PROBABILITY OF EVENTS AND PROPERTIES OF PROBABILITY

**Problem**

(a)  Two coins are tossed. What is the probability of getting at least one head?

(b)  Of the freshmen who entered Loren College last year, 12% made A's in freshman English, 8% made A's in history, and 4% made A's in both English and history. An admissions counselor would like to know what percent made A's in English or history.

**Overview**   In the previous section we computed probabilities of a single outcome from an experiment. In this section we will examine the problem of computing the probability of an event where an event is a subset of outcomes from the sample space. Because an event is a subset of the sample space, we will find that many of the concepts and much of the language of Section 5.1 on sets will be useful.

**Goals**   Illustrate the Data Analysis and Probability Standard.
Illustrate the Representation Standard.
Illustrate the Problem Solving Standard.

We now extend our definition of probability to events.

**DEFINITION**

> **Event**
>
> An *event* is a subset of a sample space.

## EXAMPLE 1

In the previous section we found a uniform sample space for the experiment of tossing a pair of fair coins. Tabulate the event "at least one coin shows a head."

**Solution**

| Sample Space | Event |
|---|---|
| {HH, HT, TH, TT} | {HH, HT, TH} |

The event consists of the possible outcomes in the sample space that include at least one occurrence of heads, circled in Figure 5.17.

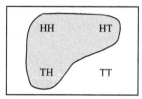

*FIGURE 5.17*

In the case of an event from a uniform sample space, the probability of an event is particularly easy to compute.

DEFINITION

**Probability of an Event**

Let $S$ represent a uniform sample space with $n(S)$ equally likely outcomes, and let $A$ be an event in $S$. If $A$ is an event consisting of $n(A)$ outcomes, then $P(A)$ is given by

$$P(A) = \frac{n(A)}{n(S)}$$

This rule is the classical definition of probability. Suppose that there are $N$ equally likely possible outcomes of an experiment. If $r$ of these outcomes have a particular characteristic so that they can be classified as a success, then the probability of a success is defined to be $r/N$.

## EXAMPLE 2

Suppose that a card is drawn from a set of 6 (numbered 1 through 6). There are 6 equally likely possible outcomes of the experiment. Let 2 of these, a 3 and a 6, represent a success $E$. Then

$$P(E) = \frac{n(E)}{n(S)} = \frac{2}{6} = \frac{1}{3}$$

*Practice Problem*

What is the probability of drawing an ace from a standard deck?

**Answer**

4/52 or 1/13

## EXAMPLE 3

Return to the experiment of tossing a pair of fair coins. What is the probability of tossing

(a) at least one head?

(b) exactly one head?

**Solution**

Using the uniform sample space

$$S = \{HH, HT, TH, TT\}$$

we can tabulate the events in question.

(a) $A = \{HH, HT, TH\}$        (b) $B = \{HT, TH\}$

Thus,

(a) $P(A) = \dfrac{n(A)}{n(S)} = \dfrac{3}{4}$        (b) $P(B) = \dfrac{n(B)}{n(S)} = \dfrac{1}{2}$

## EXAMPLE 4

A poll is taken of 500 workers to determine whether they want to go on strike. Table 5.3 indicates the results of this poll.

(a)   What is the probability that a worker selected at random is in favor of a strike?

(b)   What is the probability that such a worker has no opinion?

**Table 5.3**

| In Favor of a Strike | Against a Strike | No Opinion |
|---|---|---|
| 280 | 200 | 20 |

**Solution**

One sample space for this experiment would consist of a list of the 500 workers. To choose a worker *at random* means that each employee has the same chance of being selected; hence the sample space is uniform.

(a)   Let $E$ be the event "worker is in favor of the strike." Then

$$P(E) = \frac{n(E)}{n(S)} = \frac{280}{500} = \frac{14}{25}$$

(b)   Similarly, $P(\text{no opinion}) = \dfrac{20}{500} = \dfrac{1}{25}$

Because events are subsets, we can, of course, form unions, intersections, and complements of events. The properties that govern unions, intersections, and complements of events are very important.

**DEFINITION**

**And, Or, and Complement**

1.   The event $A \cup B$ ($A$ or $B$) is the collection of all outcomes that are in $A$ or in $B$ or in both $A$ and $B$.

2.   The event $A \cap B$ ($A$ and $B$) is the collection of all outcomes that are in both $A$ and $B$.

3.   The **complement** of an event $A$, denoted by $\overline{A}$, is the collection of all outcomes that are in the sample space and are not in $A$.

To illustrate these concepts, we consider examples involving the roll of a die.

## EXAMPLE 5

In the rolling of a fair die, what is the probability that the result will be an odd number or a 4?

**Solution**

We let $O$ represent an odd number and $F$ represent a 4, and then we seek $P(O \cup F)$. In Figure 5.18(b), using Venn diagrams, we see that

$$P(O \cup F) = \frac{n(O \cup F)}{n(S)} = \frac{4}{6} = \frac{2}{3}$$

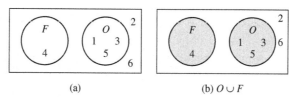

(a)                    (b) $O \cup F$

*FIGURE 5.18*

Notice in Figure 5.18(a) that

$$P(O) = \frac{3}{6} \quad \text{and} \quad P(F) = \frac{1}{6}$$

Furthermore,

$$P(O \cup F) = P(O) + P(F)$$

because

$$\frac{4}{6} = \frac{3}{6} + \frac{1}{6}$$

## EXAMPLE 6

In the rolling of the same fair die, what is the probability of getting either an even number or a multiple of 3?

**Solution**

Let $E$ represent an even number, and let $M$ represent a multiple of 3. We seek $P(E \cup M)$. In Figure 5.19(b) we see that

$$P(E \cup M) = \frac{n(E \cup M)}{n(S)} = \frac{4}{6} = \frac{2}{3}$$

Notice in Figure 5.19(a) that

$$P(E) = \frac{3}{6} \quad \text{and} \quad P(M) = \frac{2}{6}$$

Here, however,

$$P(E \cup M) \neq P(E) + P(M)$$

because

$$\frac{4}{6} \neq \frac{3}{6} + \frac{2}{6}$$

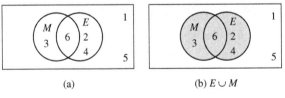

(a)                          (b) $E \cup M$

*FIGURE 5.19*

What is the difference between the problems in the two previous examples? For $P(O \cup F)$, $F$ and $O$ had no points in common. For $P(E \cup M)$, $E$ and $M$ overlapped. This discussion suggests the following definition and property of probability.

---

**Mutually Exclusive Events**

1. Events $A$ and $B$ are *mutually exclusive* if they have no outcomes in common.

2. If events $A$ and $B$ are mutually exclusive,

$$P(A \cup B) = P(A) + P(B)$$

**EXAMPLE 7**

From a standard deck of cards, you draw one card. What is the probability of your getting a spade or a red card?

**Solution**

Verify that

$$P(S) = \frac{13}{52} \quad \text{and} \quad P(R) = \frac{26}{52}$$

Because the outcome of getting a spade and the outcome of getting a red card are mutually exclusive,

$$P(S \cup R) = P(S) + P(R)$$

$$= \frac{13}{52} + \frac{26}{52} = \frac{39}{52} = \frac{3}{4}$$

Now let's return to Example 6, where we noted that $P(E \cup M) = \frac{4}{6}$, $P(E) = \frac{3}{6}$, and $P(M) = \frac{2}{6}$. The reason $P(E \cup M) \neq P(E) + P(M)$ is that the outcome 6 is in both $E$ and $M$ and thus is counted twice in $P(E) + P(M)$. The probability of getting an outcome that is in both $E$ and $M$ is

$$P(E \cap M)$$

Because $E \cap M$ is included twice in $P(E) + P(M)$, we subtract one of these and note that

$$P(E \cup M) = P(E) + P(M) - P(E \cap M)$$

or

$$\frac{4}{6} = \frac{3}{6} + \frac{2}{6} - \frac{1}{6}$$

We can generalize this concept by realizing that in set theory the number of outcomes in event $A$ *or* in event $B$ is the number in $A$ plus the number in $B$ less the number in $A \cap B$, which has been counted in both $A$ and $B$. (See Figure 5.20.) Thus,

$$n(A \cup B) = n(A) + n(B) - n(A \cap B)$$

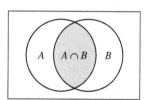

*FIGURE 5.20*

In a uniform sample space we can divide both sides of the equation by $N$, the number of elements in a sample space, to obtain probabilities

$$\frac{n(A \cup B)}{N} = \frac{n(A)}{N} + \frac{n(B)}{N} - \frac{n(A \cap B)}{N}$$

Thus,

$$P(A \cup B) = P(A) + P(B) - P(A \cap B)$$

> **Probability of A or B**
>
> For any two events A and B, the *probability of A or B* is given by
>
> $$P(A \cup B) = P(A) + P(B) - P(A \cap B)$$

We return now to the second of our introductory problems.

## EXAMPLE 8

Of the freshmen who entered Loren College last year, 12% made A's in English, 8% made A's in history, and 4% made A's in both English and history. What percent made A's in English or history?

**Solution**

$$P(E) = .12$$

$$P(H) = .08$$

$$P(E \cap H) = .04$$

$$P(E \cup H) = P(E) + P(H) - P(E \cap H)$$

$$= .12 + .08 - .04$$

$$= .16$$

16% made A's in English or history.

## EXAMPLE 9

In drawing a card from 8 cards numbered 1 through 8, what is the probability of getting an even number or a number less than 5?

**Solution**

Let A represent the event of getting a number less than 5, and let B represent the event of getting an even number. Then

$$P(A) = \frac{4}{8} = \frac{1}{2}$$

$$P(B) = \frac{4}{8} = \frac{1}{2}$$

But 2 and 4 are both even and less than 5; consequently,

$$P(A \cap B) = \frac{2}{8} = \frac{1}{4}$$

Therefore,

$$P(A \cup B) = P(A) + P(B) - P(A \cap B)$$

$$= \frac{1}{2} + \frac{1}{2} - \frac{1}{4} = \frac{3}{4}$$

*Practice Problem*

A card is drawn from a standard deck of cards. What is the probability that it is either an ace or a spade?

**Answer**

$\dfrac{4}{13}$

---

The complement of event $A$ is everything in the sample space except $A$, denoted by $\overline{A}$. Because $A \cup \overline{A} = S$, and $A \cap \overline{A} = \varnothing$, $P(A \cup \overline{A}) = 1$ and $P(A \cap \overline{A}) = 0$, then $P(A \cup \overline{A}) = P(A) + P(\overline{A}) = 1$. Therefore, $P(\overline{A}) = 1 - P(A)$.

> **Probability of the Complement of $A$**
>
> If $\overline{A}$ is the complement of $A$, then
>
> $$P(\overline{A}) = 1 - P(A).$$

## EXAMPLE 10

What is the probability of not getting an ace in the drawing of a card from a standard deck of cards?

**Solution**

$$P(\text{no ace}) = 1 - P(\text{ace}) = 1 - \frac{4}{52} = \frac{12}{13}$$

---

Sometimes probability statements are given in terms of odds, which is actually a comparison of the probability that event $E$ will occur and the probability that event $E$ will not occur (denoted by $\overline{E}$).

> **DEFINITION**  |  **Odds**
>
> The odds in favor of event $E$ equal the ratio
>
> $$P(E) : P(\overline{E}) \quad \text{or} \quad P(E) : 1 - P(E)$$
>
> and the odds against event $E$ equal
>
> $$P(\overline{E}) : P(E) \quad \text{or} \quad 1 - P(E) : P(E)$$

## EXAMPLE 11

Find the odds in favor of rolling a 6 with a single fair die.

**Solution**

$$P(6) = \frac{1}{6} \qquad 1 - P(6) = \frac{5}{6}$$

$$\text{Odds} = P(6) : 1 - P(6) \quad \text{or} \quad \frac{1}{6} \text{ to } \frac{5}{6} \quad \text{or} \quad 1 \text{ to } 5$$

Thus, the odds in favor of rolling a 6 are 1 to 5.

At times we are given the odds in favor of an event, and from the odds we obtain the probability that the event will occur.

---

**Probability from Odds**

If the odds favoring an event $E$ are $m$ to $n$, then

$$P(E) = \frac{m}{m + n} \quad \text{and} \quad P(\bar{E}) = \frac{n}{m + n}$$

---

## EXAMPLE 12

The odds that it will rain today are 1 to 3. What is the probability that it will rain?

**Solution**

For the given odds, $m$ can be taken as 1 and $n$ as 3. Therefore,

$$P(R) = \frac{1}{1 + 3} = \frac{1}{4}$$

*Practice Problem*

What are the odds against 2 heads on 2 tosses of a fair coin?

**Answer**

The odds against 2 heads are 3 to 1.

---

## Just for Fun

In a casino, a customer places quarters on numbers, and the house places silver dollars on numbers. The first to put money on any 3 squares that total 15 wins what the other has on the board.

EXAMPLE: Customer, 25¢ on 1, the house, $1 on 5; customer, 25¢ on 9, and the house $1 on 7; customer blocks with 25¢ on 3, and the house places $1 on 6; the customer blocks with 25¢ on 4 and the house places $1 on 2 and wins 4 • 25¢ = $1 because 7 + 6 + 2 = 15.

Show that this game is similar to tic-tac-toe and a 3-by-3 magic square with totals of 15. Can the house always win?

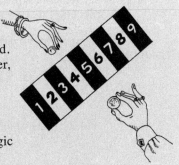

## EXERCISES

1. Use the accompanying spinner to find the probability that the needle will stop on
   (a) a 6 or 3 wedge.
   (b) a 5 or 4 wedge.

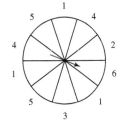

2. The letters of the word *MATHEMATICS* are placed on cards. Aaron draws a card without looking.

   (a) What is the probability that it is an M, H, or T?
   (b) What is the probability it is not an A?

3. Suppose that there is an equally likely probability that the spinner will stop at any one of the six sections for the given spinner.

(a) What is the probability of stopping on an even number?

(b) What is the probability of stopping on a multiple of 3?

(c) What is the probability of stopping on an even number or a multiple of 3?

(d) What is the probability of stopping on an even number or an odd number?

(e) What is the probability of stopping on a number less than 3?

4. A card is drawn from an ordinary deck. What is the probability of getting

(a) a heart?

(b) an ace?

(c) the jack of spades?

(d) a red card?

(e) a red ace?

5. A multiple-choice question has five possible answers. You haven't studied and hence have no idea which answer is correct, so you randomly choose one of the five answers. What is the probability that you select the correct answer? An incorrect answer?

6. A survey course in history contains 20 freshmen, 8 sophomores, 6 juniors, and 1 senior. A student is chosen at random from the class roll.

(a) Describe in words a uniform sample space for this experiment.

(b) What is the probability that the student is a freshman?

(c) What is the probability that the student is a senior?

(d) What is the probability that the student is a junior or a senior?

(e) What is the probability that the student is a freshman or a sophomore or a junior or a senior?

7. A box contains four black balls, seven white balls, and three red balls. If a ball is drawn, what is the probability of getting the following colors?

(a) Black

(b) Red

(c) White

(d) Red or white

(e) Black or white

(f) Red or white or black

8. A number $x$ is selected at random from the set of numbers $\{1, 2, 3, \ldots, 8\}$. What is the probability that

(a) $x$ is less than 5?

(b) $x$ is even?

(c) $x$ is less than 5 and even?

(d) $x$ is less than 5 or a 7?

9. An experiment consists of tossing a coin seven times. Describe in words (without using the word not) the complement of each of the following.

(a) Getting at least two heads

(b) Getting three, four, or five tails

(c) Getting one tail

(d) Getting no heads

10. If $A$ and $B$ are events with $P(A) = .6$, $P(B) = .3$, and $P(A \cap B) = .2$, find $P(A \cup B)$.

11. At Brooks College, 30% of the freshmen make A's in mathematics, 20% make A's in English, and 15% make A's in both mathematics and English. What is the probability that a freshman makes an A in mathematics or English?

12. A single card is drawn from a 52-card deck. What is the probability that it is

(a) either a heart or a club?

(b) either a heart or a king?

(c) not a jack?

(d) either red or black?

13. In the experiment of tossing a pair of fair coins compute the probabilities of these events.

(a) all heads

(b) no heads

14. In Exercise 13, compute the probability of:

(a) at least one tail

(b) exactly one tail

15. The employment status of the residents in a certain town is given in the following table.

| Gender | Employed | Unemployed |
|--------|----------|------------|
| Male   | 1000     | 40         |
| Female | 800      | 160        |

Assign a probability that each of the following is true of a randomly selected person.

(a) Person is female.

(b) Person is male.

(c) Person is unemployed.

(d) Person is employed.

16. A sociology class made a study of the relationship between an employee's age and the number of on-the-job accidents. The following table summarizes the findings.

| Age Group | Number of Accidents | | | |
|-----------|---|---|---|----------|
|           | 0 | 1 | 2 | 3 or More |
| Under 20  | 18 | 22 | 8 | 12 |
| 20–39     | 26 | 18 | 8 | 10 |
| 40–59     | 34 | 14 | 8 | 6 |
| 60 and over | 42 | 10 | 12 | 2 |

An employee is selected at random.

   (a)  What is the probability that the employee is in the 20–39 age group?
   (b)  What is the probability that the employee has had 2 accidents?
   (c)  What is the probability that the employee has had more than 2 accidents?
   (d)  What is the probability that the employee has had at least 1 accident?

17. Find the odds in favor of drawing a heart from an ordinary deck of 52 cards.

18. A fair die is rolled. What are the odds that a 2 will turn up?

19. What are the odds against selecting an ace in drawing one card from a deck of cards?

20. Leroy has a chance for 5 different summer jobs, 3 of which are at resort areas. If he selects a job at random, find the odds against its being at a resort area.

21. If $A$ and $B$ are events in a sample space such that $P(A) = .6$, $P(B) = .2$, and $P(A \cap B) = .1$, compute each of the following.

   (a)  $P(\overline{A})$
   (b)  $P(\overline{B})$
   (c)  $P(A \cup B)$
   (d)  $P(\overline{A \cap B})$

22. If $A$ and $B$ are events with $P(A \cup B) = \frac{5}{8}$, $P(A \cap B) = \frac{1}{3}$, and $P(\overline{A}) = \frac{1}{2}$, compute the following.

   (a)  $P(A)$
   (b)  $P(B)$
   (c)  $P(\overline{B})$
   (d)  $P(\overline{A} \cup B)$

23. In a survey, families with children were classified as $C$, and those without children as $\overline{C}$. At the same time, families were classified according to $D$, husband and wife divorced, and $\overline{D}$, not divorced. For 200 families surveyed, the following results were obtained.

|  | $C$ | $\overline{C}$ | Total |
|---|---|---|---|
| $D$ | 60 | 20 | 80 |
| $\overline{D}$ | 90 | 30 | 120 |
| **Total** | 150 | 50 | 200 |

What is the probability that a family selected at random

   (a)  has children?
   (b)  has husband and wife who are not divorced?
   (c)  has children or husband and wife who are divorced?
   (d)  has no divorce or no children?

24. A recent survey found that 60% of the people in a given community drink Lola Cola and 40% drink other soft drinks; 15% of the people interviewed indicated that they drink both Lola Cola and other soft drinks. What percent of the people drink either Lola Cola or other soft drinks?

25. A coin is tossed 4 times. Let event $A$ be that exactly one head appears. Compute $P(A)$ and $P(\overline{A})$, and verify that $P(A) = 1 - P(\overline{A})$.

26. After 1000 rolls of a loaded die, the following probability assignments are made. Compute the probability of the following events.

| Outcome | 1 | 2 | 3 | 4 | 5 | 6 |
|---|---|---|---|---|---|---|
| **Probability** | .15 | .3 | .3 | .05 | .05 | .15 |

   (a)  The roll is even.
   (b)  The roll is >4.
   (c)  The roll is even or divisible by 3.
   (d)  The roll is even and divisible by 3.

27. Four coins are tossed. What is the probability of getting

   (a)  4 tails?
   (b)  exactly 2 tails?
   (c)  at least 3 heads?
   (d)  exactly 1 head?

## Review Exercise

28. A family plans to have three children.

   (a)  Use a tree diagram to help list all possible outcomes of this activity by classifying each child as either a boy or a girl.
   (b)  If each of these outcomes is equally likely, what probability should be assigned to each outcome?

# Interlude: Practice Problems 5.3, Part 1

1. Suppose that you use your calculator to generate two random integers between 1 and 5 inclusive and you find the sum of those two numbers.

   (a) What is the probability that you obtain a sum of 4?

   (b) What is the probability that you obtain a sum of no more than 6?

   (c) What is the probability that you obtain a sum that is greater than 2?

   (d) What are the odds in favor of obtaining a sum of 5?

   (e) What are the odds against obtaining a sum greater than 8?

2. It has been claimed that the odds in favor of someone in the United States living in the state of their birth are 16:9. What is the probability that a randomly selected person living in the United States is living in his or her birth state?

3. A card is drawn at random from a standard 52-card deck.

   (a) What is the probability that the card drawn is either an ace or an 8?

   (b) What are the odds against drawing a card that is either red or a 2?

   (c) What are the odds in favor of drawing a face card or an ace?

   (d) What are the odds against drawing an even-numbered card?

4. A phone poll of 5000 registered voters in a city established that in a recent election, 3860 of the voters had voted on a municipal bonding proposal, 2490 of the voters had voted on school board elections, and 1495 had cast votes regarding both of those two items. Given this, what are the odds against a randomly selected voter from the city having not voted on either issue?

# Interlude: Practice Problems 5.3, Part 2

1. A pair of fair six-sided dice is rolled and the sum of the dots on the upturned faces is calculated.
   (a) What is the probability that the sum is a prime number?

   (b) What are the odds in favor of a sum that is either divisible by 3 or divisible by 4?

   (c) What are the odds against a sum that is either 7 or 11?

2. A survey of married couples in a certain state found the following:
   - The probability the husband has a college degree is 0.45
   - The probability the wife has a college degree is 0.55
   - The probability that both have college degrees is 0.3
   (a) Find the probability that neither member of the couple has a college degree.

   (b) Find the probability that the wife has a college degree and the husband does not.

3. In a standard 52-card deck of playing cards, all black cards that are not numbered are removed. A single card is drawn from what remains of the deck.

   (a) What are the odds in favor of drawing a numbered card that is red?

   (b) What are the odds against drawing either an ace or a king?

   (c) What is the probability that the card drawn is numbered less than 6?

4. A store that sells eyewear has established that 40% of their customers will purchase glasses but not contact lenses, while 25% will purchase contact lenses but not glasses. Also, 20% of the customers opt to purchase neither glasses nor contact lenses. What is the probability that a customer will purchase both glasses and contact lenses?

# Interlude: Practice Problems 5.3, Part 3

1. A bag contains five balls that are identical except that each ball is uniquely numbered 1, 2, 3, 4, or 5. An experiment consists of drawing a pair of balls (at the same time) at random from the bag and adding the numbers shown on each ball. This sum is then recorded.

   (a) Calculate the probability that the sum is at most 5.

   (b) Give the odds against obtaining a sum greater than 7.

   (c) Give the odds in favor of obtaining a sum that is an odd number.

2.  A bag contains 10 balls that are identical except for being numbered. Six of the balls are uniquely numbered 1 through 6 and the remaining balls are numbered 7. A ball is drawn randomly from the bag, replaced, and then a ball is drawn again. The sum of the numbers drawn is recorded.

(a)  What is the probability that the sum is at least 6?

(b)  What are the odds against a sum smaller than 13?

(c)  What are the odds in favor of a sum that is an even number?

# 5.4  MULTISTEP EXPERIMENTS, EXPECTED VALUE, AND SIMULATION

**Problem**  A nationwide promotion promises a first prize of $25,000, 2 second prizes of $5000, and 4 third prizes of $1000. A total of 950,000 persons enter the lottery. Is entering worth the price of a stamp required to mail the lottery form?

**Overview**  In this section we examine three important problem-solving tools that are used for resolving uncertain situations: We find a use for tree diagrams in multistep experiments; we discuss the importance of expected value in determining how much a decision maker stands to gain or lose in an uncertain situation; and we look at the process of simulation as a way of exploring probability.

**Goals**  Illustrate the Data Analysis and Probability Standard.
Illustrate the Representation Standard.
Illustrate the Connections Standard.
Illustrate the Algebra Standard.

Some experiments are performed as sequences of consecutive steps and are called *multistep experiments*. The probabilities of outcomes of multistep experiments are easily computed with tree diagrams. Consider the following example.

## EXAMPLE 1

A basket contains two red balls and three black balls. A ball is drawn, set aside, and its color is noted. Then a second ball is drawn. Find a sample space for this experiment and assign probabilities.

**Solution**

This is an example of a multistep experiment. The simple tree diagram in Figure 5.21 makes it easy to determine the sample space, but it is clear that this is not a uniform sample space. The outcome RR (red, then red) is certainly less likely than BB (black, then black).

The probability of obtaining a red ball on the first draw is 2/5, but, because the first ball is laid aside, the probability of the second draw producing a red ball is 1/4. Thus, only 1/4 of the times that the first ball is red will the second ball be red. Hence it is reasonable to assign a probability of $\frac{2}{5} \cdot \frac{1}{4} = \frac{1}{10}$ to the outcome RR. Label each branch of the tree with the probability that the outcome on that branch will occur on the next step. (See Figure 5.21.) Notice that the probability of RR we computed is just the product of the probabilities along the path RR in the tree diagram. Similarly, the probability of RB is $\frac{2}{5} \cdot \frac{3}{4} = \frac{3}{10}$, the probability of BR is $\frac{3}{5} \cdot \frac{2}{4} = \frac{3}{10}$, and BB is $\frac{3}{5} \cdot \frac{2}{4} = \frac{3}{10}$.

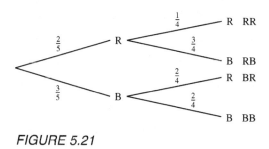

*FIGURE 5.21*

The scheme used in Example 1 to compute probabilities in a multistep experiment can be used in general. Suppose a tree diagram is used to find a sample space for a multistep experiment.

Probability Using a Tree Diagram

The probability of an outcome described by a path through a tree diagram is equal to the product of the probabilities along that path.

## EXAMPLE 2

Let us use the basket from Example 1 that contains 2 red balls and 3 black balls. This time, after we draw the first ball and record the color, we will place it back in the basket. (See Figure 5.22.) In this circumstance, what are the probabilities of the outcomes RR and RB in the sample space?

**Solution**

Do you see that the probability of RR is $\frac{2}{5} \cdot \frac{2}{5} = \frac{4}{25}$ while the probability of RB is $\frac{2}{5} \cdot \frac{3}{5} = \frac{6}{25}$?

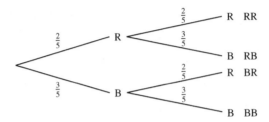

**FIGURE 5.22**

*Practice Problem*

Find the probability of BR and BB in Example 2.

**Answer**

$\frac{6}{25}, \frac{9}{25}$

## EXAMPLE 3

A card is drawn from a deck of cards. Then the card is replaced, the deck is reshuffled, and a second card is drawn. What is the probability of getting an ace on the first draw and a king on the second?

**Solution**

We do not need to draw the whole tree for the experiment of drawing two cards from a deck to compute this probability. Consider only the path of interest in the problem shown in Figure 5.23. The probability of an ace followed by a king is $\frac{4}{52} \cdot \frac{4}{52} \approx .0059$.

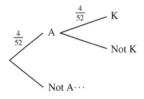

**FIGURE 5.23**

*Practice Problem*

In Example 3 compute the probability if the first card drawn is not replaced.

**Answer**

$\frac{4}{52} \cdot \frac{4}{51} = .006$

## Expected Value

An important property associated with probability is that of expectation or expected value. If we toss a fair coin 100 times, we expect to get heads approximately

$$100\left(\frac{1}{2}\right) = 50 \text{ times}$$

If we spin a fair spinner with 10 equal sectors 1000 times, we expect the spinner to stop on any given sector approximately

$$1000\left(\frac{1}{10}\right) = 100 \text{ times}$$

The concept is perhaps most easily explored in the analysis of a simple game of chance. Suppose that some poor benighted soul were persuaded to play the following game with us: A fair coin is flipped. If a head appears, we receive \$5; if a tail appears, our demented opponent receives \$2 (that is, we win −\$2).

What happens if we play the game 100 times? In 100 flips of the coin we can expect approximately 50 heads and 50 tails. Hence we can expect a payoff of approximately 50(\$5) from the heads and a payoff of (50)(−\$2) from the tails. Our net profit will thus be (50)(\$5)+ 50(−\$2) = \$150. Because the game is played 100 times, our average profit per game will be 150/100 = \$1.50.

Now observe this alternative way of computing average gain per game:

$$P(\text{winning } \$5) = P(H) = \frac{1}{2} \quad P(\text{payoff of } -\$2) = P(T) = \frac{1}{2}$$

$$(\$5) \cdot P(\text{winning } \$5) + (-\$2) \cdot P(\text{payoff of } -\$2) = \frac{1}{2} \cdot \$5 + \frac{1}{2} \cdot -\$2$$

$$= \$2.50 - \$1 = \$1.50$$

This second set of computations suggests the following definition of expected value.

---

**DEFINITION**

### Expected Value

Suppose that there are $n$ payoff values in a given experiment: $A_1, A_2, A_3, \ldots, A_n$. The *expected value* for this experiment is

$$A_1 \cdot P(A_1) + A_2 \cdot P(A_2) + \ldots + A_n \cdot P(A_n)$$

where $P(A_1) + P(A_2) + \cdots + P(A_n) = 1$.

---

## EXAMPLE 4

Let $x$ be a variable representing the number of tails that can appear in the toss of three coins. Of course, $x$ can assume the value 0, 1, 2, or 3. Tabulating the results as in Table 5.4 assists us in computing the expected value of $x$. The expected number of tails is $\frac{3}{2}$.

**Table 5.4**

| $x$ | $P(x)$ | $xP(x)$ |
|-----|--------|---------|
| 0 | $\frac{1}{8}$ | 0 |
| 1 | $\frac{3}{8}$ | $\frac{3}{8}$ |
| 2 | $\frac{3}{8}$ | $\frac{6}{8}$ |
| 3 | $\frac{1}{8}$ | $\frac{3}{8}$ |
| Total | | Expected value $= \frac{12}{8} = \frac{3}{2}$ |

One interpretation of expected value is that it is the average payoff per experiment when the experiment is performed a large number of times.

## EXAMPLE 5

A nationwide promotion promises a first prize of $25,000, two second prizes of $5000, and four third prizes of $1000. A total of 950,000 persons enter the lottery.

(a) What is the expected value if the lottery costs nothing to enter?

(b) Is it worth the stamp required to mail the lottery form?

**Solution**

(a) Because

$$P(\$25{,}000) = \frac{1}{950{,}000} \qquad P(\$5000) = \frac{2}{950{,}000} \qquad P(\$1000) = \frac{4}{950{,}000}$$

the expected value is

$$(\$25{,}000)\frac{1}{950{,}000} + (\$5000)\frac{2}{950{,}000} + (\$1000)\frac{4}{950{,}000} = \$0.04$$

(b) Hardly!

Notice again that expected value is not something to be expected, in the ordinary sense of the word. It is a long-run average of repeated experimentation.

*Practice Problem*

Alfa Car Insurance Company insures 200,000 cars each year. Records indicate that during the year the company faces the likelihood of making the liability payments shown in Table 5.5 for accidents.
   What amount should the company expect to pay per car insured?

**Table 5.5**

| Liability | Corresponding Probabilities |
|---|---|
| $500,000 | 0.0001 |
| $100,000 | 0.001 |
| $50,000 | 0.004 |
| $30,000 | 0.01 |
| $5,000 | 0.04 |
| $1,000 | 0.06 |
| $0 | 0.8849 |

**Answer**

$910

## Simulations

One of the most powerful tools available to both scientists and business analysts is the tool of simulation. Using simulations, rocket engines, marketing schemes, and weapons systems are tested before they are even built. We will certainly not test any rocket engines, but we can experience the flavor of simulations. For example, because the probability of having a boy child is roughly the same as the probability of having a girl child, we could use the flip of a coin to simulate the birth of a child. We could simulate possible outcomes for a family with 3 children by flipping a coin 3 times. By repeating this simulation 100 times, we could compute empirical probabilities for the experiment without ever having undergone the considerable difficulty of having 300 children.

A very useful tool in performing simulation is a table of random digits such as the one in Table 5.6. The numbers in Table 5.6 are collections of digits that are randomly generated by a computer. We will use this table in the following simulation.

**Table 5.6**

| | | | | | | | |
|---|---|---|---|---|---|---|---|
| 12135 | 65186 | 86886 | 72976 | 79885 | 07369 | 49031 | 45451 |
| 10724 | 95051 | 70387 | 53186 | 97116 | 32093 | 95612 | 93451 |
| 53493 | 56442 | 67121 | 70257 | 74077 | 66687 | 45394 | 33414 |
| 15685 | 73627 | 54287 | 42596 | 05544 | 76826 | 51353 | 56404 |
| 74106 | 66185 | 23145 | 46426 | 12855 | 48497 | 05532 | 36299 |
| 57126 | 99010 | 29015 | 65778 | 93911 | 37997 | 89034 | 79788 |
| 94676 | 32307 | 41283 | 42498 | 73173 | 21938 | 22024 | 76374 |
| 68251 | 71593 | *93397 | 26245 | 51668 | 47244 | 13732 | 48369 |
| 60907 | 17698 | 32865 | 24490 | 56983 | 81152 | 12448 | 00902 |
| 07263 | 16764 | 71261 | 52515 | 93269 | 61210 | 55526 | 71912 |
| 43501 | 10248 | 34219 | 83416 | 91239 | 45279 | 19382 | 82151 |
| 57365 | 84915 | 11437 | 98102 | 58168 | 61534 | 69495 | 85183 |
| 38161 | 22848 | 06673 | → 35293 | 27893 | 58461 | 10404 | 17385 |
| 26760 | 51437 | 87751 | 41523 | 10816 | 54858 | 35715 | 47947 |
| 65592 | 93388 | 36555 | 21136 | 43900 | 89837 | 78093 | 28870 |
| 48651 | 16719 | 99032 | 86292 | 40668 | 72821 | 59266 | 44970 |
| 71495 | 84760 | 35193 | 06961 | 41211 | 33548 | 40026 | 63873 |
| 81242 | 06154 | 69109 | 60926 | 62177 | 72065 | 70225 | 86018 |
| 26574 | 84854 | 38915 | 83783 | 46780 | 08735 | 38781 | 94657 |
| 07736 | 70130 | 46808 | 18940 | 14795 | 34231 | 23671 | 05856 |
| 26533 | 06561 | 09049 | 67618 | 12560 | 59539 | 41937 | 18490 |
| 36335 | 84039 | 05960 | 38850 | 62976 | 65958 | 99682 | 64250 |
| 92074 | 87770 | 31924 | 99481 | 15505 | 55099 | 42072 | 57637 |
| 00243 | 48272 | 45390 | 24171 | 96173 | 98887 | 03335 | 45965 |
| 68900 | 91374 | 18868 | 45389 | 57567 | 89557 | 56764 | 59362 |
| 57663 | 88219 | 88929 | 03419 | 28838 | 89659 | 64710 | 60768 |
| 27715 | 05262 | 06208 | 96357 | 65700 | 82054 | 28590 | 95933 |
| 91798 | 54270 | 85403 | 30110 | 00426 | 19915 | 38883 | 43423 |
| 64221 | 42325 | 55273 | 68399 | 91856 | 76729 | 25130 | 64615 |
| 10852 | 21817 | 08641 | 82759 | 75389 | 96295 | 05934 | 53697 |

## EXAMPLE 6

Simulate the experiment of tossing a pair of coins 50 times.

**Solution**

We need to use the table to determine the result of tossing each coin. Because each digit is equally likely to be even or odd, we will use the first two digits in each entry to represent the outcomes of the coins. If the first digit is even, the first coin shows a head; if the first digit is odd, the first coin shows a tail. Similarly, the second digit will determine whether the second coin shows a head or tail. We need to randomly choose a place in the table to start; close your eyes and mark a starting point with your pencil. Suppose the arrow marks the spot where to start. Because the first two digits of 35293 are odd, the outcome of our first toss is two tails. Proceeding to the next number down the column, the 4 and 1 in 41523 indicate that the next pair of tosses results in a head and a tail. Continuing, we get the results from our 50 trials as shown in Table 5.7. How does this compare to the theoretical probabilities?

**Table 5.7**

|  | Frequency | Empirical Probability |
|---|---|---|
| HH | 12 | $\frac{12}{50} = .24$ |
| HT | 12 | $\frac{12}{50} = .24$ |
| TH | 13 | $\frac{13}{50} = .26$ |
| TT | 13 | $\frac{13}{50} = .26$ |

## EXAMPLE 7

A basketball player hits 40% of her shots from three-point range. Use simulation to determine the probability that in a game in which she takes 5 long shots (from beyond the three-point line) she will hit at least 3 of them.

### Solution

We will let a single digit represent each attempted long shot. Because she hits 40% of her shots from this distance, we will let the four digits 0, 1, 2, 3 represent successful shots and the digits 4, 5, 6, 7, 8, 9 represent missed attempts. (Note that 0, 1, 2, 3 comprise 40% of the digits 0 through 9.) A five-digit entry from the random number table will represent a game in which she attempts five shots. We randomly pick a place to begin. Suppose that it is the location marked with the asterisk in Table 5.6. The first entry is 93397. Because the 3's represent successes, but 9 and 7 represent misses, in the first game she made only 2 of her 5 shots. If we simulate this 50 times, we learn that in 13 of 50 games she hit 3 or more of her 3-point shots. Thus, from the simulation we would assign a probability of .26 to this event.

## Just for Fun

The "Birthday Problem" gives a result that defies intuition. Suppose that 35 people are in a room. What is the probability that at least 2 of them have their birthdays on the same day of the year? Because there are 365 days in the year and only 35 people, it seems that the probability should be fairly small. In fact, however, it is quite large—namely, 0.814.

**The Probability $P$ That Among $n$ People at Least 2 Will Have the Same Birthday**

| $n$ | $P$ |
|---|---|
| 5 | .027 |
| 10 | .117 |
| 15 | .253 |
| 20 | .411 |
| 23 | .506 |
| 25 | .569 |
| 30 | .706 |
| 35 | .814 |
| 40 | .891 |
| 45 | .941 |
| 50 | .970 |
| 55 | .986 |
| 60 | .994 |
| 65 | .998 |
| 70 | .999 |
| 75 | .999 + |

# EXERCISES

**1.** A basket contains 4 yellow balls and 5 green balls. One ball is drawn and laid aside, and a second ball is drawn. A tree diagram showing the possible outcomes and the probabilities is shown below.

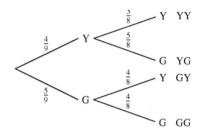

(a) What is the probability of drawing a yellow ball and then a green ball?

(b) What is the probability of drawing 2 green balls?

**2.** Consider the problem in Exercise 1 but in this case return the first ball to the basket after it is drawn and its color noted.

(a) Redraw the tree diagram from Exercise 1 with the correct probabilities.

(b) What is the probability of drawing 2 green balls?

**3.** In a lottery, 200 tickets are sold for $1 each. There are 4 prizes, worth $50, $25, $10, and $5. What is the expected value of a single ticket?

**4.** A fair die is rolled. What is the expected value of the number of dots?

**5.** The 6 letters of the word *LITTLE* are in a box. We draw 4 letters one at a time and place them in a row. We are interested in the probability that the letters spell the word *TELL*. We do not have to draw the whole tree. Indeed, the branch of the tree that spells *TELL* is our primary interest. Label this branch with the probabilities that are required and compute the probability of *TELL*.

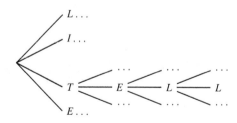

**6.** We spin the three spinners shown below. What is the probability that the resulting letters spell *NOT?*

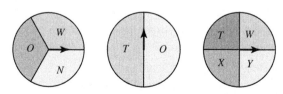

**7.** From a box containing 5 red balls and 3 white balls, 2 balls are drawn successively at random, without replacement. What is the probability that the first is white and the second is red?

**8.** John must take a 4-question true–false quiz but has failed to study. What is the probability that he will score 100% if he guesses on each question?

**9.** Suppose the quiz (see Exercise 8) that John must take consists of 4 multiple-choice questions with answers (a) through (d). What is the probability that John will make 100%?

**10.** A candy jar contains 6 pieces of peppermint, 4 pieces of chocolate, and 12 pieces of butterscotch candy. A small boy reaches into the jar, snatches a piece, and eats it rapidly. He repeats this act quickly.

(a) What is the probability that he eats a peppermint and then a chocolate?

(b) What is the probability that he eats 2 chocolates?

(c) What is the probability that he eats a chocolate and then a butterscotch?

(d) What is the probability that he eats a chocolate and a butterscotch?

**11.** Suppose that the small boy of Exercise 10 is caught by his mother immediately after he snatches his first piece. She makes him return the candy to the jar. He waits an appropriate length of time and then again snatches a piece.

(a) What is the probability that the frustrated thief snatches first a peppermint and then a chocolate?

(b) What is the probability that he gets chocolate on both tries?

**12.** Assume that two cards are drawn from a standard deck of playing cards. What is the probability that a jack is drawn, followed by a queen,

(a) if the first card is replaced before the second is drawn?

(b) if the first card is not replaced before the second is drawn?

**13.** Use Table 5.6 to simulate the experiment of flipping a coin 3 times. Use the first 3 digits of each entry in the table to determine respectively whether the first, second, and third flips show a head or a tail. Simulate the experiment 50 times and compute the appropriate empirical probabilities.

**14.** Tiny Trinkles cereal is placing 1 of 3 tiny dolls in each box of cereal. Each of the dolls is equally likely to occur in a given box of cereal. We are interested in the probability that if we open 5 boxes of cereal, we will have a collection of all 3 dolls.

(a) How could we use simulation with rolls of a die to answer this question?

(b) Use simulation to answer this question by rolling 5 dice 20 times.

15. How could we use a die or dice to simulate

    (a) flipping a coin?
    (b) flipping a coin 5 times?
    (c) 3 times at bat by a batter bating .333?
    (d) a family of 4 children, each classified as boy or girl?

16. How would we use a box containing 4 cards numbered 1 through 4 to simulate

    (a) flipping a coin?
    (b) flipping a coin 4 times?
    (c) a 3-game series by a team that wins 75% of its games?
    (d) 4 times at bat by a hitter hitting .250?

17. Ed and Jack play a match of tennis each Sunday afternoon. They play until 1 player has won 2 sets, then retire for the day. The probability that Ed will win a given set is .6 and the probability is .4 that Jack will win.

    (a) Draw a tree diagram showing the possible outcomes of a Sunday afternoon's play.
    (b) What is the probability that the match will be over after 2 sets?
    (c) What is the probability that Jack will win 2 sets?
    (d) What is the probability that Ed will win 2 sets?

18. A candy bowl contains 4 pieces of candy in identical wrappers. Two of the pieces of candy are butterscotch and two are licorice. Joyce sits down to read, absentmindedly takes a piece of candy from the bowl, and begins to eat it. Suppose she will continue to eat the candy until she pops a licorice candy in her mouth.

    (a) Draw a tree diagram that shows the possible outcomes of this eating session.
    (b) What is the probability that Joyce will place only 1 piece of candy in her mouth?
    (c) What is the probability that Joyce will place 3 pieces of candy in her mouth?

19. A baseball player is batting .300. In a typical game he will have 3 official at bats.

    (a) Draw a 3-stage tree diagram describing the possible outcomes.
    (b) What is the probability of 3 hits?
    (c) What is the probability of at least 2 hits?
    (d) What is the probability of no hits?

20. Answer the questions in Exercise 19 by simulation. Let the first 3 digits of an entry in Table 5.6 represent 3 times at bat. If the entry is 0, 1, or 2, record a hit; otherwise, record a failure. Simulate a game with 3 times at bat 50 times and compute the probabilities requested in Exercise 19.

21. A manufacturer receives a shipment of 20 articles. Unknown to him, 6 are defective. He selects 2 articles at random and inspects them. What is the probability that the first is defective and the second is satisfactory?

22. Tom and Savoy decide to play the following game for points. A single die is rolled. If it shows a non-prime, Tom receives points equal to twice the number of dots showing. If it shows a prime, Tom loses points equal to 3 times the number of dots showing. What is the expected value of the game?

23. A man who rides a bus to work each day determines that the probability that the bus will be on time is 7/16. The probability that the bus will be 5 minutes late is 3/16; 10 minutes late, 1/4; and 15 minutes late, 1/8. What is his expected waiting time if the man arrives at the bus stop at the scheduled time?

24. Two coins, each biased (or weighted) $\frac{1}{3}$ heads and $\frac{2}{3}$ tails, are tossed. The payoff is $5 for matching heads, $3 for matching tails, and $-$2 if they don't match. What is the expected value of the game?

25. The probability that a test will be positive for a person with AIDS is .95, and the probability that the test will be positive for a person who does not have AIDS is .01. Of the people in a given area of a city, 2% have AIDS. If a person is selected at random from that area, what is the probability that the test will be positive? (*Hint:* The answer will come from two branches of the accompanying tree diagram.)

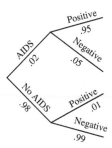

26. Use random digit simulation (using Table 5.6) to estimate the probability that two cards drawn from a standard deck with replacement will be of different suits. (*Hint:* One model could be as follows. Let 1 and 2 represent hearts, 3 and 4 spades, 5 and 6 diamonds, and 7 and 8 clubs. If either a 9 or 0 occurs, ignore that random number and move to the next one. Start at any place in Table 5.6, and consider sets of 2 digits. Record the number of times you get different suits and the number of times you get the same suit.)

27. In Exercise 1

    (a) What is the probability of drawing at least 1 yellow ball?
    (b) What is the probability of drawing 2 balls of different color?

28. In Exercise 2

    (a) What is the probability of drawing at least 1 green ball?
    (b) What is the probability of drawing 2 balls of the same color?

## Review Exercise

**29.** A bag contains 6 red balls, 4 black balls, and 3 green balls. A ball is drawn from the bag. What is the probability that the ball is

(a) red or black?

(b) blue?

(c) red or black or green?

(d) not red and not green?

(e) not black?

(f) green?

(g) not red or not black?

(h) not green?

**30.** A couple wishes to have either 3 children or 4 children. Further, they want exactly 2 girls. Is it more likely that they will get their wish with 3 children or with 4? Answer this question by using Table 5.6 to do the necessary simulations.

(a) Choose a starting point in Table 5.6 and simulate the experiment of having 3 children 50 times. You may use the first 3 digits of each entry in the table to represent the 3 children.

(b) Choose a starting point in Table 5.6 and simulate the experiment of having 4 children 50 times. You may use the last 4 digits in each number to represent the 4 children.

(c) On the basis of your simulations, answer the young couple's question.

# Interlude: Practice Problems 5.4

1. A deck of flash cards has cards numbered 3 through 20. A card is drawn randomly and the number on the card is noted as either even or odd. That card is set aside and a second card is drawn randomly and the number is noted as prime or not prime. Resulting pair of observations is recorded.

   (a) Draw a probability tree that models the outcomes of the experiment.

   (b) If the first number drawn is odd, then what is the probability that the second number drawn is prime?

   (c) What is the probability that the second number drawn is not prime?

2. A box contains five charged batteries and two dead batteries. Batteries are selected from the box (without replacement) until a charged battery is selected. What is the probability that it will take three selections to get a charged battery?

3. Kathy and Robin agree to a game of chance. Three white balls numbered 2, 4, and 7 and one red ball numbered 4 are placed in a bag. Kathy will draw one ball from the bag, replace that ball, and draw a ball from the bag again. The sum of the numbers on the balls that were drawn is noted. If the sum is odd, Kathy loses the game, and any other sum results in Kathy winning the game. If Kathy loses, she will give Robin $20. This game is considered to be a **fair game** if Kathy's expected winnings are $0. How much must Robin agree to pay Kathy if Kathy wins so that the game is fair?

4. A game at a carnival consists of drawing balls, one at a time (without replacement) from a bag containing two black and four red balls. The game proceeds until a black ball is drawn. It costs $5 to play, and the player receives $2 for each ball drawn. Given this, what are the expected winnings for someone who plays this game?

# 5.5   THE FUNDAMENTAL PRINCIPLE OF COUNTING AND PERMUTATIONS

**Problem**   The Hardy College Bulldogs are purchasing uniforms. Members can purchase red or white shorts. They can choose red, white, or striped shirts. How many possible choices are there for the uniforms?

**Overview**   In a uniform sample space the probability of an event is computed by counting the outcomes in the event and the outcomes in the sample space and then dividing. In some cases it is easy to list all the outcomes and then count them. In other cases, it is difficult to list all the outcomes. Hence, we need to learn some procedures that will allow us to count elements in sets without listing all of their elements.

**Goals**   Illustrate the Data Analysis and Probability Standard.
Illustrate the Problem Solving Standard.

As we have seen earlier, tree diagrams are useful in listing all the possible outcomes of many experiments. Observe carefully the tree diagrams in the next two examples, and you may discover a hint about a powerful counting technique.

## EXAMPLE 1

The college chorale is planning a concert tour with performances in Dallas, St. Louis, and New Orleans. In how many ways can its itinerary be arranged?

### Solution

If there is no restriction on the order of the performances, any one of the three cities can be chosen as the first stop. After the first city is selected, either of the other cities can be second, and the remaining city is then the last stop. (See Figure 5.24.)

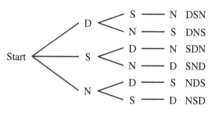

*FIGURE 5.24*

Did you notice that there were three choices for the first stop, two choices for the second stop, and one choice for the third stop for a total of $3 \cdot 2 \cdot 1 = 6$ possible itineraries?

## EXAMPLE 2

The members of the chorale in Example 1 decided to sing first in New Orleans, next in Dallas, and finally in St. Louis. Now, they must decide on their modes of transportation. They can travel from the campus to New Orleans by bus or plane; from New Orleans to Dallas by bus, plane, or train; and from Dallas to St. Louis by bus or train. The tree diagram in Figure 5.25 indicates the chorale's options. The first part of the trip can be made in 2 ways, the second part in 3 ways, and the last part in 2 ways. Notice that the total number of ways the transportation can be chosen is $2 \cdot 3 \cdot 2 = 12$ ways.

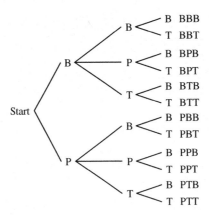

*FIGURE 5.25*

These examples suggest the following principle, called the *Fundamental Principle of Counting*.

---

**Fundamental Principle of Counting**

1. If an experiment consists of two steps, performed in order, with $n_1$ possible outcomes of the first step and $n_2$ possible outcomes of the second step, then there are

$$n_1 \cdot n_2$$

possible outcomes of the experiment.

2. In general, if $k$ steps are performed in order, with possible number of outcomes $n_1, n_2, n_3, \ldots, n_k$, respectively, then there are

$$n_1 \cdot n_2 \cdot n_3 \cdot \ldots \cdot n_k$$

possible outcomes of the experiment.

---

## EXAMPLE 3

A coin is tossed 5 times. If we classify each outcome as either a head or a tail, how many outcomes are in the sample space?

**Solution**

Because there are 5 steps, each with 2 possible outcomes, there are $2 \cdot 2 \cdot 2 \cdot 2 \cdot 2 = 32$ different outcomes.

***Practice Problem***

A die is rolled 3 times. How many different outcomes are there?

**Answer**

216

---

The Fundamental Principle of Counting is helpful in solving problems such as the following.

## EXAMPLE 4

In the state of Georgia, automobile license plates contain an arrangement of 3 letters followed by 3 digits. If all letters and digits may be used repeatedly, how many different arrangements are available?

**Solution**

There are 26 letters to choose from for each of the 3 letter places, and there are 10 digits to choose from for the digit places. By the Fundamental Principle of Counting, the number of arrangements is

$$26 \cdot 26 \cdot 26 \cdot 10 \cdot 10 \cdot 10 = 17{,}576{,}000$$

## EXAMPLE 5

If the letters and numbers on a license plate in Georgia (see Example 4) are assigned randomly, what is the probability that you will receive a license plate on which the letters read *DOG?*

**Solution**

In Example 4 we learned that the sample space contains 17,576,000 outcomes, all of which are equally likely. By the counting principle there are

$$1 \cdot 1 \cdot 1 \cdot 10 \cdot 10 \cdot 10 = 1000$$

possible license plates that begin with *DOG* (because we have only 1 choice for each of the letters). Thus the probability of a license plate with the word *DOG* is

$$\frac{1000}{7{,}576{,}000} = \frac{1}{17{,}576}$$

## EXAMPLE 6

An urn contains 5 red balls and 7 white balls. A ball is drawn, its color is noted, but the ball is not replaced. A second ball is drawn. What is the probability of drawing a red ball followed by a white ball?

**Solution 1**

By the counting principle, there are 12 ways of drawing the first ball and 11 ways of drawing the second. Therefore, there are

$$12 \cdot 11 = 132$$

ways of drawing the 2 balls. All of these ways are equally likely. At the same time, there are 5 ways of drawing the red ball on the first draw and 7 ways of drawing the white ball on the second draw. Therefore, the number of ways of drawing a red ball and then a white ball is

$$5 \cdot 7 = 35$$

Thus,

$$P(R \text{ followed by } W) = \frac{35}{132}$$

**Solution 2**

We, of course, recognize that this problem could have been solved with a tree diagram as discussed in the previous section. Consider the tree diagram in Figure 5.26 and then observe that

$$P(R \text{ followed by } W) = \frac{5}{12} \cdot \frac{7}{11} = \frac{35}{132}$$

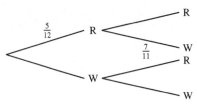

*FIGURE 5.26*

## EXAMPLE 7

In how many ways can 6 students line up outside Wheeler's office to complain about their grades?

**Solution**

As a first step, we can choose a person to be first in line; there are 6 ways to do this. Then we can choose a second person; only 5 persons are available after the first person is chosen. Similarly, the third place must be filled by one of 4 persons and so on.

| First | Second | Third | Fourth | Fifth | Sixth |
|:-----:|:------:|:-----:|:------:|:-----:|:-----:|
|   6   |   5    |   4   |   3    |   2   |   1   |

By the Fundamental Principle of Counting, there are 6 • 5 • 4 • 3 • 2 • 1 or 720 ways to accomplish this task.

## Permutations

Notice in Example 7 that each of the 720 possible lineups is a different ordered arrangement of the set of 6 persons. Ordered arrangements of sets of objects are called ***permutations*** of the sets.

## EXAMPLE 8

Count the number of different permutations of the letters A, B, and C, and then verify the count with a tree diagram.

**Solution**

Because there are 3 ways to choose a first element, then only 2 ways to choose a second element, and finally only 1 way to choose a third, there are

$$3 • 2 • 1 = 6$$

different permutations. This fact is demonstrated with the tree diagram in Figure 5.27.

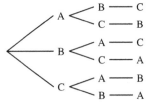

*FIGURE 5.27*

The 6 permutations are listed below:

$$
\begin{array}{lll}
\text{ABC} & \text{BAC} & \text{CAB} \\
\text{ACB} & \text{BCA} & \text{CBA}
\end{array}
$$

In the previous 2 examples, we observed that a set of 6 distinct elements has $6 \cdot 5 \cdot 4 \cdot 3 \cdot 2 \cdot 1$ permutations, whereas a set of 3 distinct elements has $3 \cdot 2 \cdot 1$ permutations. In general,

---

### Number of Permutations of $n$ Objects

The number of permutations of $n$ distinct objects is denoted $P(n, n)$ and

$$
P(n, n) = n(n - 1)(n - 2) \cdot \ldots \cdot 3 \cdot 2 \cdot 1
$$

which can be written as $n!$ (read as $n$ factorial).

---

The preceding discussion suggests that $n!$ is the product of positive integers 1 to $n$, inclusive. The product $6 \cdot 5 \cdot 4 \cdot 3 \cdot 2 \cdot 1$ may be denoted by $6!$, called $6$ *factorial*. We define both $1!$ and $0!$ to be 1. The statement that $0! = 1$ may seem surprising, but you will learn later in your work with factorials that this definition is reasonable and consistent with the idea of factorials for positive integers.

## Calculator Hint

On many calculators, there is a key $\boxed{!}$, $\boxed{x!}$, or $\boxed{n!}$ that can be used to compute factorials. We use the $\boxed{\text{PRB}}$ key to locate a menu on which is located a $!$ entry. For example $6!$ can be found as follows:

$6$ $\boxed{\text{PRB}}$ (underline $!$) $\boxed{=}$ $\boxed{=}$ to get 720. Verify that $10! = 3,628,800$.

---

## EXAMPLE 9

A first-year class is to elect a president, a vice-president, a secretary, and a treasurer from among 6 class members who qualify. In how many ways can the class officers be selected?

### Solution

If we consider the order as president, vice-president, secretary, and treasurer, then (Maria, Tom, Jim, Tomoko) is certainly different from (Tomoko, Tom, Maria, Jim). Thus, each selection of officers is a permutation, not of the whole set, but of a subset of 4 chosen from the whole set. Using the Fundamental Principle of Counting, you can see that the position of president can be filled in 6 ways. After this occurs, the position of vice-president can be filled in 5 ways, so the 2 positions can be filled in $6 \cdot 5$ ways. Then the secretary can be selected in 4 ways, so the 3 positions in $6 \cdot 5 \cdot 4$ ways. Finally, only 3 people remain as candidates for treasurer. Hence, the number of ways that all 4 positions can be filled is $6 \cdot 5 \cdot 4 \cdot 3$.

---

An ordered arrangement of 4 things chosen from 6 things is called a *permutation of 6 things taken 4 at a time*. The number of permutations of 6 things taken 4 at a time is denoted $P(6, 4)$. In the previous example, we determined that

$$
P(6, 4) = 6 \cdot 5 \cdot 4 \cdot 3
$$

If we wished to express $P(6, 4)$ using factorials, we could observe that

$$
P(6, 4) = 6 \cdot 5 \cdot 4 \cdot 3 = \frac{6 \cdot 5 \cdot 4 \cdot 3 \cdot 2 \cdot 1}{2 \cdot 1} = \frac{6!}{2!}
$$

Thus,

$$
P(6, 4) = \frac{6!}{2!} = \frac{6!}{(6 - 4)!}
$$

By reasoning in the same way, we find the following.

---

**Number of Permutations**

The *number of permutations* of n things taken r at a time is given by

$$P(n, r) = \frac{n!}{(n - r)!} \quad 1 \le r \le n$$

---

### Calculator Hint

The number of permutations, $P(n, r)$ can be computed using the factorial key of a calculator and the formula for $P(n, r)$. On some calculators there are specific keys for computing $P(n, r)$; sometimes these keys are named *nPr*. We compute the number of permutations by going to the [PRB] menu and selecting *nPr*. First we compute $P(6, 4)$ by using the definition involving factorials, and then we compute P(6,4) by selecting the *nPr* entry.

6 [PRB] (underline *!*) [=] [÷] 2 [PRB] (underline *!*)[=][=] , and

6 [PRB] (underline *nPr*) [=] 4 [=]. Did you get 360 in both cases?

---

## EXAMPLE 10

Professor Wheeler is asked to judge the Homecoming Pageant. Overcome by the sheer beauty and talent of the 10 contestants, he decides to randomly assign his rankings of first, second, and third. What is the probability that he will award Lori first prize, Jodi Lyn second prize, and Joy third prize?

### Solution

The sample space would consist of all permutations of 3 chosen from the 10 contestants. Hence,

$$n(S) = P(10, 3) = \frac{10!}{7!} = 10 \cdot 9 \cdot 8 = 720$$

Only one of the 720 outcomes ranks Lori first, Jodi Lyn second, and Joy third. Hence, the probability of this result is $\frac{1}{720}$.

---

## Just for Fun

Move only one glass so that empty glasses alternate with full glasses.

# EXERCISES

1. A student plans a trip from Atlanta to Boston to London. From Atlanta to Boston, he can travel by bus, train, or airplane. However, from Boston to London, he can travel only by ship or airplane.

   (a) In how many ways can the trip be made?
   (b) Verify your answer by drawing an appropriate tree diagram and counting the routes.

2. A sociology quiz contains a true–false question and 2 multiple-choice questions with possible answers (a), (b), (c), and (d).

   (a) In how many possible ways can the test be answered?
   (b) Draw a tree diagram and count the options to check your answer to (a).
   (c) If Jodi guesses on each problem, what is the probability that she will get all 3 correct?

3. There are 6 roads from $A$ to $B$ and 4 roads between $B$ and $C$.

   (a) In how many ways can Joy drive from $A$ to $C$ by way of $B$?
   (b) In how many ways can Joy drive round trip from $A$ to $C$ through $B$?

4. Kate wants to buy an automobile. She has a choice of 2 body styles (standard or sports model) and 4 colors (green, red, black, or blue). In how many ways can she select the automobile?

5. (a) In how many ways can 2 speakers be arranged on a program?
   (b) In how many ways can 3 speakers be arranged on a program?
   (c) In how many ways can 4 speakers be arranged on a program?

6. Compute the following without a calculator and then check your answers with a calculator

   (a) 4!
   (b) 7!
   (c) 0!
   (d) 1!

7. (a) How many permutations are there of the set $P, Q, R$?
   (b) Write down all of the permutations of $P, Q, R$.

8. (a) How many permutations taken 2 at a time are there of the set $W, X, Y, Z$?
   (b) List all of the permutations in (a).

9. A die is tossed and a chip is drawn from a box containing 3 chips numbered 1, 2, and 3. How many possible outcomes can be obtained from this experiment? Verify your answer with a tree diagram.

10. Evaluate each of the following without a calculator and then check your answers with a calculator:

    (a) $P(5, 3)$         (b) $P(6, 5)$
    (c) $P(8, 1)$         (d) $P(9, 2)$
    (e) $P(7, 2)$         (f) $P(8, 7)$

11. Write a simple expression for each of the following:

    (a) $P(r, 1)$         (b) $P(k, 2)$
    (c) $P(r, r - 1)$     (d) $P(k, k - 2)$
    (e) $P(k, 3)$         (f) $P(k, k - 3)$

12. The license plates for a certain state display 3 letters followed by 3 numbers (examples: MFT-986, APT-098). How many different license plates can be manufactured if no repetitions of letters or digits are allowed?

13. Employee ID numbers at a large factory consist of 4-digit numbers such as 0133, 4499, and 0000.

    (a) How many possible ID numbers are there?
    (b) How many possible ID numbers are there in which all 4 digits are different?

14. (a) How many 3-digit numbers are there? (Remember, a 3-digit number cannot begin with 0.)
    (b) How many 3-digit numbers are there that end in 3, 6, or 9?
    (c) If you are randomly assigned a 3-digit number as an ID number, what is the probability that it will end in 3, 6, or 9?

15. An ice chest contains 5 cans of cola, 7 cans of ginger ale, and 3 cans of root beer. Al randomly selects a can, and then Sheila takes one. Compute each of the following probabilities in two ways. Use a uniform sample space with 15 • 14 elements. Use portions of a tree diagram as in Section 3.

    (a) Al gets a cola and Sheila a root beer.
    (b) Al gets a ginger ale and Sheila a cola.
    (c) Both get root beers.
    (d) Neither gets a cola.

16. Employee ID numbers at a large factory consist of 4-digit numbers possibly beginning with 0. What is the probability that, if a number is chosen at random from the list of ID numbers, all 4 of its digits will be different?

17. A typical social security number is 413-22-9802. If a social security number is chosen at random, what is the probability that all the digits will be the same? (Social security numbers may begin with 0.)

18. Consider the license plates in Exercise 12. What is the probability that a citizen will receive a license plate whose first 3 letters read *WHY*?

## Review Exercises

19. A single ball is drawn from a basket that contains balls numbered 1 through 10. Find the probability that the number of the ball that is chosen is:

    (a) even.
    (b) greater than 7.
    (c) even and greater than 7.
    (d) even or greater than 7.
    (e) What property of probability can be illustrated with the facts in (a) through (d)?

20. Two cards are drawn from a standard deck of playing cards. What is the probability that a king is drawn, followed by an ace,

    (a) if the first card is replaced before the second is drawn?
    (b) if the first card is not replaced before the second is drawn?

21. A desperate student offers to play the following game with her professor. A single die is rolled. If the roll is odd, the student receives points equal to twice the number of dots showing. If the roll is even, the student loses points equal to 3 times the number of dots showing. What is the expected value of this game for the student?

22. Two cards are drawn without replacement from a standard deck. What is the probability that

    (a) the first is red and the second is black?
    (b) both are red?
    (c) both are black?

## PCR EXCURSION

23. In this section we have been concerned with counting the ordered subsets of a set (called *permutations*). In the next section we will be interested in counting the ordinary unordered subsets of a set, which sometimes are called *combinations*. In this excursion, we do some preparatory work for the next section and discover interesting patterns that have fascinated mathematicians for centuries.

    **A.** The set $\{a, b\}$ has 4 subsets. $\{\ \}$, $\{a\}$, $\{b\}$, and $\{a, b\}$. (Remember that the null set, $\{\ \}$, is a subset of all sets.) Count the number of subsets for each of sets $\{a, b, c\}$, $\{a, b, c, d\}$, $\{a, b, c, d, e\}$ by completing the table:

| Set | Number of Elements | Subsets | Number of Subsets |
|---|---|---|---|
| $\{\ \}$ | 0 | $\{\ \}$ | 1 |
| $\{a\}$ | 1 | $\{\ \}, \{a\}$ | 2 |
| $\{a, b\}$ | 2 | $\{\ \}, \{a\}, \{b\}, \{a, b\}$ | 4 |
| $\{a, b, c\}$ | | | |
| $\{a, b, c, d\}$ | | | |
| $\{a, b, c, d, e\}$ | | | |

**B.** Do you see a pattern in the cases you examined in A? How many subsets does a set with 6 elements have? A set with 10 elements? A set with *n* elements?

**C.** Let us look more carefully at the subsets of each set. If we classify the subsets of $\{a, b\}$ by the number of elements in each subset, we observe that there are 1 subset with no elements, 2 subsets with 1 element, and 1 subset with 2 elements. Notice how this fact is recorded in row 3 of the following triangular chart. Similarly, the set $\{a, b, c\}$ has 1 subset with no elements, 3 subsets with 1 element, 3 subsets with 2 elements, and 1 subset with 3 elements (row 4 of the following chart). By looking at your work from A, complete the rows in the following chart for sets with 4 elements (fifth row) and sets with 5 elements (sixth row).

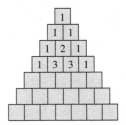

**D.** To analyze the subsets of a set with 6 elements, we would like to have the entries in row 7 of our triangular chart. Look at the pattern formed by the previous rows and complete the last row. Describe what the entries in row 7 tell us about the subsets of a set with 6 elements.

**E.** The triangular table you created is called *Pascal's Triangle*. It contains a wealth of information, but, among other things, it counts for us the unordered subsets (combinations) of various sizes in a set. Complete 2 more rows of Pascal's Triangle and then look for as many patterns as can be found. (*Hint:* What is the sum of the entries in a row? What can be said about the second entry in each row?)

# Interlude: Practice Problems 5.5, Part 1

1. A serial number is formed by three letters followed by two digits. The letters are chosen from the first 10 letters of the alphabet and the digits are chosen from the integers 1-9 (inclusive).

   (a) If letters and digits cannot be repeated, then how many serial numbers are possible?

   (b) If neither letters nor digits can be repeated, then how many serial numbers are possible?

   (c) If letters and digits can be repeated and a serial number is generated at random, then what is the probability that the number does not contain the digit 6?

   (d) If letters and digits can be repeated and a serial number is generated at random, then what is the probability that the number does not contain repeated letters?

2. Suppose that five sci-fi novels and eight mystery novels are to be arranged on a shelf. How many arrangements are possible if novels of the same genre must be grouped together?

3. Five people are seated around a circular table. How many **distinct** seating arrangements are possible?

4. How many batting orders are possible for a team of nine baseball players if the pitcher always bats last and the catcher bats in either the second or third spot?

5. Two cards are drawn at random without replacement from a standard deck of cards.
   (a) What is the probability that exactly two spades are drawn?

   (b) What is the probability that the cards are different colors?

# Interlude: Practice Problems 5.5, Part 2

1. Suppose that a six-digit number is selected at random from the collection of all six-digit numbers that contain the digits 1, 2, 3, 4, 5, and 6 exactly once each. What is the probability that the block 654 appears in the number selected?

2. A campus club consisting of 20 members needs to select a leadership group consisting of a president, vice president, and treasurer. What is the probability that Larry, Curly, and Moe wind up on the leadership group if the leadership positions are assigned randomly?

3. Suppose that I have five copies or *War and Peace*, three copies of *Moby-Dick*, and two copies of *Fifty Shades of Grey*. How many distinct different arrangements of these books can I have on a bookshelf (assume that the books with multiple copies are indistinguishable)?

4.  A 5-digit ZIP code cannot start with a zero. How many 5-digit ZIP codes are possible in which the product of the digits is even?

5.  At a social gathering, everyone shakes hands with everyone else. If 182 handshakes took place, then how many people were at the gathering?

# 5.6   COUNTING AND COMBINATIONS

**Problem**   Eight students each submit an essay for competition. In how many ways can 3 essays be chosen to receive a certificate of merit?

**Overview**   In the previous section we developed the ability to count the number of ordered subsets or permutations of a set. However, there are many circumstances in which we need to count the number of unordered subsets of some set. In the context of counting, unordered subsets of a sample space are called **combinations**.

**Goals**   Illustrate the Data Analysis and Probability Standard.
Illustrate the Representation Standard.
Illustrate the Connections Standard.
Illustrate the Algebra Standard.

Write all subsets with 3 elements that can be chosen from the set $\{P, Q, R, S\}$. Did you get

$$\{P, Q, R\}, \{P, Q, S\}, \{P, R, S\}, \{Q, R, S\}?$$

Notice that there are 4 subsets with 3 elements. In the context of counting, we say that there are 4 combinations of 3 objects that can be chosen from a set of 4 objects. In symbols, we write $C(4, 3) = 4$.

---

### Combinations

In general, a subset of $r$ elements chosen from a set $S$ with $n$ elements is called *an r combination of S*. The number of $r$ combinations that can be chosen from a set of $n$ elements is denoted $C(n, r)$. $C(n, r)$ is sometimes called the *number of combinations of n things taken r at a time*.

---

Notice that each of the 3 combinations of $\{P, Q, R, S\}$ can be ordered in $3! = 6$ ways. For instance,

$$PQR, PRQ, QPR, QRP, RPQ, RQP$$

are all the different permutations of the combination (subset) $\{P, Q, R\}$. Thus, the number of permutations taken 3 at a time from a set with 4 elements is $3! \cdot C(4, 3)$; that is,

$$3! \cdot C(4, 3) = P(4, 3) \quad \text{or}$$

$$C(4, 3) = \frac{P(4, 3)}{3!} \quad \text{or because } P(4, 3) = \frac{4!}{(4 - 3)!}$$

$$C(4, 3) = \frac{4!}{3!1!}$$

This reasoning generalizes as follows:

---

### Combination of $n$ Things Taken $r$ at a Time

The number of ways of selecting $r$ objects from $n$ objects without regard to order (the number of combinations of $n$ things taken $r$ at a time) is

$$C(n, r) = \frac{n!}{r!(n - r)!} \quad 1 \leq r \leq n$$

---

### Calculator Hint

Some books use $_nC_r$ or $\binom{n}{r}$ instead of $C(n, r)$, and some calculators have a $\boxed{nCr}$ key. We use a key found under the $\boxed{\text{PRB}}$ menu, where we underline $nCr$. For the preceding example $C(4, 3)$ is found by 4 $\boxed{\text{PRB}}$ (underline $nCr$) $\boxed{=}$ 3 $\boxed{=}$. Did you get 4?

---

## EXAMPLE 1

Compute the number of unordered subsets with 4 elements that can be chosen from a set with 9 elements.

**Solution**

$$C(9, 4) = \frac{9!}{4!(9-4)!} = \frac{9 \cdot 8 \cdot 7 \cdot 6 \cdot 5 \cdot 4 \cdot 3 \cdot 2 \cdot 1}{4 \cdot 3 \cdot 2 \cdot 1 \cdot 5 \cdot 4 \cdot 3 \cdot 2 \cdot 1} = \frac{9 \cdot \overset{2}{8} \cdot 7 \cdot 6}{4 \cdot 3 \cdot 2 \cdot 1} = 126$$

Notice that to compute $C(n, r)$ easily we first write the answer in factorial notation, expand, divide out common factors, and multiply.

### Practice Problem

Compute $C(7, 4)$ without a calculator. Then check your answer with a calculator.

**Answer**

35

It is important to learn to distinguish between counting problems in which you are counting permutations and counting problems in which you are counting combinations. Consider the problem with which we began this section.

## EXAMPLE 2

Eight students each submit an essay for competition. In how many ways can 3 essays be chosen to receive a certificate of merit?

**Solution**

In this problem we are counting the number of unordered subsets with 3 elements that can be chosen from a set of 8 elements. Hence there are

$$C(8, 3) = \frac{8!}{3!5!} = \frac{8 \cdot 7 \cdot 6 \cdot 5 \cdot 4 \cdot 3 \cdot 2 \cdot 1}{3 \cdot 2 \cdot 1 \cdot 5 \cdot 4 \cdot 3 \cdot 2 \cdot 1} = 56$$

possible outcomes.

## EXAMPLE 3

Eight students each submit an essay for competition. In how many ways can first, second, and third prizes be awarded?

**Solution**

Because we are not only choosing 3 essays but also placing them in order by awarding first, second, and third prizes, we are counting permutations. Thus there are

$$P(8, 3) = \frac{8!}{(8-3)!} = \frac{8 \cdot 7 \cdot 6 \cdot 5 \cdot 4 \cdot 3 \cdot 2 \cdot 1}{5 \cdot 4 \cdot 3 \cdot 2 \cdot 1} = 8 \cdot 7 \cdot 6 = 336$$

possible outcomes.

## EXAMPLE 4

The president of the Student Government Association wishes to appoint a committee of senators consisting of 3 men and 4 women. Currently there are 12 male senators and 8 female senators. In how many ways can this committee be formed?

**Solution**

There are $C(12, 3)$ ways of choosing the male members of the committee and $C(8, 4)$ ways of choosing the female members of the committee. By the Fundamental Principle of Counting, the number of ways of appointing this committee is

$$C(12, 3) \bullet C(8, 4) = 220 \bullet 70 = 15,400$$

*Practice Problem*

A 5-card hand is chosen from a 52-card deck. How many hands contain 3 aces and 2 kings?

**Answer**

$C(4, 3) \bullet C(4, 2) = 24$

---

The ability to count combinations allows us to solve many interesting probability problems.

## EXAMPLE 5

A basket contains 5 red marbles and 6 black marbles. A handful of 4 marbles is chosen from the basket.

(a) How many outcomes are in the sample space for this experiment?
(b) How many outcomes are in the event "2 red, 2 black"?
(c) What is the probability of this event?

**Solution**

(a) The outcomes are all possible combinations of 4 chosen from the 11 marbles so the sample space contains $C(11, 4) = 330$ outcomes.
(b) There are $C(5, 2)$ ways to choose the red marbles and $C(6, 2)$ ways to choose the black marbles, so the event contains $C(5, 2) \bullet C(6, 2) = 150$ outcomes.
(c) The probability is $\dfrac{150}{330} = \dfrac{5}{11}$

## EXAMPLE 6

Find the number of 5-card hands that can be drawn from an ordinary deck of cards. What is the probability of getting all hearts in a given hand?

**Solution**

There are 52 cards in a deck of cards. We will be selecting 5 cards at a time, without regard to the order in which the cards are drawn. The outcome is a combination of 52 things taken 5 at a time:

$$C(52, 5) = \frac{52!}{(52 - 5)!5!} = \frac{52!}{47! \bullet 5!}$$

$$= \frac{52 \bullet 51 \bullet 50 \bullet 49 \bullet 48}{1 \bullet 2 \bullet 3 \bullet 4 \bullet 5}$$

$$= 2,598,960$$

There are 13 hearts. The number of ways in which 5 cards can be drawn from these 13 is $C(13, 5) = 1287$. Therefore,

$$P(\text{all hearts}) = \frac{1287}{2,598,960} \approx .0005$$

## Calculator Hint

To check the computation in Example 6 with a calculator try

52 [PRB] (underline *nCr*) 5 [=] 2,598,960.

# EXERCISES

1. Consider the set of 4 objects {*W, X, Y, Z*}.

   (a) How many combinations of 2 objects can be chosen from this set? List them.
   (b) How many permutations of 2 objects can be chosen from this set? List them.
   (c) Verify that $P(4, 2) = (2!)C(4, 2)$.
   (d) How many combinations of 3 objects can be chosen from this set? List them.
   (e) How many permutations of 3 objects can be chosen from this set? List them.
   (f) Verify that $P(4, 3) = 3!C(4, 3)$.

*In Exercises 2 through 7, determine whether the problem is counting permutations or combinations and compute the correct answer.*

2. In how many ways can a student select 3 books from a reading list of 10 books?

3. In how many ways can 4 books be chosen from 12 books and arranged on a shelf?

4. In how many ways can a starting 5 be chosen from a team of 12 basketball players if one disregards position?

5. In how many ways can a starting 5 be chosen from a team of 12 basketball players if each of the 5 players is assigned a position?

6. In how many ways can a chair and vice-chair be chosen from a committee of 6 members?

7. In how many ways can a coach select 3 cocaptains from a team of 22 baseball players?

8. Three students who are seniors in mathematics and 3 mathematics faculty members are to be placed on a student activities committee for the department. If there are 6 mathematics faculty members and 8 senior mathematics majors, in how many ways can this committee be formed?

9. How many different hands consisting of 7 cards can be drawn from an ordinary deck of cards?

10. A special committee of 3 persons must be selected from a 12-person board of directors. In how many ways can the committee be selected? Check your answer with a calculator.

11. From a standard deck of cards, how many different 7-card hands can be drawn consisting of

    (a) 7 spades?
    (b) 5 clubs and 2 hearts?
    (c) 4 clubs, 1 spade, and 2 hearts?
    (d) 3 clubs, 2 hearts, and 2 diamonds?

12. Ten cities are competing to be selected as the site of a new Jupiter automobile assembly plant. In how many ways can the CEO of Jupiter Inc. select 4 cities as finalists for this new industrial development?

13. A firm buys material from 3 local companies and 5 out-of-state companies. If 4 orders are submitted at once, in how many ways can 2 orders be submitted to a local firm and 2 to an out-of-state firm?

14. A bowl contains 8 red marbles and 14 black marbles. At random, 3 marbles are selected from the bowl without replacement.

    (a) What is the probability that all 3 are black?
    (b) What is the probability that 1 is red and 2 are black?

15. A hat contains 20 slips of paper numbered 1 to 20. If 3 are drawn without replacement, what is the probability that all are numbered less than 10?

16. (a) How many 5-card hands contain 2 aces and 3 kings?
    (b) How many 5-card hands can be drawn from a standard deck?
    (c) What is the probability that a 5-card hand chosen at random contains exactly 2 aces and 3 kings?

17. A hospital ward contains 12 patients, of whom 6 have heart disease. If an intern randomly selects 4 of the patients to examine, what is the probability that all of them will have heart disease?

18. If a 5-card hand is selected from a standard deck, what is the probability of

    (a) 4 aces?          (c) no aces?
    (b) exactly 1 ace?   (d) at least 1 ace?

## Review Exercises

19. A bag contains 6 red balls, 4 black balls, and 3 green balls. Two balls are drawn in succession from the bag without replacement. What is the probability of getting

    (a) 2 red balls?          (c) 2 black balls?
    (b) a red ball followed by a green ball?

20. Two dice are tossed. What is the probability that the sum of the 2 numbers is

    (a) greater than 10?      (c) equal to 7?
    (b) equal to 9?

21. In a certain college, 30% of the students failed mathematics, 20% failed English, and 15% failed both mathematics and English. What is the probability that a student failed mathematics or English?

# Interlude: Practice Problems 5.6, Part 1

1. Suppose that the Mathematics Department at a university is forming a committee of six members to study the possibility of making Finite Mathematics a two-semester course. The department has 12 tenured faculty members and 8 non-tenured faculty members. How many committees are possible if

   (a) the committee must have an equal number of tenured and non-tenured members?

   (b) the committee must have at least two tenured members?

   (c) the committee must have exactly 4 tenured members?

2. A city government is considering bids submitted by seven different firms for each of three separate contracts. How many ways can the city award the contracts if no firm can receive more than two contracts?

3. A group of 12 faculty at a university have applied for three available travel grants. In how many ways can these grants be awarded if

   (a) no preference is given to any applicant?

   (b) one particular applicant must be awarded a grant?

   (c) the group of applicants consists of five men and seven women and at least one man must be awarded a grant?

4. A restaurant offers a choice of three side dishes with each meal. The side dishes are chosen from a list of 15 options and duplicate choices are allowed (so, a customer can order two sides of pickled beets and a side of French fries, if desired). How many possible choices of three side dishes are there?

# Interlude: Practice Problems 5.6, Part 2

1. A bag contains 3 red marbles, 3 blue marbles, 3 green marbles, and 2 yellow marbles, all of which are distinguishable from each other. Four marbles are drawn from the bag (without replacement).

   (a) How many possible draws are there?

   (b) How many draws contain four different-colored marbles?

   (c) How many draws contain at least two red marbles?

   (d) How many draws contain at least one green marble and no red marbles?

2. Suppose that a hand of five cards is dealt from a standard 52-card deck.

   (a) How many 5-card hands are possible?

   (b) How many 5-card hands consist of exactly three aces and two 8s?

   (c) A "full house" is a 5-card hand with three cards of one denomination and two cards of another denomination (see (b) for an example of such a hand). How many different full house hands are possible?

   (d) How many hands contain exactly three cards of one denomination, one of another denomination, and one of a third denomination?

# Interlude: Practice Problems 5.6, Part 3

1. A box contains 30 calculator batteries, of which 10 are defective. Suppose that four are selected at random.

   (a) Calculate the probability that all four are defective.

   (b) Calculate the probability that exactly three are defective.

   (c) Calculate the probability that at least three are defective.

   (d) Calculate the probability that none are defective.

2. A quiz consists of 10 true/false questions. What is the probability of getting exactly six questions correct just by guessing?

3.  The DMV administers five different written driver's examinations and one is selected at random for each driver's license applicant. Suppose that one day a group of six people, two women and four men, take a written exam.

    (a) Calculate the probability that exactly two of the six people take the same test.

    (b) Calculate the probability that the two women take the same test.

4.  Four dry-erase markers are chosen at random from a box containing eight black, six blue, three red, and two green markers.

    (a) Calculate the probability that exactly two blue markers are chosen.

    (b) Calculate the probability that each marker is a different color.

    (c) Calculate the probability that at least two black markers are chosen.

# 5.7   CONDITIONAL PROBABILITY

Earlier, in Example 1 of Section 5.4, we encountered a probability calculation that seemed quite intuitive, when considering probabilities of drawing certain colored balls on the second draw. What is noteworthy there is that we used knowledge of what happened on the first draw to make the subsequent probability assessments. It is that situation that we will explore more thoroughly now.

## Conditional Probability

Suppose that you are attempting to answer a multiple-choice question with five choices, but you have no idea what the correct answer is. If you simply guess, we would say that you have a probability of $\frac{1}{5}$ of choosing the correct answer. However, suppose that you know one of the choices cannot possibly be the correct answer. Then your chances of selecting the correct answer by guessing improve—you have a probability of $\frac{1}{4}$ of guessing correctly. In this case, we say that the $\frac{1}{4}$ is a *conditional probability*. In essence, what that additional knowledge has done is reduce the sample space for this experiment from five outcomes to four outcomes. We would say here "the probability of guessing the correct answer **given** that one of the choices is known to be incorrect is $\frac{1}{4}$" to indicate that the probability was assigned based on knowledge that another situation was present that modified the sample space (and hence the probability assigned to the outcome in question, guessing correctly).

In general, suppose that event $E$ in sample space $S$ has probability $P(E)$. Suppose also that we know some other event $F$ in $S$ has occurred, giving us reason to modify the sample space and, subsequently, reassess the likelihood that event E occurs. This newly appraised probability will be denoted $P(E \mid F)$, and will be called the ***conditional probability of event E, given event F.***

## EXAMPLE 1   CONDITIONAL PROBABILITIES

Consider an experiment that consists of flipping a fair coin two times and recording the sequence of results for each flip ("heads" (H) or "tails" (T)). Then the sample space for this experiment would be

$$S = \{HH, HT, TH, TT\}$$

Since the coin is fair, the sample space is uniform and a probability of $\frac{1}{4}$ would be assigned to each of the outcomes. However, suppose that you are charged with the task of assigning probabilities to the outcomes, but somebody else has actually performed the experiment and informs you that one of the flips was "tails". Immediately, this removes the outcome $HH$ from consideration, resulting in a "reduced" sample space of

$$\{HT, TH, TT\}$$

Armed with this information and knowing that these three outcomes are all equally likely, you would then assign to each of them a (conditional) probability of $\frac{1}{3}$.

Now, let's consider another relatively simple situation. Suppose that in a group of 500 college students, 90 of them are enrolled in a mathematics course and the group contains 280 females. Of the females, 60 of them are enrolled in a mathematics course. A person from the group is chosen at random and we are told that the person is a female. Then the probability that the person chosen is enrolled in a mathematics course, given that the person is a female is $\frac{60}{280} = \frac{3}{14}$ (this is another conditional probability). Now, let $S$ denote the entire group of college students, let $E$ be the event "a person chosen at random is enrolled in a mathematics course", and let $F$ be the event "a person chosen at random is a female." Then we would write $P(E \mid F) = \frac{3}{14}$.

Consider Figure 5.28. It demonstrates that when calculating $P(E \mid F)$ we only consider the ratio of the number of outcomes in both event $E$ and event $F$ to the number of outcomes that are in $F$. Since $n(E \cap F) = 60$ and $n(F) = 280$, we have

$$P(E \mid F) = \frac{n(E \cap F)}{n(F)} = \frac{60}{280} = \frac{3}{14}.$$

But, notice also

$$P(E|F) = \frac{n(E \cap F)/n(S)}{n(F)/n(S)} = \frac{P(E \cap F)}{P(F)}$$

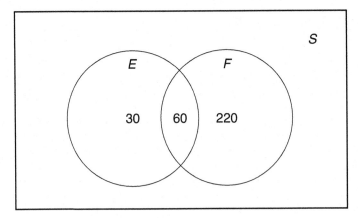

FIGURE 5.28

This suggests a general definition for conditional probability.

---

**Conditional Probability**

Let $E$ and $F$ be events in a sample space $S$ and suppose that $P(F) > 0$. The **conditional probability of event $E$, given event $F$,** denoted $P(E|F)$, is defined as

$$P(E|F) = \frac{P(E \cap F)}{P(F)}$$

Although the situations we have encountered so far involve uniform sample spaces, this definition is valid even when outcomes in the sample space are not equally likely.

## EXAMPLE 2   CALCULATING A CONDITIONAL PROBABILITY

Suppose that an irregular six-sided die is thrown and the number of dots on the upturned face is noted. Since the die is not fair, the probability assignment on the sample space will not be uniform. The following table shows the probabilities assigned to the six outcomes.

| # of dots | 1 | 2 | 3 | 4 | 5 | 6 |
|-----------|-----|-----|-----|-----|------|------|
| **Probability** | 0.2 | 0.1 | 0.1 | 0.4 | 0.15 | 0.05 |

What is the probability that the number of dots showing is a prime number, given that the number of dots showing is odd?

### Solution

Let $E$ be the event "the number of dots showing is a prime number" and let $F$ be the event "the number of dots showing is odd." Then $E = \{2, 3, 5\}$, $F = \{1, 3, 5\}$, and $E \cap F = \{3, 5\}$. Consequently,

$$P(E|F) = \frac{P(E \cap F)}{P(F)} = \frac{0.1}{0.45} = \frac{10}{45} = \frac{2}{9}$$

Notice in Example 2 that if we would not have had the additional condition that the number of dots showing was odd, but had simply wanted to know the probability that the number of dots showing was a prime number, we would have had a probability

$$P(E) = 0.1 + 0.1 + 0.15 = 0.35$$

## Intersection of Events, Revisited

We now revisit intersections of events. Suppose that $E$ and $F$ are events in a sample space with $P(F) > 0$. Then we have $P(E|F) = \dfrac{P(E \cap F)}{P(F)}$. Notice that we can multiply both sides of the equation by $P(F)$ to obtain

$$P(E \cap F) = P(E|F) \bullet P(F) = P(F) \bullet P(E|F)$$

This new equation gives what is called the **product rule** for probability.

---

**Product Rule**

Let $E$ and $F$ be events in a sample space $S$ and suppose that both have nonzero probabilities. Then

$$P(E \cap F) = P(F) \bullet P(E|F)$$

Since $E \cap F = F \cap E$, we also have

$$P(E \cap F) = P(E) \bullet P(F|E)$$

---

We now return to a problem you have encountered earlier.

## EXAMPLE 3   USING THE PRODUCT RULE

Suppose that two cards are drawn consecutively, at random, without replacement, from a standard deck of playing cards. What is the probability that the first card drawn is an ace and the second card drawn is a king?

### Solution

Let $E$ be the event "the first card drawn is an ace" and let $F$ be the event "the second card drawn is a king." Then $E \cap F$ is the event "the first card drawn is an ace and the second card drawn is a king." We seek $P(E \cap F)$. Notice that since there are 52 cards in the deck, $P(E) = \frac{4}{52} = \frac{1}{13}$. Also, if event $E$ happens, then there are only 51 cards remaining, so $P(F|E) = \frac{4}{51}$. By the Product Rule,

$$P(E \cap F) = P(F \cap E) = P(E) \bullet P(F|E) = \frac{1}{13} \bullet \frac{4}{51} = \frac{4}{663}$$

In general, $P(E|F) \neq P(F|E)$, even though $P(E \cap F) = P(F \cap E)$, as we will see in the next example.

## EXAMPLE 4   COMPARING CONDITIONAL PROBABILITIES

Suppose that $E$ and $F$ are events in a sample space such that $P(E) = 0.6$, $P(F) = 0.7$, and $P(E \cap F) = 0.3$. Then

$$P(E|F) = \frac{P(E \cap F)}{P(F)} = \frac{0.3}{0.7} = \frac{3}{7}$$

and

$$P(F|E) = \frac{P(F \cap E)}{P(E)} = \frac{0.3}{0.6} = \frac{3}{6} = \frac{1}{2}$$

### Probability Trees, Revisited

In Section 5.4, the following statement was made.

> The probability of an outcome described by a path through a tree diagram is equal to the product of the probabilities along that path.

What we see now, then, is that that statement is merely asserting the Product Rule (perhaps being used more than once along a path), as the probabilities being assigned along the path are just conditional probabilities, since each probability assignment is based on what happened previously.

## EXAMPLE 5   USING A PROBABILITY TREE AND PRODUCT RULE

Suppose that a fair six-sided die is rolled three times and the number of dots showing on the upturned face is noted each time. What is the probability that the three rolls are 1 dot, an even number of dots, and 6 dots, in that order?

**Solution**

We will start by looking at a tree diagram modeling this three-step experiment (Figure 5.29). We will not put in complete paths that do not relate to the outcome in question, however, so the tree is not "complete."

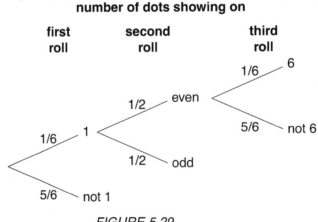

FIGURE 5.29

Define three events as follows.

Let $O$ be the event "one dot showing on first roll"
Let $E$ be the event "even number of dots showing on second roll"
Let $S$ be the event "six dots showing on third roll"

Then what we seek is $P(O \cap E \cap S)$. Notice that $O \cap E \cap S = (O \cap E) \cap S$, so what we will actually calculate is $P((O \cap E) \cap S)$.

What we see from the tree are these probabilities.

$$P(O) = \tfrac{1}{6} \quad \text{and} \quad P(E \mid O) = \tfrac{1}{2}$$

Consequently,

$$P(O \cap E) = P(O) \bullet P(E \mid O) = \tfrac{1}{6} \bullet \tfrac{1}{2}$$

Notice also from the tree that

$$P(S \mid (O \cap E)) = \tfrac{1}{6}$$

(i.e., the probability that the third roll shows six dots given that the first roll showed one dot and the second roll showed an even number of dots is $\tfrac{1}{6}$)

So,

$$P((O \cap E) \cap S) = P(O \cap E) \bullet P(S \mid (O \cap E)) = \tfrac{1}{6} \bullet \tfrac{1}{2} \bullet \tfrac{1}{6} = \tfrac{1}{72}$$

While the notation may seem a bit cumbersome, what this last example illustrates is just what was proposed earlier, which is that the probability we sought was nothing more than the product of the probabilities along the path on the tree leading to the indicated outcome.

# Exercises

*In Exercises 1–5, use the given information to calculate $P(E \mid F)$ and $P(F \mid E)$.*

1. $P(E) = 0.6$, $P(F) = 0.7$, $P(E \cap F) = 0.3$

2. $P(E) = 0.4$, $P(F) = 0.5$, $P(E \cap F) = 0.2$

3. $P(E) = \tfrac{2}{7}$, $P(F) = \tfrac{4}{7}$, $P(E \cap F) = \tfrac{1}{7}$

4. $P(E) = \tfrac{3}{8}$, $P(F) = \tfrac{1}{8}$, $P(E \cap F) = \tfrac{1}{8}$

5. $P(E) = 0.68$, $P(F) = 0.33$, $P(E \cap F) = 0.16$

6. A Mathematics instructor gave the same quiz to two different sections of a class. Section A has 32 students, and 16 passed the quiz. Section B has 26 students, and 15 passed the quiz. Suppose that one quiz paper is selected at random.

   (a) What is the probability that the quiz paper is a passing paper?

   (b) If the paper is known to be from Section A, then what is the probability it is a passing paper?

   (c) If the paper is a passing paper, then what is the probability that it is from Section B?

7. A wallet contains seven $1 bills, three $5 bills, and five $10 bills. A bill is selected at random from the wallet. Find the probability that the bill is

   (a) a $5 bill, given that it is not a $1 bill

   (b) a $1 bill, given that it is not a $10 bill

8. A class has 18 males and 16 females. Kristyn is one of the females in class. One student is selected at random.

   (a) Find the probability that Kristyn was selected, given that the student is female.

   (b) Find the probability that a female was selected, given that Kristyn was the student selected.

9. In a group of 60 children, 28 are enrolled in a summer swimming program, 20 are signed up for soccer, and 6 are signed up for both. Suppose that a child is selected at random.

   (a) What is the probability that the child is enrolled in swimming?

   (b) What is the probability that the child is enrolled in swimming if the child selected is signed up for soccer?

   (c) What is the probability that the child is signed up for soccer if the child selected is enrolled in swimming?

10. Two cards are drawn from a standard deck of playing cards without replacement. Find the probability that the first is a 6 and the second is a face card.

**11.** A dining center on a university campus surveyed the students who ate breakfast there for their coffee preferences. The survey results are given in the accompanying table.

| | Don't Drink Coffee | Prefer Regular Coffee | Prefer Decaf Coffee | Total |
|---|---|---|---|---|
| Female | 23 | 145 | 69 | 237 |
| Male | 18 | 196 | 46 | 260 |
| Total | 41 | 341 | 115 | 497 |

A student is selected at random from this group. Find the probability that

**(a)** the student does not drink coffee.

**(b)** the student is male.

**(c)** the student is a female who prefers regular coffee.

**(d)** the student prefers decaf coffee, given that the student is a male.

**(e)** the student is a male, given that the student prefers decaf coffee.

**(f)** the student is female, given that the student prefers regular coffee or does not drink coffee.

**12.** Two cards are drawn from a standard 52-card deck of playing cards without replacement. Find the probability that the first is a 6 and the second is a king.

**13.** A card is drawn at random from a standard 52-card deck of playing cards. It is replaced, and another card is drawn.

**(a)** Find the probability that the first card is a 3 and the second card is a 9.

**(b)** Find the probability that both are hearts.

**(c)** Find the probability that both are black cards.

**14.** A person draws three balls in succession without replacement from a box containing three black balls, five red balls, and two white balls. Find the probability that the balls drawn are black, red, and white (in that order).

**15.** A fair coin is flipped 10 times.

**(a)** What is the probability of flipping "heads" on the 10th flip, given that the previous nine flips were all "heads"?

**(b)** What is the probability of flipping "heads" 10 consecutive times?

**16.** A circular spinner has five sectors, each numbered with an integer from 1 to 5, but not all of the sectors have the same area. The probability assigned to a pointer landing on a sector numbered with a particular integer is the ratio of the area of the sector to the area of the entire circle, and is given in the following table.

| Pointer lands on | 1 | 2 | 3 | 4 | 5 |
|---|---|---|---|---|---|
| Probability | 0.25 | 0.10 | 0.20 | 0.30 | 0.15 |

Define two events as follows.

Let $E$ be the event "pointer lands on an odd number."
Let $F$ be the event "pointer lands on a number greater than 2."

**(a)** Calculate $P(E \mid F)$.

**(b)** Calculate $P(F \mid E)$.

**17.** Refer to the tree diagram in Figure 5.30 to find the following probabilities.

**(a)** $P(A \cap D)$

**(b)** $P(B \cap C)$

**(c)** $P(C)$

**(d)** $P(D)$

**(e)** $P(D \mid \overline{A})$

**(f)** $P(B \mid C)$

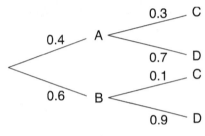

**FIGURE 5.30**

**18.** Refer to the Venn diagram in Figure 5.31 to find the following probabilities.

**(a)** $P(B \mid A)$

**(b)** $P(A \mid B)$

**(c)** $P(A)$

**(d)** $P(A \cup B)$

**(e)** $P(\overline{A})$

**(f)** $P(A \cap B)$

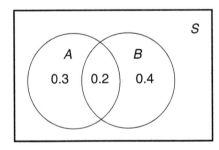

**FIGURE 5.31**

**19.** A bag contains three white, two green, and 1 yellow ball. Two balls are drawn from the bag without replacement. Find the probability that one white and one yellow ball are drawn.

**20.** In a small town in Pennsylvania, it is known that 20% of the families have no children, 30% have one child, 20% have 2 children, 16% have three children, 8% have four children, and 6% have more than four children. Find the probability that a family has more than two children if it is known that the family has at least one child.

21. The following table summarizes the graduating class at a particular university.

| | Arts and Sciences | Education | Business | Total |
|---|---|---|---|---|
| Male | 342 | 424 | 682 | 1448 |
| Female | 324 | 102 | 144 | 570 |
| Total | 666 | 526 | 826 | 2018 |

One student is selected at random from the graduating class. Find the probability that the student

(a) is male.
(b) is receiving an Arts and Sciences degree
(c) is a female, given that the student is receiving an Education degree.
(d) is receiving an Arts and Sciences degree, given that the student is a male
(e) is not receiving a business degree and is a male.
(f) is female, given that an education degree is not received.

22. In a group of 80 professionals, 25 are accountants, 15 are engineers, 30 are teachers, and 10 are nurses. Among the females in this group, 11 are accountants, six are engineers, 24 are teachers, and nine are nurses. A person is selected at random from the group. Find the probability that this person is

(a) a female.
(b) an engineer.
(c) a female engineer.

23. A box contains two white balls and one red ball. Two balls are drawn in succession without replacement. Find the probability of drawing

(a) two white balls.
(b) a white ball followed by a red ball.
(c) a red ball and a white ball, in no particular order.

24. Of a group of children, 30% are boys and 70% are girls. Of the boys, 40% have brown eyes; of the girls, 50% have brown eyes. A child is selected at random from the group. Find the probability that the child

(a) has brown eyes, given that the child is a boy.
(b) is a boy and has brown eyes.
(c) is a girl and has brown eyes.
(d) has brown eyes.

25. A box contains five balls numbered 1 through 5. Three balls are selected in succession without replacement. Find the probability of getting

(a) numbers 1, 2, and 3, in that order.
(b) numbers 1, 2, and 3, in any order.
(c) three balls, the sum of whose numbers is 4.
(d) three balls, the sum of whose numbers is 6.
(e) three odd-numbered balls.
(f) two odd-numbered balls and one even-numbered ball.

26. A box contains two red, three white, and four blue balls. Two balls are drawn in succession without replacement. What is the probability that both balls are the same color?

27. A fair six-sided die is rolled twice.

(a) What is the probability of getting an even number on each roll?
(b) What is the probability of getting an even number on either the first or second roll?

# Interlude: Practice Problems 5.7, Part 1

1. Samples of a new flavor of yogurt were given to customers visiting a warehouse store on three separate days and the customer preference was noted, as recorded in the table.

| | Liked Flavor | Disliked Flavor | Indifferent |
|---|---|---|---|
| Day 1 | 155 | 60 | 35 |
| Day 2 | 140 | 80 | 10 |
| Day 3 | 52 | 98 | 30 |

Find the probability that a customer selected at random

(a) sampled the yogurt on Day 2.

(b) liked the flavor.

(c) disliked the flavor, given the sample took place on Day 1.

(d) sampled on Day 3, given that the customer was indifferent to the flavor.

(e) was indifferent and sampled the yogurt one of the first two days.

(f) sampled the yogurt one of the first two days, given that the customer liked the flavor.

2.  A bag contains 8 red marbles, 6 black marbles, and 4 green marbles. Suppose that two marbles are drawn successively from the bag without replacement.

(a)  What is the probability that the second marble drawn is black, given that the first marble drawn was red?

(b)  What is the probability that the second marble drawn is green?

(c)  What is the probability that two marbles of the same color are drawn?

(d)  What is the probability that the second marble drawn is black, given that the first marble drawn was not black?

# Interlude: Practice Problems 5.7, Part 2

Suppose that $E$ and $F$ are events in a sample space. We will say that $E$ and $F$ are **independent** if

$$P(E \mid F) = P(E) \quad \text{or} \quad P(F \mid E) = P(F)$$

Otherwise, we say that $E$ and $F$ are **dependent**.

In essence, events $E$ and $F$ are independent if the occurrence of one does not affect the probability that the other will occur.

1. A circular spinner has five sectors, each numbered with an integer from 1 to 5, but not all of the sectors have the same area. The probability assigned to a pointer landing on a sector numbered with a particular integer is the ratio of the area of the sector to the area of the entire circle, and is given in the following table.

| Pointer lands on | 1 | 2 | 3 | 4 | 5 |
|---|---|---|---|---|---|
| **Probability** | 0.4 | 0.1 | 0.3 | 0.05 | 0.15 |

(a) Define two events as follows.

Let $E$ be the event "pointer lands on an even number."

Let $F$ be the event "pointer lands on a number less than 4."

Determine if $E$ and $F$ are independent events.

(b) Define two events as follows.

Let $O$ be the event "pointer lands on an odd number."

Let $P$ be the event "pointer lands on a prime number."

Determine if $O$ and $P$ are independent events.

2. An experiment consists of flipping a fair coin and then tossing a fair six-sided die. Determine if the events "flip heads" and "roll three" are independent.

If events $E$ and $F$ are independent, then we have the following result (called the **Multiplication Rule for Independent Events**), which follows directly from the definition of independent events:

Two events $E$ and $F$ are independent if, and only if, $P(E \cap F) = P(E) \cdot P(F)$

This means that we have another way of testing two events for independence, and, if it is known that two events are independent, a way of calculating the probability that both events occur.

Sometimes, it is intuitively obvious whether or not two events are independent, but other times it is not so clear, and so testing is required.

3.  A single card is drawn from a standard 52-card deck. Determine if the following events are independent, but try to guess the result before you test.

(a)  the events "card drawn is a heart" and "card drawn is a face card"

(b)  the events "card drawn is a spade" and "card drawn is a diamond"

4.  Many employers screen job applicants for illicit drug use. Suppose that one company uses a drug test that is 95% effective—the test will come up "positive" 95% of the time if the person being tested is using drugs, and will come up "negative" 95% of the time if the person is not using drugs. Suppose that in the population, 10% of the people are using drugs illicitly, and that a potential employee takes this drug test.

(a)  Determine if the events "person uses drugs" and "test comes up positive" are independent.

(b)  Calculate the probability that the person uses drugs, given that the test comes up positive.

# Interlude: Practice Problems 5.7, Part 3

1. In a certain city, census information indicates that 20% of the households have no children living within, 30% have one child, 20% have two children, 16% have three children, 8% have four children, and 6% have five or more children. Calculate the probability that a household has more than two children living there if it is known that at least one child lives there.

2. Suppose that for a certain type of car battery, 80% of the time the battery will last 6 years or more and 15% of the time the battery will last 10 years or more. Given that a battery has lasted 6 years, calculate the probability it will last 10 years or more.

3. The Food and Drug Administration requires testing of a proposed red food dye for cancer-causing potential. After experimentation on laboratory rats, 6% of the rats that consumed the red dye developed cancer. Also, 2% of the rats that did not consume the dye developed cancer. Exactly half of the rats utilized in the experiment consumed the red dye. Determine if the events "consumed the red dye" and "developed cancer" are independent.

4. Suppose that $A, B, C, D$, and $E$ are events in some sample space and the following *probability table* describes probabilities for these events. To read the table, the totals at the end of rows or bottom of columns give the probability of the event labeling the row or column, while the intersection of rows and columns gives the probability of the intersection of the corresponding events. For instance, $P(A) = 0.28$ and $P(A \cap D) = 0.24$.

|        | A    | B    | C    | Totals |
|--------|------|------|------|--------|
| D      | 0.24 | 0.34 | 0.09 | 0.67   |
| E      | 0.04 | 0.22 | 0.07 | 0.33   |
| Totals | 0.28 | 0.56 | 0.16 | 1.00   |

Calculate these probabilities. Leave your results in fraction form (i.e., $p/q$).

(a) $P(C)$

(b) $P(E)$

(c) $P(D\,|\,B)$

(d) $P(C\,|\,E)$

(e) $P(B\,|\,D)$

(f) $P(E\,|\,C)$

(g) $P(D\,|\,C)$

(h) $P(E\,|\,A)$

(i) $P(A\,|\,C)$

(j) $P(B\,|\,A)$

Solution to Introductory Problem

**Understand the problem.** A couple wants to have either 3 or 4 children, including exactly 2 girls. We are to find the probability of exactly 2 girls with 3 children and the probability of exactly 2 girls with 4 children.

**Devise a plan.** Use a tree diagram to describe all possible arrangements of children in order of birth for a family with 3 children and a family with 4 children. Let B represent boy and G represent girl. Count the number of possible arrangements and the number of arrangements with 2 girls.

| **Three Children** | | **Four Children** | | |
|---|---|---|---|---|
| BBB | BGG | BBBB | BGGB | GBBG |
| BBG | GBG | BBBG | BBGG | GBBB |
| BGB | GGB | BBGB | BGBG | GBGG |
| GBB | GGG | BGBB | GBGB | GGBG |
|     |     | BGGG | GGBB | GGGB |
|     |     |     |     | GGGG |

**Carry out the plan.** The resulting sample spaces are tabulated in the table. Because the probabilities of a girl and a boy are approximately the same, this is a uniform sample space. Hence,

$$P(\text{exactly 2 girls from 3 children}) = \frac{3}{8}$$

$$P(\text{exactly 2 girls from 4 children}) = \frac{6}{16} = \frac{3}{8}$$

Are you surprised?

**Look back.** The probability of exactly 2 girls is the same whether the family has 3 children or 4 children. We can verify this result with simulation. For help with this verification, see the PCR Excursion at the end of Section 3 of this chapter.

# SUMMARY AND REVIEW

## Sample Space

(a) Any result of an experiment is called an *outcome*.

(b) If each outcome has the same chance of occurring as any other, the outcomes are equally likely.

(c) The outcomes of a sample space satisfy the following criteria:

   (i) The categories do not overlap.
   (ii) No result is classified more than once.
   (iii) The list exhausts all possibilities.

(d) An *event* is a subset of sample space.

(e) If each outcome of a sample space is equally likely to occur, the sample space is a *uniform* sample space.

(f) A tree diagram is useful in identifying the outcomes in a multistep experiment.

## Counting Techniques

(a) *Fundamental Principle of Counting:* If two steps are performed in order, with $n_1$ possible outcomes of the first step and $n_2$ possible outcomes of the second step, then there are $n_1 \bullet n_2$ combined possible outcomes of the first step followed by the second.

(b) $n! = n(n - 1) \bullet (n - 2) \bullet \ldots \bullet 1$

(c) The ordered arrangements of $r$ objects selected from a set of $n$ different objects ($r \leq n$) are called *permutations* of $n$ things taken $r$ at a time. The number of permutations of $n$ things taken $r$ at a time is

$$P(n, r) = \frac{n!}{(n - r)!}$$

(d) An $r$ combination is a subset (unordered) with $r$ elements. The number of $r$ combinations from a set with $n$ elements is

$$C(n, r) = \frac{n!}{r!(n - r)!}$$

where $1 \leq r \leq n$.

## Probability

(a) In a uniform sample space with $m$ outcomes, each outcome has a probability of $1/m$.

(b) In a uniform sample space the probability of event $A$ is given by

$$P(A) = \frac{n(A)}{n(S)}$$

(c) If an experiment is performed $N$ times, where $N$ is a large number, then

$$P(A) = \frac{\text{number of times } A \text{ occurs}}{N}$$

(d) If $A$ and $B$ are mutually exclusive (have no outcomes in common), then the probability of $A$ or $B$ is given by $P(A \cup B) = P(A) + P(B)$.

(e) For any two events $A$ and $B$,

$$P(A \cup B) = P(A) + P(B) - P(A \cap B)$$

(f) The complement of event $A$ is everything in the sample space except $A$, denoted by $\overline{A}$.

$$P(\overline{A}) = 1 - P(A)$$

(g) In a multistep experiment the probability of an outcome described by a path in a tree diagram can be computed by multiplying the probabilities along that path.

## Odds

(a) The odds in favor of event $E$ are equal to the ratio $P(E)/P(\overline{E})$.

(b) The odds against event $E$ are equal to $P(\overline{E})/P(E)$.

## Expected Value

Suppose that there are $n$ payoff values in an experiment: $A_1, A_2, A_3, \ldots A_n$. Then the expected value of the experiment is

$$E = A_1 P(A_1) + A_2 P(A_2) + \ldots + A_n P(A_n)$$

where $P(A_1) + P(A_2) + \ldots + P(A_n) = 1$.

# CHAPTER TEST

1. What is the probability of getting heads when a 2-headed coin is tossed?

2. In how many ways can 6 books be arranged on a shelf?

3. A jar contains 3 red balls, 2 green balls, and 1 yellow ball. Tabulate a sample space for the following experiments.

    (a) A single ball is drawn (a sample space with 3 outcomes).
    (b) A ball is drawn and pocketed. A second ball is drawn. (A tree diagram might be helpful.)
    (c) A ball is drawn, its color is recorded, and it is replaced in the jar. A second ball is drawn. (Try a tree diagram.)

4. Suppose that $P(A) = .35$, $P(B) = .51$, and $P(A \cap B) = .17$. Compute the following.

    (a) $P(\overline{A})$      (b) $P(A \text{ or } B)$

5. A box contains 3 red balls and 4 white balls. What is the probability of drawing 2 white balls

    (a) if the first ball is replaced before the second one is drawn?
    (b) if the first ball is not replaced?

6. From a group of 7 people, how many committees of 4 can be selected?

7. In a certain college, 15% of the students failed mathematics, 20% failed English, and 75% passed both mathematics and English. What is the probability that a student passed English or passed mathematics?

8. A file contains 20 good sales contracts and 5 canceled contracts. In how many ways can 4 good contracts and 2 canceled contracts be selected?

9. A card is drawn from a standard deck. What is the probability that it is a king or a spade?

10. The city jail of Rocky Hill has 5 cells, numbered 1 to 5. One evening, 4 drunks were arrested for disturbing the peace. In how many ways can they be assigned to separate cells?

11. A red die and a green die are tossed.

    (a) Describe a uniform sample space for this experiment.
    (b) What is the probability that the red die shows a 6 while the green die shows a 1?
    (c) What is the probability that the sum of the 2 numbers is 7?
    (d) What is the probability that the red die shows an even number?

12. A box contains 6 red and 4 black balls. Three balls are drawn at random. What is the probability of getting 2 red balls and 1 black ball?

13. In how many ways can a chairman, a treasurer, and a secretary be selected from a board of 12 persons?

14. Consider a family of 3 children. Find the probability that all 3 children are of the same sex.

15. A new pod is being formed for a middle school. The principal has available 3 mathematics specialists, 4 social studies specialists, 2 science specialists, and 4 language arts specialists. In how many ways can the principal choose 1 teacher from each area to form a team to teach in the pod?

# Chapter 5 Progress Check I

1. The senior class at Hipster High is holding elections for class president. There are 10 people interested in the position and each student is asked to fill out a ballot by indicating their first, second, and third choices for president, in order of preference.

   (a) How many possible ways can a ballot be filled out?

   (b) If the instructions on the ballot were just to list three candidates without order of preference, then how many groups of three names could appear on a ballot?

2. In my sock drawer, I have six different pairs of socks. If I randomly draw two socks from the drawer, what is the probability that I have drawn a matching pair?

3. A fair coin is flipped five times. What is the probability that heads was flipped at least three times?

4. Determine the number of **distinguishable** ways that the letters of the word "lugubriousness" can be arranged.

5. A bag contains three white balls that are identical except that they are numbered 1, 3, and 8. The bag also contains a red ball that is numbered 1. A person randomly picks a ball, puts the ball back in the bag, and randomly picks a ball again. The sum of the numbers on the two selected balls is recorded.

   (a) Calculate the expected sum.

   (b) Calculate the odds in favor of the sum being either at most 4 or greater than 11.

6. A bag contains four nickles, five dimes, and three quarters. An experiment consists of drawing two coins randomly from the bag. What is the expected monetary value of the pair of coins drawn?

# Chapter 5  Progress Check II

1. An urn contains 10 red, 20 white, and 30 blue balls. Six balls are drawn from the bag without replacement.

   (a) What is the probability that one ball is red, two balls are white, and three balls are blue?

   (b) What is the probability that at least one ball is red?

   (c) What is the probability that the balls are all the same color?

2. At a particular university, all students in Finite Mathematics are assigned a seven-digit identification number, where digits may be repeated. Suppose that these numbers are assigned randomly to students.

   (a) What is the probability of being assigned an identification number that alternates even and odd digits, starting with an odd digit?

   (b) What is the probability of being assigned an identification number that is either all the same digits or all different digits?

3. A bag contains three white balls that are identical except that they are numbered 2, 3, and 7. The bag also contains a red ball that is numbered 3. A person randomly picks a ball, puts the ball back in the bag, and randomly picks a ball again. The sum of the numbers on the two selected balls is recorded.

   (a) Calculate the probability that the sum is odd, given that the sum is greater than 5.

   (b) Calculate the probability that the sum is odd, given that the red ball was selected on the first draw.

   (c) Calculate the probability that the sum is even, given that the red ball was selected at least once.

   (d) Calculate the probability that the sum is even and the red ball was not drawn.

# Chapter 5  Progress Check III

1. Two fair six-sided dice (one red and one green) are rolled and the sum of the numbers on the upturned faces is observed. Calculate the following probabilities.

   (a)  The sum is 5, given the green die is not a 1

   (b)  The sum is 6, given the green die is either a 3 or a 4

   (c)  The red die is 1, given the sum is 6

   (d)  The red die is 4, given the green die is also 4

   (e)  The sum is 7, given that one die is even and one die is odd

2. Everyone's Everyday Electronics uses microprocessors in its desktop computers that come from one of three production plants, located in Clitherall, Vining, and Henning. The Clitherall plant provides 40% of the microprocessors, whereas the Vining and Henning plants provide 50% and 10%, respectively, of the microprocessors. The percentage of defective microprocessors produced by the Clitherall, Vining, and Henning plants are 2%, 4%, and 1%, respectively.

   (a) What is the probability that Everyone's Everyday Electronics receives a defective microprocessor from one of the production plants?

   (b) What is the probability that a microprocessor is defective, given that it came from the Vining plant?

   (c) What is the probability that a microprocessor came from the Henning plant, given that it is defective?

# Chapter 5 Progress Check IV

1. Percephone recently returned from the annual spring meeting of Math Geeks Anonymous (MGA). There were 500 people in attendance. Percephone noticed that 200 of the attendees were wearing pocket protectors, 150 were wearing taped glasses, and 25 were wearing both of those items.

    (a) How many of the people at the MGA meeting weren't wearing either of the aforementioned items?

    (b) How many of the people at the MGA meeting were wearing only taped glasses?

2. Iggy's Ice Cream is a national chain that sells 42 flavors of ice cream. Lynsey is trying to decide what flavors to order on a three-scoop ice cream cone.

    (a) If Lynsey wants only two flavors and doesn't want two consecutive scoops to be the same flavor of ice cream, then how many different three-scoop cones could she order?

    (b) Suppose Lynsey wants exactly two scoops of vanilla ice cream. If she insists that the two scoops of vanilla be put on the cone consecutively, then how many different three-scoop cones could she order?

3. The musical group Massive Ego Attack is planning a short club tour in the area to support their most recent album, *The Largest Noise to Talent Ratio Ever Recorded*. They will perform in Winnepeg, Green Bay, Duluth, Minneapolis, Moorhead, Bismarck, and Sioux Falls.

   (a) How many travel itineraries for the group are possible if the three performances in Minnesota must be given consecutively?

   (b) If the group chooses cities at random to form the itinerary, what is the probability that the tour is bookended by performances in the Dakotas?

4. Suppose that a five-card hand has been randomly dealt from a standard deck of playing cards.

   (a) Calculate the probability that you are dealt a full house (recall that a full house consists of 3 cards of one denomination and 2 of another).

   (b) Suppose that the hand you are dealt contains three aces, the queen of hearts, and the jack of spades. You discard the two mismatched cards (without replacing in the deck) and are dealt two new cards. What is the probability that your new hand has either four aces or is a full house?

# ANSWERS TO SELECTED EXERCISES

## Section 5.1

**1. (a)** True
   **(b)** False; 81 is not less than 16
   **(c)** False; $5 \in A$. {5} is a set and is not an element of this set.
   **(d)** False; {1, 2, 3, . . . 15} is a subset.
   **(e)** True
   **(f)** False; 0 is not a counting number.

**2. (a)** {1, 2, 3, . . . , 16}   **(c)** { }

**3. (a)** {$x \mid x$ is a counting number less than or equal to 16}
   **(b)** {$x \mid x$ is an even counting number}
   **(c)** {$x \mid x$ is a woman president of the United States}

**4. (a)** (i), (v)   **(c)** (i), (iii), (v)

**5. (a)** True   **(b)** False   **(c)** True
   **(d)** False   **(e)** True

**6. (a)** No   **(c)** Yes

**7. (a)** $\in$   **(b)** $\subseteq$ or $\subset$
   **(c)** $\subseteq$   **(d)** $\notin$

**8. (a)** $R \cup T = \{5, 10, 15, 20\}$   $R \cap T = \{15\}$
   **(c)** $A \cup B = \{0, 10, 100, 1000\}$ $A \cap B = \{10, 100\}$
   **(e)** $A \cup B = \{x, y, z, r, s, t\}$   $A \cap B = \{x, y\}$

**9. (a)** Set of all elements in A but not in $B(A - B)$
   **(b)** Set of all elements in $A$ and $B(A \cap B)$
   **(c)** Set of all elements in $B$ but not in $A(B - A)$
   **(d)** Set of all elements not in $A$ and not in $B(\overline{A} \cap \overline{B})$

**10. (a)** {$a, c, e, g$}   **(c)** {$b, f, h$}   **(e)** {$c$}

**11. (a)** $A \cap D$   **(b)** $B \cap (C \cap D)$
   **(c)** $B \cap (\overline{C} \cap \overline{D})$   **(d)** $A \cap (C \cup D)$

**12. (a)** 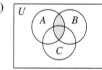   **(c)**

**13. (a)**    **(b)**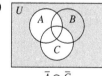
$A \cup B$   $\overline{A} \cap \overline{C}$

   **(c)** 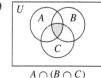   **(d)**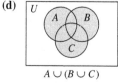
$A \cap (B \cap C)$   $A \cup (B \cup C)$

**14. (a)** Yes; {$\varnothing$} is a set with one element; $\varnothing$ is the empty set.
   **(c)** No   **(e)** No

**15. (a)**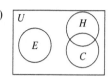

E: elementary teachers
H: high-school teachers
C: college teachers
U: all teachers in Ourtown

   **(b)**

M: math teachers
O: teachers with overhead projector
C: teachers with cassette recorder
U: all teachers at Ourtown High
CK: teachers with chalkboards

**16. (a)** $c, f, g, h$

**17. (a)** {$a$}, {$b$}, {$a, b$}, $\varnothing$
   **(b)** {$a$}, {$b$}, {$c$}, {$a, b$}, {$b, c$}, {$a, c$}, {$a, b, c$}, $\varnothing$
   **(c)** {$a$}, {$b$}, {$d$}, {$a, b$}, {$a, d$}, {$b, d$}, $\varnothing$

**18. (a)** $B - A = \{7, 8\}$

**19. (a)** 8   **(b)** 16   **(c)** 32   **(d)** $2^n$

**20. (a)** $A \cap (\overline{B} \cap \overline{C})$ or $A \cap (\overline{B \cup C})$

**21. (a)**
$A \cap B$   $\overline{A \cap B}$

$\overline{A}$   $\overline{B}$

$\overline{A} \cup \overline{B} = \overline{A \cap B}$

   **(b)** $\overline{A \cup B} = \overline{A} \cap \overline{B}$
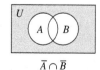
$\overline{A \cup B}$   $\overline{A} \cap \overline{B}$

**23.**

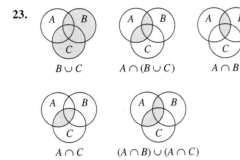

$B \cup C$     $A \cap (B \cup C)$     $A \cap B$

$A \cap C$     $(A \cap B) \cup (A \cap C)$

**24. (a)** $\{a, b\} \cup (\{b, c, d, e\} \cup \{d, e, f\})$
$= (\{a, b\} \cup \{b, c, d, e\}) \cup \{d, e, f\}$
as $\{a, b\} \cup \{b, c, d, e, f\}$
$= \{a, b, c, d, e\} \cup \{d, e, f\}$
as $\{a, b, c, d, e, f\} = \{a, b, c, d, e, f\}$

**25. (a)** Yes, $A \cap B \subseteq A \cup B$
**(b)** Not necessarily; $x$ could be an element in either $A$ or $B$ but not in $A \cap B$.

**27.** $\{A, B, C\}, \{A, B, D\}, \{A, B, E\}, \{A, C, D\}, \{A, C, E\},$
$\{A, D, E\}, \{B, C, D\}, \{B, C, E\}, \{B, D, E\}, \{C, D, E\},$
$\{A, B, C, D\}, \{A, B, C, E\}, \{A, B, D, E\}, \{A, C, D, E\},$
$\{B, C, D, E\}, \{A, B, C, D, E\}$

**28. (a)** $\{A, B\} \{A, C\} \{B, C\} \{A, B, C\} \{A, B, D\} \{A, B, E\}$
$\{A, C, D\} \{A, C, E\} \{B, C, D\} \{B, C, E\} \{A, B, C, D\}$
$\{A, B, C, E\} \{A, B, D, E\} \{A, C, D, E\} \{B, C, D, E\}$
$\{A, B, C, D, E\}$

**29. (a)** $\overline{A} \cap \overline{B}$ or $\overline{A \cup B}$     **(b)** $B \cap \overline{A}$ or $B - A$
**(c)** $A \cap \overline{B}$ or $A - B$

## Section 5.2

**1. (a)** 4     **(b)** {blue, black, yellow, green}
**(c)** $\dfrac{1}{4}$     **(d)** $\dfrac{1}{4}$
**(e)** Yes

**2. (a)** {H, O, P, E, F, U, L}     **(c)** 1/7

**3.** $\dfrac{1}{4}$

**5. (a)** 6     **(b)** {A, B,C, D, E, F}
**(c)** $\dfrac{11}{60}$     **(d)** $\dfrac{1}{6}$
**(e)** Empirical probability only approximates theoretical probability.
**(f)** Possibly. As the number of trials increases, empirical probability should more closely approximate theoretical probability.

**6. (a)** $\dfrac{19}{40}$
**(c)** The answer in (a) is an approximation. The accuracy of this answer will increase with additional tosses of the coin.

**7. (a)** $\{A, B, C\}$     **(b)** $\{R, G\}$
**(c)** $\{1, 2, 3, 4, 5, 6\}$     **(d)** $\{P, R\}$

**8. (a)** $\dfrac{1}{3}$     **(c)** $\dfrac{1}{6}$

**9.** $\dfrac{2}{2}$ or 1

**11. (a)** 5000
**(b)** No, because accuracy increases with the number of tosses

**13.** $\{1, 2, 3, 4, 5, 6\}$
$$P(1) = .181, \quad P(2) = .152,$$
$$P(3) = .144, \quad P(4) = .138,$$
$$P(5) = .156, \quad P(6) = .229$$

**15.** {bass, bream, carp, catfish}
$$P(\text{bass}) = .235; \quad P(\text{bream}) = .327$$
$$P(\text{carp}) = .144; \quad P(\text{catfish}) = .294$$

**16. (a)** No, probability cannot be negative.
**(c)** Yes
**(e)** No, probability cannot be larger than 1.
**(g)** No, probability cannot be larger than 1.

**17. (a)** {HHH, HHT, HTH, HTT, THH, THT, TTH, TTT}
**(b)** 1/8

**18. (a)** $\{(1, 1), (1, 2), (1, 3), (1, 4), (2, 1), (2, 2), (2, 3), (2, 4),$
$(3, 1), (3, 2), (3, 3), (3, 4), (4, 1), (4, 2), (4, 3), (4, 4)\}$

**19.** {H1, H2, H3, H4, H5, H6, T1, T2, T3, T4, T5, T6}

**20. (a)** $\{R, B\}$

**21. (a)** No; neglects 0 heads as an outcome
**(b)** No; neglects 3 heads as an outcome
**(c)** No, 4 is not a possible outcome.
**(d)** Yes

**22. (a)** Yes, but the odd number must be 1 or 3.

**23.** $\dfrac{1}{2}$

**21. (a)**

**(b)** $\overline{A \cup B} = \overline{A} \cap \overline{B}$

 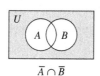

$\overline{A \cup B}$ $\overline{A} \cap \overline{B}$

**23.**

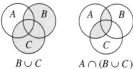 

$B \cup C$ $A \cap (B \cup C)$ $A \cap B$

$A \cap C$ $(A \cap B) \cup (A \cap C)$

**24. (a)** $\{a, b\} \cup (\{b, c, d, e\} \cup \{d, e, f\})$
= $(\{a, b\} \cup \{b, c, d, e\}) \cup \{d, e, f\}$
as $\{a, b\} \cup \{b, c, d, e, f\}$
= $\{a, b, c, d, e\} \cup \{d, e, f\}$
as $\{a, b, c, d, e, f\} = \{a, b, c, d, e, f\}$

**25. (a)** Yes, $A \cap B \subseteq A \cup B$
   **(b)** Not necessarily; $x$ could be an element in either $A$ or $B$ but not in $A \cap B$.

**27.** $\{A, B, C\}, \{A, B, D\}, \{A, B, E\}, \{A, C, D\}, \{A, C, E\},$
$\{A, D, E\}, \{B, C, D\}, \{B, C, E\}, \{B, D, E\}, \{C, D, E\},$
$\{A, B, C, D\}, \{A, B, C, E\}, \{A, B, D, E\}, \{A, C, D, E\},$
$\{B, C, D, E\}, \{A, B, C, D, E\}$

**28. (a)** $\{A, B\} \{A, C\} \{B, C\} \{A, B, C\} \{A, B, D\} \{A, B, E\}$
$\{A, C, D\} \{A, C, E\} \{B, C, D\} \{B, C, E\} \{A, B, C, D\}$
$\{A, B, C, E\} \{A, B, D, E\} \{A, C, D, E\} \{B, C, D, E\}$
$\{A, B, C, D, E\}$

**29. (a)** $\overline{A} \cap \overline{B}$ or $\overline{A \cup B}$   **(b)** $B \cap \overline{A}$ or $B - A$
   **(c)** $A \cap \overline{B}$ or $A - B$

## Section 5.3

**1. (a)** $\dfrac{1}{5}$   **(b)** $\dfrac{2}{5}$

**2. (a)** $\dfrac{5}{11}$

**3. (a)** $\dfrac{1}{2}$   **(b)** $\dfrac{1}{3}$
   **(c)** $\dfrac{2}{3}$   **(d)** $1$
   **(e)** $\dfrac{1}{3}$

**4. (a)** $\dfrac{1}{4}$   **(c)** $\dfrac{1}{52}$   **(e)** $\dfrac{1}{26}$

**5.** $\dfrac{1}{5}, \dfrac{4}{5}$

**6. (a)** Any 1 of the 35 students would be an equally likely outcome.
   **(c)** $\dfrac{1}{35}$   **(e)** $1$

**7. (a)** $\dfrac{2}{7}$   **(b)** $\dfrac{3}{14}$
   **(c)** $\dfrac{1}{2}$   **(d)** $\dfrac{5}{7}$
   **(e)** $\dfrac{11}{14}$   **(f)** $1$

**8. (a)** $\dfrac{1}{2}$   **(c)** $\dfrac{1}{4}$

**9. (a)** Getting less than 2 heads (0 or 1)
   **(b)** Getting 0, 1, 2, 6, or 7 tails
   **(c)** Getting 0, 2, 3, 4, 5, 6, or 7 tails
   **(d)** Getting at least 1 head

**11.** .35

**12. (a)** $\dfrac{1}{2}$   **(c)** $\dfrac{12}{13}$

**13. (a)** $\dfrac{1}{4}$   **(b)** $\dfrac{1}{4}$

**14. (a)** $\dfrac{3}{4}$

**15. (a)** .48   **(b)** .52   **(c)** .1   **(d)** .9

**16. (a)** $\dfrac{31}{125}$

**17.** 1 to 3
**19.** 12 to 1
**21. (a)** .4   **(b)** .8   **(c)** .7   **(d)** .9

**22. (a)** $\dfrac{1}{2}$

**23. (a)** $\dfrac{3}{4}$   **(b)** $\dfrac{3}{5}$
   **(c)** $\dfrac{17}{20}$   **(d)** $\dfrac{7}{10}$

**25.** $P(A) = \dfrac{1}{4}$   $P(\overline{A}) = \dfrac{12}{16} = \dfrac{3}{4}$
   $P(A) = 1 - P(\overline{A}) = 1 - \dfrac{12}{16} = \dfrac{1}{4}$

**26. (a)** .5

**27. (a)** $\dfrac{1}{16}$   **(b)** $\dfrac{6}{16} = \dfrac{3}{8}$
   **(c)** $\dfrac{5}{16}$   **(d)** $\dfrac{4}{16} = \dfrac{1}{4}$

**28. a.**

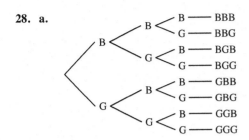

## Section 5.4

**1. (a)** $\dfrac{5}{18}$          **(b)** $\dfrac{5}{18}$

**2. a.**

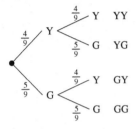

**3.** $-\$.55$

**5.** $\dfrac{2}{6}$ T $\dfrac{1}{5}$ E $\dfrac{2}{4}$ L $\dfrac{1}{3}$ L

  $P(TELL) = \dfrac{1}{90}$

**7.** $\dfrac{15}{56}$

**9.** $\dfrac{1}{256}$

**10. (a)** $\dfrac{4}{77}$

**11. (a)** $\dfrac{6}{121}$          **(b)** $\dfrac{4}{121}$

**12. (a)** $\dfrac{1}{169}$

**13.** Starting with the first entry in Table 5.7 and using 0–4 as a head and 5–9 as a tail:

|  | Frequency | Empirical Probability |
|---|---|---|
| HHH | 4 | .08 |
| HHT | 4 | .08 |
| HTH | 4 | .08 |
| HTT | 10 | .20 |
| THH | 9 | .18 |
| THT | 8 | .16 |
| TTH | 8 | .16 |
| TTT | 3 | .06 |

**14. (a)** One or 2 represent doll 1; 3 or 4 represent doll 2; 5 or 6 represent doll 3. Roll 5 times.

**15. (a)** If an even number is rolled on the die, the coin shows "heads." If an odd number is rolled, the coin shows "tails."
  **(b)** Roll the die 5 times, letting an even number represent a head and an odd number a tail.
  **(c)** Let 1 and 2 represent a hit and 3, 4, 5, and 6 a miss; roll the die 3 times for the 3 times at bat.
  **(d)** Let an odd number represent a boy and an even number a girl; roll the die 4 times for 4 children.

**16. (a)** If an even-numbered card is drawn, the coin shows "heads"; if an odd-numbered card is drawn, the coin shows "tails."

**17. a.**

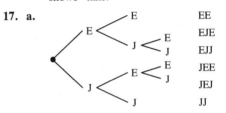

  **(b)** .52   **(c)** .352   **(d)** .648

**18. a.**

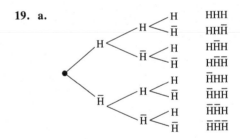

**19. a.**

H < H̄ < H HHH / HH̄ etc.

  **(b)** .027   **(c)** .216   **(d)** .343

**21.** $\dfrac{21}{95}$

**23.** $5\dfrac{5}{16}$ min.

**25.** .0288

**27. (a)** $\dfrac{13}{18}$          **(b)** $\dfrac{5}{9}$

**28. (a)** $\dfrac{65}{81}$

**29. (a)** $\dfrac{10}{13}$   **(b)** 0   **(c)** 1

  **(d)** $\dfrac{4}{13}$   **(e)** $\dfrac{9}{13}$   **(f)** $\dfrac{3}{13}$

  **(g)** 1   **(h)** $\dfrac{10}{13}$

## Section 5.5

**1.** **(a)** 6 ways

**(b)**

| | Atlanta | Boston | London |
|---|---|---|---|

**2.** **(a)** $2 \cdot 4 \cdot 4 = 32$ ways **(c)** $\dfrac{1}{32}$

**3.** **(a)** 24 **(b)** 576

**5.** **(a)** 2 **(b)** 6 **(c)** 24

**6.** **(a)** 24 **(c)** 1

**7.** **(a)** 6 **(b)** *PQR; PRQ; QPR; QRP; RPQ; RQP*

**8.** **(a)** 12

**9.** 18

**10.** **(a)** 60 **(c)** 8 **(e)** 42

**11.** **(a)** $r$ **(b)** $k^2 - k$

**(c)** $r!$ **(d)** $\dfrac{k!}{2}$

**(e)** $k(k - 1)(k - 2)$ **(f)** $\dfrac{k!}{6}$

**13.** **(a)** 10,000 **(b)** 5040

**14.** **(a)** 900

**15.** **(a)** $\dfrac{1}{14}$ **(b)** $\dfrac{1}{6}$

**(c)** $\dfrac{1}{35}$ **(d)** $\dfrac{3}{7}$

**17.** $\dfrac{1}{10^8}$

**19.** **(a)** $\dfrac{1}{2}$ **(b)** $\dfrac{3}{10}$

**(c)** $\dfrac{1}{5}$ **(d)** $\dfrac{3}{5}$

**(e)** $P(A \cup B) = P(A) + P(B) - P(A \cap B)$

**20.** **(a)** $\dfrac{1}{169}$

**21.** $-3$ points

**22.** **(a)** $\dfrac{13}{51}$

## Section 5.6

**1.** **(a)** 6: *WX,WY,WZ, XY, XZ, YZ*

**(b)** 12: *WX, WY, WZ, XW, XY, XZ, YW, YX, YZ, ZW, ZX, ZY*

**(c)** $12 = 2 \cdot 6$

**(d)** 4: *WXY, WXZ, WYZ, XYZ*

**(e)** 24: *WXY, WYX, WXZ, WZX, WYZ, WZY, XWY, XYW, XWZ, XZW, XYZ, XZY, YWX, YXW, YWZ, YZW, YXZ, YZX, ZWX, ZXW, ZWY, ZYW, ZXY, ZYX*

**(f)** $24 = 6 \cdot 4$

**3.** $P(12, 4) = 11,880$

**5.** $P(12, 5) = 95,040$

**7.** $C(22, 3) = 1540$

**9.** 133,784,560

**11.** **(a)** 1716 **(b)** 100,386

**(c)** 725,010 **(d)** 1,740,024

**13.** 30

**14.** **(a)** $\dfrac{13}{55}$

**15.** $\dfrac{7}{95}$

**16.** **(a)** 24

**17.** $\dfrac{1}{33}$

**18.** **(a)** $\dfrac{1}{54,145} = .0000185$

**19.** **(a)** $\dfrac{5}{26}$ **(b)** $\dfrac{3}{26}$

**(c)** $\dfrac{1}{13}$

**20.** **(a)** $\dfrac{1}{12}$

**21.** $\dfrac{7}{20}$

## Section 5.7

**1.** $P(E \mid F) = \frac{3}{7}, P(F \mid E) = \frac{1}{2}$

**3.** $P(E \mid F) = \frac{1}{4}, P(F \mid E) = \frac{1}{2}$

**5.** $P(E \mid F) = \frac{16}{33}, P(F \mid E) = \frac{4}{17}$

**7.** **(a)** $\frac{3}{8}$ **(b)** $\frac{7}{10}$

**9.** **(a)** $\frac{7}{15}$ **(b)** $\frac{3}{10}$

**(c)** $\frac{3}{14}$

**11.** **(a)** 0.0825 **(b)** 0.523

**(c)** 0.292 **(d)** 0.177

**(e)** 0.4 **(f)** 0.44

**13.** **(a)** $\frac{1}{169}$ **(b)** $\frac{1}{16}$

**(c)** $\frac{1}{4}$

**15.** **(a)** $\frac{1}{2}$ **(b)** $\frac{1}{1024}$

**17.** **(a)** 0.28 **(b)** 0.06

**(c)** 0.18 **(d)** 0.82

**(e)** 0.9 **(f)** $\frac{1}{3}$

19. $\frac{1}{5}$

21. (a) $\frac{724}{1009}$    (b) $\frac{333}{1009}$

    (c) $\frac{51}{263}$    (d) $\frac{171}{724}$

    (e) $\frac{383}{1009}$    (f) $\frac{117}{333}$

23. (a) $\frac{1}{3}$    (b) $\frac{1}{3}$

    (c) $\frac{2}{3}$

25. (a) $\frac{1}{60}$    (b) $\frac{1}{10}$

    (c) $0$    (d) $\frac{1}{10}$

    (e) $\frac{1}{10}$    (f) $\frac{3}{5}$

27. (a) $\frac{1}{4}$    (b) $\frac{3}{4}$

# The Uses and Misuses of Statistics

The management of Acme Manufacturing published an average annual salary for 50 employees of $19,280. The Trainsters Union, trying to organize the Acme plant, published an average annual salary of $14,000. Some of the employees made an informal survey and published an average annual salary of $16,000. The interesting point is that all were correct. How can this be?

When Aunt Jane asserts that smoking is not harmful to a person's health because Uncle Joe lived to be 88 years of age and smoked 2 packs of cigarettes every day of his adult life, Aunt Jane is using statistical thinking; that is, she has organized the data of her experience (Uncle Joe) and then made a statement on the basis of her data. However, she lacks an understanding of how many data are needed, how the data should be organized, and what conclusions are appropriate or inappropriate relative to the data. This chapter should help you avoid making the types of mistakes that Aunt Jane made.

Statistics can be divided into two subdivisions: in descriptive statistics, techniques are used to summarize and describe the characteristics of a set of data, and inferential statistics, generalizations or conclusions are made about the data in a large group (called a population) from a small portion (called a sample).

In this chapter, we focus on descriptive statistics: understanding a set of data by forming frequency distributions, drawing associated graphs, finding measures of central tendency (mean, median, and mode), and finding measures of the scattering of data.

## 6.1 FREQUENCY DISTRIBUTIONS AND GRAPHICAL REPRESENTATIONS

**Problem**  Line graphs are often used to make predictions about the future. Based on the trend for the years 1980 to 1985, estimate the world population in the year 2000.

**Overview**  We are immersed daily in a torrent of numerical data flowing in bubbling splendor from our televisions, our radios, our newspapers, our hair stylist, and our favorite Uncle Al. Although such information is part of our daily lifestyle, we are often unsure of how to organize, interpret, or understand the messages it conveys. In this section, we consider how to organize and summarize data for a better understanding of statistics.

First, we discuss some basic techniques for classifying and summarizing a set of observed measurements (data). Then we represent the data with graphs.

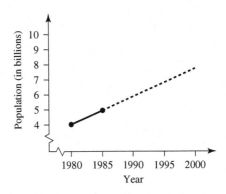

**Goals**    Collect, organize, and describe data.
Construct, read, and interpret displays of data. (Standard 11, K–4)
Construct, read, and interpret tables, charts, and graphs. (Standard 10, 5–8)
    Representations of data should include various types of graphs, including bar, line, circle, . . . as well as line plots, stem and leaf plots, box plots, histograms, and scatter plots. (*Teaching Standards,* p. 136)

## Organizing Data

The first objective of a statistician is to develop a plan to collect data for study. After the data have been collected, the second objective is to make sense of this large mass of information. Suppose that you have collected the numbers shown in Table 6.1. The table contains your tabulation of the number of colds experienced during one winter by each of a group of 30 elementary-school children.

**Table 6.1**

| 7 | 1 | 1 | 0 | 3 | 4 | 5 | 5 | 3 | 2 | 3 | 3 | 6 | 6 | 2 |
| 4 | 2 | 1 | 0 | 0 | 3 | 4 | 5 | 6 | 3 | 1 | 4 | 1 | 3 | 4 |

A quick glance at Table 6.1 reveals little about what the numbers imply about the group of children represented in the data. Closer observation indicates that the largest number of colds experienced was 7 and the smallest number experienced was 0. The difference between the largest and smallest entries in the data is called the **range** of the data. In this case, the range is $7 - 0 = 7$.

Table 6.1 allows us to make only general observations because the numerical values have not been organized. Using a **frequency distribution table** to help organize the information enables us to uncover more meaning. To summarize data with a frequency distribution, we record the number of students who reported each number of colds. By the **frequency** of a number, we mean the number of times a number occurs in a set of data.

## EXAMPLE 1

Make a frequency distribution for the data in Table 6.1.

**Solution**

From the summary in Table 6.2, it is obvious that 3 colds was the number most often reported. The summary also shows how the number of colds was distributed among the 30 students.

**Table 6.2**

| Number of Colds | Tally | Frequency |
|:---:|:---:|:---:|
| 0 | ||| | 3 |
| 1 | Ж | 5 |
| 2 | ||| | 3 |
| 3 | Ж || | 7 |
| 4 | Ж | 5 |
| 5 | ||| | 3 |
| 6 | ||| | 3 |
| 7 | | | 1 |
| | | Total   30 |

When many numerical values are involved, a frequency distribution may become cumbersome. In this case we choose **intervals** (or **classes**) of data and construct a grouped frequency distribution table for them.

In a **grouped frequency distribution,** we cover the range of data by intervals of equal length and record the number of values that fall into each interval. Constructing a grouped frequency distribution involves four main steps.

---

Steps in Making a Grouped Frequency Distribution

1. Find the range.
2. Choose the number and size of the classes into which the information should be grouped.
3. Find class limits.
4. Count the values in each class. This number is the frequency of the class.

---

## EXAMPLE 2

Table 6.3 gives the lengths of engagements (in months) of 30 newly married students. Construct a grouped frequency distribution for these values.

**Solution**

The range of the data is $18 - 1 = 17$. We arbitrarily select six classes for our grouping. Since $17 \div 6$ is 2.833, the length of the classes (if the classes are to be of equal length) must be more than 2.833 in order to include all the data in six classes. Whenever feasible, classes should be of equal integral length. Thus, we arbitrarily select the following class limits: 1–3, 4–6, 7–9, 10–12, and so on. The grouped frequency distribution is found in Table 6.4.

**Table 6.3**

| Lengths of Engagements | | | | |
|:---:|:---:|:---:|:---:|:---:|
| 10 | 2 | 9 | 6 | 11 |
| 17 | 4 | 10 | 7 | 3 |
| 1 | 4 | 11 | 6 | 3 |
| 8 | 15 | 12 | 9 | 12 |
| 8 | 18 | 12 | 6 | 10 |
| 8 | 18 | 12 | 6 | 9 |

**Table 6.4**

| Class Intervals | Tallies | Frequency |
|:---:|:---:|:---:|
| 1–3 | |||| | 4 |
| 4–6 | Ж | | 6 |
| 7–9 | Ж || | 7 |
| 10–12 | Ж |||| | 9 |
| 13–15 | | | 1 |
| 16–18 | ||| | 3 |

In Table 6.4, the class interval 7–9 includes all measurements between 6.5 and 9.5. Thus, an engagement of 6.7 months would be placed in the class 7–9, but an engagement of 9.6 months would go in the class 10–12. The numbers that fall halfway between class limits (.5, 3.5, 6.5, 9.5, 12.5, 15.5, and 18.5) are called **class boundaries.**

The **length** of a class interval can be found by taking the difference between the boundaries of a class. In Table 6.4 the length of the class intervals is $3.5 - .5 = 3$. The middle value of a class interval is called a **class mark.** It can be computed as the mean of the class boundaries of the class interval. Thus the class marks for the lengths of engagements in Table 6.4 are 2, 5, 8, 11, 14, and 17.

## Stem and Leaf Plots

A grouped frequency distribution allows us to summarize the information in a set of data by replacing many values by a few representative values. The **stem and leaf plot** provides a way of summarizing data that does not lose the individual values in the data set. To illustrate the stem and leaf plot, let's consider the grades obtained by two classes who were tested on the material. These grades are given in Table 6.5.

In these stem and leaf plots, the first digit serves as the stem, and the second digit as the leaf. For example, the stem of the 46 in Class I is 4, and the leaf is 6. Likewise, 56 and 58 have stems of 5 and leaves of 6 and 8, respectively. The data for Class I are listed with six stems—4, 5, 6, 7, 8, and 9—and with appropriate leaves in Table 6.6. In Table 6.7, the same leaves are arranged in increasing order.

Now let's compare Class I and Class II, using the same stems. In Table 6.8, the leaves of Class I increase from left to right, and the leaves of Class II increase from right to left.

A quick inspection of how the leaves increase in Class I and in Class II suggests that the students in Class II did better on this test.

**Table 6.5**

| Class I | Class II |
|---|---|
| 56, 64, 72, 73, 84, | 99, 81, 50, 64, 76, |
| 98, 80, 86, 75, 68, | 63, 71, 78, 81, 92, |
| 46, 78, 75, 91, 63, | 87, 79, 74, 60, 68, |
| 84, 79, 69, 76, 58 | 92, 84, 86, 65, 78 |

**Table 6.6**

| | Class I |
|---|---|
| Stems | Leaves |
| 4 | 6 |
| 5 | 6, 8 |
| 6 | 4, 8, 3, 9 |
| 7 | 2, 3, 5, 8, 5, 9, 6 |
| 8 | 4, 0, 6, 4 |
| 9 | 8, 1 |

**Table 6.7**

| | Class I |
|---|---|
| Stems | Leaves |
| 4 | 6 |
| 5 | 6, 8 |
| 6 | 3, 4, 8, 9 |
| 7 | 2, 3, 5, 5, 6, 8, 9 |
| 8 | 0, 4, 4, 6 |
| 9 | 1, 8 |

**Table 6.8**

| Class II | | Class I |
|---|---|---|
| Leaves | Stems | Leaves |
| | 4 | 6 |
| 0 | 5 | 6, 8 |
| 8, 5, 4, 3, 0 | 6 | 3, 4, 8, 9 |
| 9, 8, 8, 6, 4, 1 | 7 | 2, 3, 5, 5, 6, 8, 9 |
| 7, 6, 4, 1, 1 | 8 | 0, 4, 4, 6 |
| 9, 2, 2 | 9 | 1, 8 |

---

Steps in Making Stem and Leaf Plot

1. Decide on the number of digits in the data to be listed under stems (1-digit, 2-digit, or 3-digit numbers).

2. List the stems in a column, from least to greatest.

3. List the remaining digits in each data entry as leaves. (You may wish to order these values from smallest to largest.)

It is always a good idea to describe or explain a stem and leaf plot with a heading. Likewise, it is a good idea to explain the notation. For example, 17 6 in Table 6.9 represents a weight of 176.

**Table 6.9** Weights of Members of the Basketball Team

| Stems | Leaves |
|-------|--------|
| 15 | 8 |
| 16 | 4, 7 |
| 17 | 5, 6, 8 |
| 18 | 3, 4, 6, 8 |
| 19 | 4, 5 |
| 20 | 3 |

17 | 6 represent 176 lb.

## Bar Graphs

We have seen throughout this text that we can greatly improve our understanding and our problem-solving ability if we can draw a graph, picture, or diagram. There are several ways to represent graphically a conglomeration of data. One such representation is a **bar graph.** To construct a bar graph, first construct a frequency distribution or a grouped frequency distribution, whichever is appropriate. Then plot the frequencies on the vertical axis and the data values or intervals on the horizontal axis. Finally, draw a bar to show the relationship between the values and the frequencies.

## EXAMPLE 3

Draw a bar graph that represents the number of graduates each May as shown in Table 6.10.

**Table 6.10**

| May of Year | 1980 | 1981 | 1982 | 1983 | 1984 | 1985 | 1986 | 1987 | 1988 | 1989 | 1990 | 1991 |
|-------------|------|------|------|------|------|------|------|------|------|------|------|------|
| Number of Graduates | 152 | 163 | 197 | 185 | 201 | 196 | 210 | 189 | 195 | 205 | 200 | 180 |

**Solution**

Figure 6.1 is a bar graph of the data in Table 6.10. Notice that the number of graduates is measured on the vertical axis and that the years are given on the horizontal axis. The break in the vertical axis, denoted by ⌇, indicates that the scale is not accurate from 0 to 150. The height of each bar represents the number of students who graduated in a given year. To determine from the bar graph the number of students who graduated in 1983, locate the bar labeled 1983 and draw a horizontal line from the top of the bar to the vertical axis. The point where this horizontal line meets the vertical axis identifies the number of graduating students. Thus, about 185 students graduated in 1983.

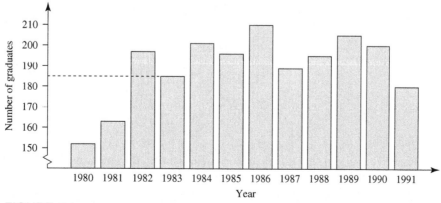

*FIGURE 6.1*

A **line graph** does a better job of showing fluctuations and emphasizing changes in the data than does a bar graph. The line graph in Figure 6.2 represents the distance (in meters) run in 6 min. by a group of freshmen in a physical education class. Looking at this graph, you can readily see the variations in the numbers who ran given distances in 6 min.

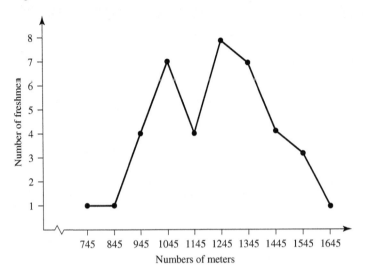

*FIGURE 6.2*

## Histogram

A bar graph representing a grouped frequency distribution is called a **histogram.** To construct a histogram, we first construct a grouped frequency distribution. Then we represent each interval with a bar or a rectangle. The height of the rectangle indicates the frequency of the interval. To label each rectangle, we use the interval that the rectangle represents or the midpoint of the interval, called the *class mark.* Usually, the rectangles touch at a point halfway between each pair of class limits so that there are no gaps in the histogram. In Figure 6.3(c) the rectangles touch at 19.5, 24.5, 29.5, and 34.5. These are termed the *class boundaries*.

**EXAMPLE 4**

Draw a histogram of the data in Table 6.11.

**Table 6.11**

| Class Intervals | Tallies | Frequency |
|---|---|---|
| 15–19 | IIII | 4 |
| 20–24 | IIII II | 7 |
| 25–29 | IIII | 5 |
| 30–34 | II | 2 |
| 35–39 | II | 2 |

**Solution**

Figures 6.3(a), (b), and (c) are identical representations of the histogram for these data except that in (a) each rectangle is labeled according to its class interval, in (b) each rectangle is labeled according to the midpoint of its class interval (the class mark), and in (c) each rectangle is labeled by class boundaries.

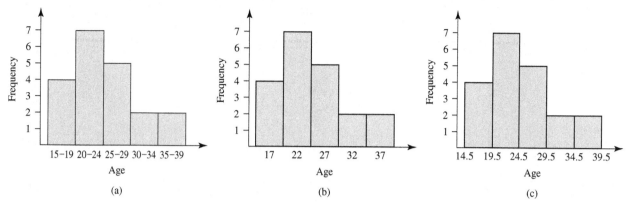

*FIGURE 6.3*

## Frequency Polygon

When a line graph is used to represent a grouped frequency distribution, it is sometimes called a **frequency polygon.** To draw a frequency polygon, we plot the midpoints (class marks) of the intervals versus the frequency of the intervals, and then we connect the resulting points with straight-line segments. Finally, we connect the first and last class marks to points on the horizontal axis that are located one interval beyond these marks.

## EXAMPLE 5

Draw a frequency polygon for the data in Table 6.11.

**Solution**

Figure 6.4 presents a frequency polygon for the grouped frequency distribution of Table 6.11.

*Practice Problem*

Draw a frequency polygon for the data in Table 6.12.

**Answer**

Figure 6.5 shows the required polygon.

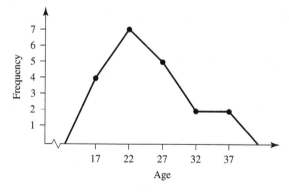

*FIGURE 6.4*

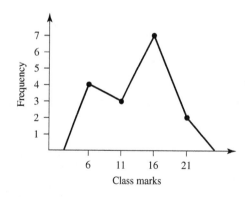

*FIGURE 6.5*

**Table 6.12**

| Class | Frequency |
|-------|-----------|
| 4–8   | 4         |
| 9–13  | 3         |
| 14–18 | 7         |
| 19–23 | 2         |

## Circle Graph

One of the simplest types of graphs is the **circle graph,** sometimes called a **pie chart.** It consists of a circle partitioned into sectors, each of which represents a percentage of the whole.

## EXAMPLE 6

Table 6.13 records examination grades in a class. Represent the data with a circle graph.

**Table 6.13**

| Final Examination Grades | Frequency |
|--------------------------|-----------|
| A                        | 4         |
| B                        | 15        |
| C                        | 36        |
| D                        | 3         |
| F                        | 2         |
|                    Total | 60        |

**Solution**

The circle graph in Figure 6.6 is a visual representation of the data. The largest percentage of the class made C's; in fact, more than half of the class made C's.

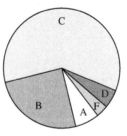

Final examination grades

*FIGURE 6.6*

In constructing the circle graph shown in Figure 6.6, we use a protractor to obtain angle measurements. Because 36 of 60 or $\frac{36}{60}$ or 60% of the students made C's, the sector representing C's has an angle of $0.60(360°) = 216°$. Similarly, because $\frac{1}{15}$ of the grades were A's, the section representing A's encompasses $(\frac{1}{15})(360°) = 24°$. The remaining sections are constructed in the same way.

## Misuses of Statistics

Throughout this chapter, we shall discuss the misuses of statistics. In fact, it is at least as important for you to be able to recognize the incorrect use of statistical concepts as it is to be able to use them correctly yourself. Suppose that Senator Cloghorn has announced that 70% of the people in his state oppose a 10¢ per gallon increase in the tax on gasoline. Here are some of the questions we should ask before accepting the validity of the senator's statistic:

1. How many people were surveyed?
2. How were those who participated in the survey selected?
3. Did Senator Cloghorn do a random survey? That is, did each adult in the state have an equal opportunity to be selected as part of the survey?
4. What proportion of people asked to participate actually responded to the survey?
5. How was the question stated?

Suppose that those surveyed were limited to individuals who had contributed to Senator Cloghorn's campaign. The thinking of this group would tend to be similar to that of the senator and in any case would not necessarily reflect the thinking of the population as a whole. Suppose that only 30% responded to the survey. Then you might be getting a sample of only those who felt strongly about the issue. Before accepting a statistic as completely accurate, ask some questions about the procedure used to obtain the statistic.

Many statistical ideas are communicated with graphs and charts. You need to be able to recognize when these visual representations have been constructed to misrepresent and to confuse. One way that erroneous conclusions can be suggested by grouped frequency distributions is through the use of unequal intervals. Table 6.14 is a retabulation of the data from Table 6.3 in Example 2. This table suggests that the most common length of engagement is in the interval 6–9. This is misleading. Why?

**Table 6.14**

| Class | Tallies | Frequency |
|-------|---------|-----------|
| 1–3   | IIII    | 4         |
| 4–5   | II      | 2         |
| 6–9   | JHT JHT I | 11      |
| 10–12 | JHT IIII | 9        |
| 13–15 | I       | 1         |
| 16–18 | III     | 3         |

Another misuse of statistics involves graphs. By changing scales on graphs, it is possible to create false impressions. For instance, although Figures 6.7(a) and (b) represent exactly the same data, (b) has been drawn to conceal the fact that the second interval has at least twice the frequency of the other intervals.

There are several ways to manipulate a line graph to produce a misleading impression. Compare the graph shown in Figure 6.8(a) with the graphs shown in (b) and (c). In (b), the vertical axis has been stretched, making the graph appear steeper. In (c), the horizontal axis has been stretched so that the graph appears flatter. The effect on the graph is especially dramatic if one of the axes is stretched while the other is compressed.

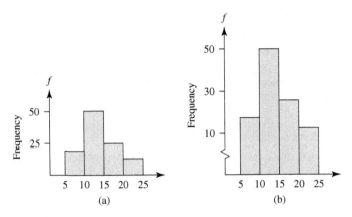

FIGURE 6.7

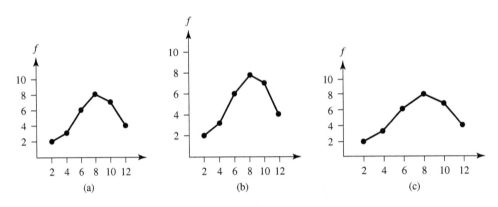

FIGURE 6.8

# EXERCISES

1. In the given pie chart or circle graph,

    (a) what percent are professionals?

    (b) what percent are crafts workers?

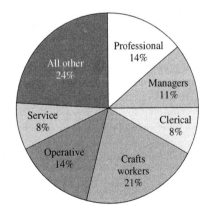

(c) what percent are managers or clerical workers?

(d) what percent are neither managers nor professionals?

2. In a transportation survey, bus riders on the Friday evening run were asked how many times they had ridden the bus that week. Summarize the data in a frequency distribution.

| | | | |
|---|---|---|---|
| 4 | 8 | 6 | 4 |
| 7 | 2 | 2 | 8 |
| 2 | 5 | 8 | 1 |
| 7 | 9 | 8 | 3 |
| 8 | 2 | 4 | 8 |
| 10 | 3 | 3 | 9 |

3. The accompanying bar graph shows changes in population for the four largest cities in the United States.

    (a) Which city increased in population between 1970 and 1984?

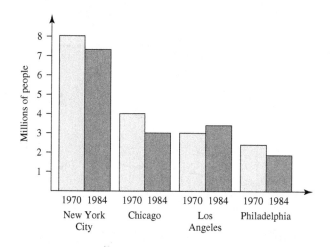

**(b)** In which city did the population decrease the most?

**(c)** What was the city with the second largest population in 1984?

**4.** Following is a tabulation of the ages of mothers of the first babies born in Morningside Hospital in 1994:

| Class | Tally | Frequency |
|-------|-------|-----------|
| 15–19 | IIII | 4 |
| 20–24 | ЖІ II | 7 |
| 25–29 | ЖІ | 5 |
| 30–34 | II | 2 |
| 35–39 | II | 2 |

**(a)** What is the number of mothers in the tabulation?

**(b)** What is the number of mothers who were younger than 30?

**(c)** What is the number of mothers who were at least 20 years of age?

**5.** The following table lists how many students had a specific number of absences in a given semester.

| Number of Absences | Frequency |
|--------------------|-----------|
| 0 | 25 |
| 1 | 18 |
| 2 | 20 |
| 3 | 31 |
| 4 | 34 |
| 5 | 14 |
| 6 | 13 |
| 7 | 12 |
| 8 | 8 |
| 9 | 3 |
| 10 | 1 |

**(a)** Display these data with a bar graph.

**(b)** Represent these data with a line graph.

**6.** The following stem and leaf plot records the distances in feet that 22 children in a recreation program could throw a softball.

**(a)** Write the distances represented in the stem and leaf plot.

**(b)** What is the shortest throw?

**(c)** How many of the throws traveled more than 60 ft?

Lengths of Softball Throws

| | |
|---|---|
| 3 | 1, 7 |
| 4 | 1, 2, 6 |
| 5 | 1, 2, 3, 3, 7, 8, 9 |
| 6 | 2, 8, 8, 9 |
| 7 | 1, 3, 6 |
| 8 | 2, 2 |
| 9 | 6 |

**7.** Alabaster College has 1426 students. The accompanying table classifies them by age.

**(a)** Find the class marks.

**(b)** Present the data as a frequency polygon.

**(c)** Present the data as a histogram.

| Age | Number of Students |
|-----|--------------------|
| 15–19 | 562 |
| 20–24 | 450 |
| 25–29 | 350 |
| 30–34 | 58 |
| 35–39 | 6 |

**8.** The accompanying circle graphs record the contributions to federal candidates for office in the late 1970s.

To Democratic candidates

$21.5 Million = 100%

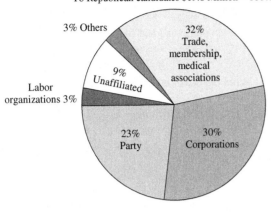

To Republican candidates $19.8 Million = 100%

(a) Compare the percent of contributions to Democratic and Republican candidates that came from labor organizations; repeat the comparison for contributions from corporations.

(b) What dollar amount of support for Democratic candidates came from the party?

(c) What dollar amount of support for Republican candidates came from corporations?

9. Comment on the following misuses of statistics.

(a) More people die in hospitals than at home. There are severe deficiencies in our medical care system.

(b) More people died in accidents in the United States last year than died in fighting in World War II. Therefore, it is more dangerous to live in the United States today than it was to fight in World War II.

(c) Professor Jones gave more A's last semester than Professor Smith. Therefore, you should enroll next semester in Professor Jones's class.

(d) Most automobile accidents occur near home. Therefore, you are much safer on a long trip.

10. (a) Make a bar graph comparing the number of students in a freshman mathematics course majoring in various academic fields, as recorded in the accompanying table.

| Academic Fields | Number of Students |
|---|---|
| Business administration | 110 |
| Social sciences | 100 |
| Life sciences | 60 |
| Humanities | 30 |
| Physical sciences | 60 |
| Elementary education | 140 |

(b) Display the data given in part (a), using a circle graph.

(c) Draw a line graph for these data.

*Exercises 11 through 14 refer to the following line graph.*

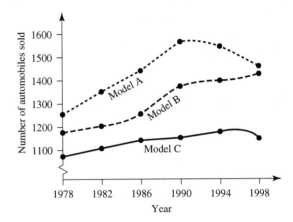

11. In 1998, how many more model A cars were sold than model B cars?

12. In which 4-year period did model C have the greatest decrease in sales? What was the decrease?

13. In which 4-year period did model B have the greatest increase in sales? What was the increase?

14. Compare the increase in sales of the 3 models from 1982 to 1998.

15. Park officials sought to analyze public use of a municipal park. One evening, 36 people were interviewed and their ages recorded.

| | | | | | |
|---|---|---|---|---|---|
| 7 | 18 | 35 | 73 | 18 | 28 |
| 15 | 19 | 41 | 61 | 16 | 24 |
| 51 | 65 | 12 | 65 | 61 | 26 |
| 16 | 62 | 14 | 73 | 72 | 48 |
| 17 | 59 | 16 | 62 | 43 | 68 |
| 21 | 16 | 17 | 19 | 32 | 72 |

Summarize the resulting data in a stem and leaf plot. Arrange the numbers in increasing order on each stem. What general observations can you make about the persons who use the park?

16. Summarize the data from Exercise 15 in a grouped frequency distribution with seven intervals. Let the first interval be 5 to 14.

17. Use a protractor to construct a pie chart showing the percent of women who work in various occupations. The percents are as follows:

| | |
|---|---|
| Professional | 16% |
| Managers | 4% |
| Clerical | 35% |
| Crafts | 2% |
| Operative | 14% |
| Service | 17% |
| All other | 12% |

**18.** A small country exports 20 main products each year, ranging from iron ore to toy medical kits to surgical instruments. The value of each export (in millions of dollars) is given in the accompanying table. Group the values into 6 intervals of minimum whole-number length, starting the first class at 60.

| | | | | |
|---|---|---|---|---|
| 86 | 62 | 239 | 290 | 207 |
| 285 | 232 | 214 | 131 | 195 |
| 424 | 343 | 476 | 140 | 398 |
| 363 | 348 | 156 | 222 | 370 |

**19.** The amounts (rounded to the nearest $1) that a sample of 50 freshmen spent on textbooks per class during a fall semester are listed in the accompanying table.

| | | | | | | | | | |
|---|---|---|---|---|---|---|---|---|---|
| 43 | 51 | 45 | 63 | 52 | 57 | 51 | 41 | 48 | 47 |
| 40 | 48 | 47 | 43 | 51 | 45 | 52 | 60 | 51 | 48 |
| 49 | 52 | 51 | 50 | 50 | 48 | 47 | 51 | 55 | 58 |
| 45 | 46 | 45 | 48 | 43 | 59 | 50 | 50 | 57 | 48 |
| 47 | 48 | 47 | 44 | 45 | 54 | 54 | 56 | 50 | 49 |

(a) Find the range.
(b) Create a grouped frequency distribution for this information using 6 intervals of minimum whole-number length, with the first class beginning at 40.
(c) Find the class marks.
(d) Draw a frequency polygon.

**20.** Make a stem and leaf plot for the data in Exercise 19.

**21.** Given the following histogram, tabulate a frequency distribution and find the class marks.

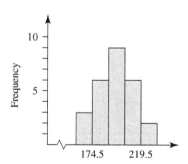

**22.** Given the following frequency polygon, tabulate a grouped frequency distribution.

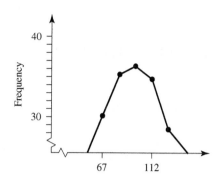

**23.** The new vice-president of the Seashore Oil Company claims that production has doubled during the first 12 months of his tenure. To present this fact to the board of directors, he has the following graph prepared in which both the height and width of the barrel are doubled. What is misleading about this graph? (*Hint:* Think in terms of volume.)

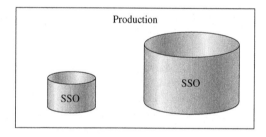

**24.** Two treatments are given for a disease, both to men and to women. The fraction cured, calculated as (number cured)/(number treated), is written as a decimal in the accompanying table.

| | Treatment $X$ | Treatment $Y$ |
|---|---|---|
| **Men** | $\frac{20}{100} = 0.20$ | $\frac{50}{210} = 0.24$ |
| **Women** | $\frac{40}{60} = 0.67$ | $\frac{15}{20} = 0.75$ |

Glancing at the table, would you say that Treatment $Y$ is better than Treatment $X$? Be careful! Find the total number cured by each treatment divided by the number treated. Now which treatment is better? Explain!

# 6.2 WHAT IS AVERAGE?

**Problem** Sandra receives grades of 69, 71, 78, 82, and 73 on her 5 tests in Math 102.
    She gives her average grade as 74.6, but her friend Sam claims that Sandra's average is 73. Which average is correct?

**Overview** In this section, we learn that both averages are correct. Sandra found the mean, and Sam the median.

In the preceding section, we used stem and leaf plots, histograms, frequency polygons, circle graphs, and other graphs to summarize and explain sets of data. Sometimes, however, we need a more concise procedure for characterizing a set of data. Specifically, we want a simple number that estimates the location of the center of the set of data. Therefore, in this section we introduce averages, which are measures of central tendency. Three measures are in general use: the arithmetic mean, the median, and the mode. The fact that these three (as well as others) exist often invites misuses of statistics. One measure may be quoted, but the reader may automatically assume another. When a measure of central tendency is quoted, immediately ask the question, "Which one?"

**Goals**  Key statistical concepts include measures of central tendency, measures of variation . . . , and general distributions. (*Teaching Standards*, p. 136)

## Finding Means

The most widely used measure of central tendency is the *arithmetic mean* (sometimes called the *arithmetic average*). The arithmetic mean of a set of $n$ measurements is the sum of the measurements divided by $n$.

**DEFINITION**

**Arithmetic Mean**

Consider $n$ measurements $x_1, x_2, x_3, \ldots, x_n$. The formula for the *arithmetic mean*, denoted by $\bar{x}$, is

$$\bar{x} = \frac{x_1 + x_2 + x_3 + \ldots + x_n}{n}$$

Mathematicians have developed a very useful notation to use in expressing complicated sums. When the Greek letter sigma, $\Sigma$, occurs in a mathematical expression, it means "add the indicated terms." For instance, the sum $x_1 + x_2 + x_3 + \ldots + x_n$ can be represented as $\Sigma_{i=1}^{n} x_i$. The index $i = 1$ at the bottom of $\Sigma$ and the $n$ at the top indicate that the $x$ terms should be added starting with the first one and stopping at the $n$th one. Using this notation, the formula for the mean can be expressed as follows:

$$\bar{x} = \frac{\sum\limits_{i=1}^{n} x_i}{n} \quad \text{or} \quad \bar{x} = \frac{\Sigma x}{n}$$

## EXAMPLE 1

Find the arithmetic mean of 8, 16, 4, 12, and 10.

**Solution**

$$\bar{x} = \frac{8 + 16 + 4 + 12 + 10}{5} = 10$$

## EXAMPLE 2

Find the arithmetic mean of 25, 25, 25, 25, 30, 30, 30, 40, 40, 40, 40, and 50.

**Solution**

$$\bar{x} = \frac{25 + 25 + 25 + 25 + 30 + 30 + 30 + 40 + 40 + 40 + 40 + 50}{12}$$

$$= \frac{25(4) + 30(3) + 40(4) + 50(1)}{4 + 3 + 4 + 1} = \frac{400}{12} = 33\frac{1}{3}$$

In the preceding example, observe that the 4, 3, 4, and 1 are the frequencies of 25, 30, 40, and 50, respectively. The mean is obtained by multiplying each value by its frequency of occurrence and then dividing the sum of these products by the sum of the frequencies. Let us now generalize the formula for finding the arithmetic mean to include the frequencies of the observations.

**DEFINITION**

> **Arithmetic Mean, Frequency Distribution**
>
> Let $x_1, x_2, \ldots, x_m$ be different measurements. Then the formula for the arithmetic mean $\bar{x}$ is
>
> $$\bar{x} = \frac{x_1 f_1 + x_2 f_2 + x_3 f_3 + \ldots + x_m f_m}{f_1 + f_2 + f_3 + \ldots + f_m}$$
>
> where $f_i$ is the frequency of $x_i$ for $i = 1, 2, 3, \ldots, m$.

Using the Greek letter $\Sigma$, the formula for the arithmetic mean of a grouped frequency distribution can be written as

$$\bar{x} = \frac{\sum\limits_{i=1}^{m} x_i f_i}{\sum\limits_{i=1}^{m} f_i} \quad \text{or} \quad \bar{x} = \frac{\Sigma x f}{n}$$

## EXAMPLE 3

Find the arithmetic mean of the following set of data:

| $x$ | 4 | 14 | 24 | 34 |
|---|---|---|---|---|
| $f$ | 2 | 8 | 20 | 10 |

**Solution**

$$\bar{x} = \frac{4(2) + 14(8) + 24(20) + 34(10)}{2 + 8 + 20 + 10} = \frac{940}{40} = 23.5$$

If data are presented in a grouped frequency table, we may have no way of knowing the distribution of the data within a class. We therefore must assume either that the data are uniformly distributed within a class interval around the class mark or that all the data within a class interval are located at the class mark. Thus, to compute the mean we use the formula for $\bar{x}$, where $x$ represents the class mark, $f$ the frequency of each class, and $m$ the number of class intervals.

## EXAMPLE 4

Find the arithmetic mean of the data in Table 6.1.

**Solution**

From Table 6.1,

$$\bar{x} = \frac{495}{20} = 24.75$$

**Table 6.1**

| $x$ (Classmark) | $f$ | $xf$ |
|---|---|---|
| 17 | 4 | 68 |
| 22 | 7 | 154 |
| 27 | 5 | 135 |
| 32 | 2 | 64 |
| 37 | 2 | 74 |
| Total | 20 | 495 |

*Practice Problem*

Find the arithmetic mean of the following set of data:

**Answer**

123.5

| $x$ | 104 | 114 | 124 | 134 |
|-----|-----|-----|-----|-----|
| $f$ | 2 | 8 | 20 | 10 |

A small company has four employees with annual salaries of \$25,500, \$26,053, \$27,144, and \$27,553. The president of the company has an annual salary of \$55,000. The mean of the 5 salaries is \$32,250. This is a true average, as given by the arithmetic mean, but most people would not accept it as a meaningful measure of central tendency. The median salary is more representative, because the one large salary tends to weight the mean upward to a misleading extent.

## Finding Medians

The median of a set of observations is the middle number when the observations are ranked according to size.

**DEFINITION**

> ### Median
>
> If $x_1, x_2, x_3, \ldots, x_n$ is a set of data placed in increasing or decreasing order, the *median* is the middle entry, if $n$ is odd. If $n$ is even, the median is the mean of the two middle entries.

## EXAMPLE 5

Consider the set of five measurements 7, 1, 2, 1, and 3. Arranged in increasing order, they may be written as

$$1, 1, 2, 3, 7$$
$$\downarrow$$
$$\text{median}$$

Hence, the median is 2.

## EXAMPLE 6

The array

$$25, 2, 5, 6, 5, 23, 7, 10, 22, 15, 21, 23$$

can be arranged in decreasing order as

$$25, 23, 23, 22, 21, 15, 10, 7, 6, 5, 5, 2$$
$$\searrow \swarrow$$
$$\text{median}$$

So the median is $\dfrac{15 + 10}{2} = 12.5$

## Finding Modes

The third measure of central tendency is called the *mode*. It refers to the measurement that appears most often in a given set of data.

DEFINITION

> ## Mode
>
> The *mode* of a set of measurements is the observation that occurs most often. If every measurement occurs only once, then there is no mode. If the two most common measurements occur with the same frequency, the set of data is *bimodal*. It may be the case that there are three or more modes.

## EXAMPLE 7

Baseball caps with the following head sizes were sold in a week by the Glo-Slo Sporting Goods Store: $7, 7\frac{1}{2}, 8, 6, 7\frac{1}{2}, 7, 6\frac{1}{2}, 8\frac{1}{2}, 7\frac{1}{2}, 8$, and $7\frac{1}{2}$. Find the mode of the head sizes.

**Solution**

The mode is $7\frac{1}{2}$; it occurs 4 times, more times than any other size.

*Practice Problem*

Find the mode of 21, 23, 24, 22, 24, 20, 22, 24, 25, 20, 22, and 21.

**Answer**

There are two modes: 22 and 24.

---

The decision about which measure of central tendency to use in a given situation is not always easy. The mean is a good average of magnitudes, such as weights, test scores, and prices, provided that no extreme values are present to distort the data. When extraordinarily large or small values are included in the data set, the median is usually better than the mean. However, the mean is the average most often used because it gives equal weight to the value of each measurement. The median is a positional average. The mode is used when the "most common" measurement is desired. The most appropriate measure for the price of pizzas in town would be the arithmetic mean. However, to select the best-tasting pizza in town, you could make a survey and use the mode (if a large number of persons were involved). Unfortunately, people with a stake in the outcome tend to use the average that best suits the objectives they hope to accomplish, and they quote the result as an accomplished (and exclusive) fact. This, of course, leads to a widespread mistrust of statistics.

## EXAMPLE 8

In one group of games against the Dodgers, the Reds won 6 of 7 games by the scores given in Table 6.2. Find the mean, median, and mode of these scores.

**Table 6.2**

| Dodgers | 2 | 6 | 1 | 15 | 4 | 2 | 2 |
|---------|---|---|---|----|---|---|---|
| Reds    | 4 | 7 | 2 | 1  | 5 | 3 | 3 |

**Solution**

When the mean scores are computed, the following results are obtained:

$$\text{Dodgers' mean score} = 4.57$$

$$\text{Reds' mean score} = 3.57$$

Although the Reds dominated the series, the Dodgers' mean score was substantially higher. In this case, the mean is not a good average to use because the Dodgers' extraordinarily high score in one game biased the mean. In such cases, it is often better to use the median. Thus,

|                                    |     |     |     |     |     |     |     |
|------------------------------------|-----|-----|-----|-----|-----|-----|-----|
| Reds' scores (placed in order):    | 1   | 2   | 3   | ③   | 4   | 5   | 7   |

↑
median
↓

|                                    |     |     |     |     |     |     |     |
|------------------------------------|-----|-----|-----|-----|-----|-----|-----|
| Dodgers' scores (placed in order): | 1   | 2   | 2   | ②   | 4   | 6   | 15  |

In this case, the median offers a better measure for comparing the scores. Coincidentally, the mode score for the Reds is 3, and the mode score for the Dodgers is 2.

## Percentiles

One way of reporting a person's relative performance on a test is to identify the percentage of people taking the test who scored lower than the person under consideration. For example, someone who scores higher than 70% of those who take a test is said to be in the 70th percentile. Conversely, a percentile score of 85 means that the person scored higher than 85% of those in the sample. In the next several examples, we will practice using percentiles.

**DEFINITION**

> ### Percentile
>
> Let $x_1, x_2, x_3, \ldots, x_n$ be a set of $n$ measurements arranged in order of magnitude. The *Pth percentile* is the value of $x$ such that $p\%$ of the measurements are less than the value of $x$ and $(100 - p)\%$ are greater than $x$. For small data sets, some percentiles cannot be computed.

## EXAMPLE 9

Monica scored in the 68th percentile on the mathematics portion of the Iowa Test of Basic Skills. Can she conclude that she got 68% of the questions correct on this test?

### Solution

No, Monica has no information about how many problems she got correct. She does know that she scored higher than 68% of those who took the mathematics portion of the test. We say that the *percentile rank* of Monica's score is 68.

Because percentiles are of practical value only for very large data sets, we are concerned with how to interpret them, not how to compute them. However, there are three percentiles to which we wish to give closer attention. The 25th percentile is also called the **first quartile,** the 50th percentile is the **second or middle quartile,** and the 75th percentile is the **third or upper quartile.** The median is approximately equal to the second quartile (the 50th percentile).

For small data sets we can get an easy approximation of the first quartile by computing the median of the scores below the median. Similarly, we can approximate the upper quartile by computing the median of the scores above the median.

## EXAMPLE 10

Find the median and approximate the first and third quartiles of the following data.

$$8, 14, 12, 64, 7, 9, 42, 84, 76, 92, 41, 15, 17, 26, 47, 16, 21, 22, 23, 24$$

### Solution

Arrange the data in order of increasing magnitude:

$$7, 8, 9, 12, 14, 15, 16, 17, 21, 22, 23, 24, 26, 41, 42, 47, 64, 76, 84, 92$$

There are 20 observations, so the median is the average of the 10th and 11th scores. The median is 22.5, the average of 22 and 23. There are 10 scores below the median, and their median is 14.5. This score of 14.5 is the first quartile. Likewise there are 10 scores above the median and their median is 44.5. This score of 44.5 is the third quartile.

## Just for Fun

The average salary of 6 office workers in the XYZ Corporation is $14,000. John remembers five of the six salaries: $10,000, $13,000, $19,000, $16,000, and $12,000. If the average is the median, describe the missing salary for John. Do the same if the average salary is the mean.

## EXERCISES

1. Compute the mean, the median, and the mode (or modes, if appropriate) for the given sets of data.

   (a) 3, 4, 5, 8, 10
   (b) 4, 6, 6, 8, 9, 12
   (c) 3, 6, 2, 6, 5, 6, 4, 1, 1
   (d) 7, 1, 3, 1, 4, 6, 5, 2
   (e) 21, 13, 12, 6, 23, 23, 20, 19
   (f) 18, 13, 12, 14, 12, 11, 16, 15, 21

2. An elevator has a maximum capacity of 15 people and a load limit of 2250 lb. What is the mean weight of the passengers if the elevator is loaded to capacity with people and weight?

3. Find the mean of the given distribution.

   | x | Frequency |
   |---|---|
   | 10 | 2 |
   | 20 | 6 |
   | 30 | 8 |
   | 40 | 4 |

4. Make up a set of data with 4 or more measurements, not all of which are equal, with each of the following characteristics.

   (a) The mean and median are equal.
   (b) The mean and mode are equal.
   (c) The mean, median, and mode are equal.
   (d) The mean and median have values of 8.
   (e) The mean and mode have values of 6.
   (f) The mean, median, and mode have values of 10.

5. The mean of a set of 8 scores is 65. What is the sum of the 8 scores?

6. The mean of 9 of 10 scores is 81. The tenth score is 100. What is the mean of the 10 scores?

7. Which of the three averages should be used for the following data?

   (a) The average salary of four salesmen and the owner of a small store.

   (b) The average height of all male students in W. R. Berry High School.
   (c) The average dress size sold at Acme Apparel.

8. At the initial meeting of an athletic club, the weights of the members were found to be 220, 275, 199, 246, 302, 333, 401, 190, 286, 254, 302, 323, and 221.

   (a) Compute the mean, median, and mode of the data.
   (b) Which measure is most representative of the data?

9. The weights (in kg) of the members of the Laramy High School football squad are as follows: 75, 60, 62, 94, 78, 80, 72, 74, 76, 89, 95, 98, 97, 80, 98, 91, 96, 90, 84, 73, 80, 92, 94, 96, 99, 84, 60, 68, 74, 80, 92, 96, 88, 74, 84, 94, 72, 76, 64, and 80.

   (a) What is the mean weight of the football squad?
   (b) What is the median weight?
   (c) What is the modal weight?
   (d) If you were a sportswriter assigned to do a story on this squad, how would you describe the (average) weight?

10. The following grades were recorded for a test on Chapter 9. Find the arithmetic mean.

    | Score | Frequency |
    |---|---|
    | 100 | 3 |
    | 90 | 5 |
    | 80 | 7 |
    | 70 | 15 |
    | 60 | 14 |
    | 50 | 3 |
    | 40 | 3 |

11. For each of the given sets of observations, approximate the 3 quartiles.

    (a) 16, 14, 12, 13, 15, 18, 24, 8, 10, 4
    (b) 18, 47, 64, 32, 41, 92, 84, 27, 14, 12

12. In a class of 80 students, Jodi scored in the 80th percentile. How many students scored lower than Jodi?

**13.** Aaron scored in the 98th percentile of the verbal portion of the SAT. Interpret this result.

**14.** For the following data, find the median. 17, 26, 34, 41, 52, 14, 13, 18, 27, 31, 39, 43, 44, 47, 49.

*For the histograms shown in Exercises 15 and 16, find the mean of the distributions.*

**15.**

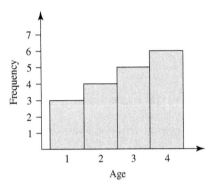

**16.**

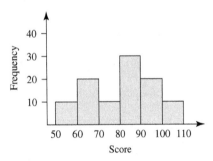

**17.** The president of Wargo Furniture Factory draws a salary of $100,000 per year. Four supervisors have salaries of $20,000 each. Twenty workers have salaries of $10,000 each. Discuss each of the following.

(a) The president says the average salary is $15,200.

(b) The union says the average salary is $10,000.

(c) Which average is more representative of the factory salaries?

**18.** If 99 people have a mean income of $8000, by how much does the mean income increase when an employee is added who has an income of $150,000?

**19.** A student has a mean average grade of 89 on 9 tests. What must she make on the tenth test to have an average of 90?

**20.** An interesting property of the mean is that the sum of the differences in the mean and each observation (deviations of each score from the mean, considered as signed numbers) is 0. Show that this statement is true for the following data: 5, 8, 10, 12, 15.

**21.** The accompanying table shows the distribution of scores on a test administered to freshmen at Laneville College. Find the arithmetic mean.

| Score | Frequency |
|-------|-----------|
| 140–149 | 3 |
| 130–139 | 4 |
| 120–129 | 8 |
| 110–119 | 13 |
| 100–109 | 4 |
| 90–99 | 2 |
| 80–89 | 0 |
| 70–79 | 1 |

**22.** Find the mean price–earnings ratio of 100 common stocks listed on the New York Stock Exchange, where the distribution is as follows.

| Interval | Frequency |
|----------|-----------|
| 0–4 | 6 |
| 5–9 | 46 |
| 10–14 | 30 |
| 15–19 | 10 |
| 20–24 | 4 |
| 25–29 | 2 |
| 30–34 | 2 |

**23.** The following frequency distribution gives the weekly salaries, by title, of the employees of the Glascow Light Bulb Factory.

| Title | Number | Weekly Salary |
|-------|--------|---------------|
| Manager | 1 | $550 |
| Supervisors | 3 | 450 |
| Inspectors | 3 | 350 |
| Line workers | 21 | 250 |
| Clerks | 5 | 150 |

Frank examined this list and concluded that the mean salary is

$$\frac{550 + 450 + 350 + 250 + 150}{5} = \$350$$

(a) Is he correct?

(b) If he is not correct, what should the mean be?

**24.** The data in the following table have been collected on the expenses (excluding travel and lodging) for 6 trips made by teachers in the mathematics department at Snelling College.

| Number of Days on Trip | Total Expenses | Expenses Per Day |
|------------------------|----------------|------------------|
| 0.5 | $18.00 | $36.00 |
| 2.5 | 75.00 | 30.00 |
| 3 | 60.00 | 20.00 |
| 1 | 19.50 | 19.50 |
| 8 | 132.00 | 16.50 |
| 5 | 160,00 | 32.00 |
| 20 | $464.50 | $154.00 |

Let     $\bar{x} = \dfrac{\$154.00}{20}$

          $= \$7.70$ average expense per day

or      $\bar{x} = \dfrac{\$464.50}{20}$

          $= \$23.23$ average expense per day

or      $\bar{x} = \dfrac{\$154.00}{6}$

          $= \$25.67$ average expense per day

Which average is realistic?

**25.** A study of the number of oil spills into the nation's waterways in recent years gives the following number of spills of various sizes. Find the arithmetic mean.

| Millions of Gallons of Oil | Number of Spills |
|---|---|
| 1–3 | 6 |
| 4–6 | 9 |
| 7–9 | 13 |
| 10–12 | 10 |
| 13–15 | 7 |
| 16–18 | 3 |

**26.** A company employs 50 men and 50 women. The table below summarizes the salary information of the company.

| Years of Service | Men | | Women | |
|---|---|---|---|---|
| | Number | Average Salary | Number | Average Salary |
| Less than 5 years | 10 | $10,000 | 40 | $12,500 |
| More than 5 years | 40 | $17,400 | 10 | $20,000 |

**(a)** What is the average (mean) salary for men in the company?

**(b)** What is the average (mean) salary for women in the company?

**(c)** Does the information from this table indicate that the company discriminates on the basis of gender? Explain.

## Review Exercises

**27.** Group the following test scores, which were received by 24 students, into 10 classes (95–99, 90–94, 85–89, 80–84, 75–79, 70–74, 65–69, 60–64, 55–59, and 50–54):

63, 71, 85, 96, 94, 90, 75, 72, 77, 71, 62, 84,

81, 76, 61, 54, 87, 94, 62, 81, 94, 77, 63, 60

Construct (a) a histogram and (b) a frequency polygon.

# 6.3   HOW TO MEASURE SCATTERING

**Problem**   The fact that the mean salary of management of the Doran Company exceeds the mean salary of management of the Wargo Company does not imply that the salaries of the Doran Company are superior to those of the Wargo Company. Compare four monthly salaries: $4000, $4100, $4100, and $21,000, with a mean of $8,300; and four salaries: $6,800, $6,900, $7,000, and $7,200, with a mean of $6,975. The mean salary of the first company exceeds the mean salary of the second company, but the lowest salary of the second company is better than all but the largest salary of the first company.

**Overview**   The preceding example indicates the need for a measure of dispersion or scattering of data. In other words, we need a measure that indicates whether the entries in a set of data are close or not close to the mean. Thus, in this section we consider the following measures of scattering: range, variance, standard deviation, and interquartile range.

**Goals**   Key statistical concepts include measures of central tendency, measures of variation (range, standard deviation, interquartile range, and outliers). (*Teaching Standards*, p. 136)

## Measures of Dispersion (Scattering)

There are several ways to measure dispersion of data. The one that is easiest to calculate is the range, introduced in Section 1.

**EXAMPLE 1**

For the set of data, 7, 3, 1, 15, 41, 74, and 35, the range is 73 because $74 - 1 = 73$.

Although the range is easy to obtain, it is not always a good measure of dispersion because it can be so radically affected by a single extreme value. For example, suppose that the 74 in the set of observations listed previously was miscopied and listed as 24 instead. As a result, the range would change from 73 to 40.

Because the range is significantly affected by extreme values, other measures of scattering or dispersion are preferable. In this section, we consider variance, denoted by $s^2$, and standard deviation, s.

**DEFINITION**

## Variance

*Variance* for a set of data can be obtained in four steps:

1.  Compute the mean $\bar{x}$.
2.  Compute the difference between each observation and the arithmetic mean, $x - \bar{x}$.
3.  Square each difference $(x - \bar{x})^2$.
4.  Divide the sum of the differences squared by $n$, where $n$ is the number of observations.

In other words, the variance is the mean of the squares of the differences of the data from the mean, or the mean of the squared deviations.

## EXAMPLE 2

Find the variance for the data 5, 7, 1, 2, 3, and 6.

**Solution**

1.  Compute the mean of the data. (See Table 6.1.) In this case, we have

$$\bar{x} = \frac{24}{6} = 4$$

2.  Determine the difference between each $x$ and $\bar{x} = 4$. (See the second column of Table 6.1.)
3.  Compute the square of each of these differences; that is, compute $(x - \bar{x})^2$ (the third column of Table 6.1).
4.  Sum the squares of differences (that is, sum the third column of Table 6.1), and divide by $n = 6$.

$$s^2 = \frac{28}{6} = 4.67$$

The variance is 4.67.

**Table 6.1**

| $x$ | $x - \bar{x}$ | $(x - \bar{x})^2$ |
|---|---|---|
| 5 | 1 | 1 |
| 7 | 3 | 9 |
| 1 | −3 | 9 |
| 2 | −2 | 4 |
| 3 | −1 | 1 |
| 6 | 2 | 4 |
| Total  24 | | 28 |

The preceding four-step calculation is equivalent to the following formula for variance.

> **Formula for Variance**
>
> Variance, denoted by $s^2$, is calculated as
>
> $$s^2 = \frac{(x_1 - \bar{x})^2 + (x_2 - \bar{x})^2 + (x_3 - \bar{x})^2 + \ldots + (x_n - \bar{x})^2}{n}$$
>
> where $\bar{x}$ is the mean of the observations.

This formula is sometimes written in summation notation as

$$s^2 = \frac{\sum_{i=1}^{n}(x_i - \bar{x})^2}{n} \quad \text{or} \quad \frac{\sum(x - \bar{x})^2}{n}$$

Because the variance is the average of the squared deviations from the mean, it is clear that it measures dispersion. However, the variance represents different units than the original data. If the original data was in inches, the variance is measured in square inches. Hence, the most often used measure of dispersion is the **standard deviation.** Standard deviation has the advantage of being expressed in the same units as the original data.

To compute standard deviation, remember that standard deviation is simply the square root of the variance.

$$s = \sqrt{s^2} = \sqrt{\frac{\sum_{i=1}^{n}(x_i - \bar{x})^2}{n}}$$

## EXAMPLE 3

Find the standard deviation of the data given in Example 2.

**Solution**

$$s = \sqrt{s^2} = \sqrt{4.67} \quad s^2 = 4.67 \text{ in Example 2}$$
$$= 2.16$$

*Practice Problem*

Compute the standard deviation of 112, 108, 114, 100, and 116.

**Answer**

5.66

It is interesting to observe that in inferential statistics, when the standard deviation of a sample is being computed in order to estimate the standard deviation of the population, one divides by $(n - 1)$ instead of $n$ when computing the standard deviation of the sample. Because we are primarily concerned with descriptive statistics in this text, we will not need to be concerned with this subtlety.

## Calculator Hint

Some calculators have keys for standard deviation. If you are using such a calculator, you must determine whether the formula in the calculator divides by $n$ or by $n - 1$.

These formulas for variance and standard deviation are useful because they make it clear that they measure variation of the data from the arithmetic mean. However, a different formula is more accessible for calculator computations.

---

Computational Formula for Variance

$$s^2 = \frac{x_1^2 + x_2^2 + \ldots + x_n^2 - n\bar{x}^2}{n} = \frac{\sum\limits_{i=1}^{n} x_i^2 - n\bar{x}^2}{n}$$

---

## EXAMPLE 4

Use the computational formula for variance to find the variance of the data given in Example 2.

**Solution**

The first column of Table 6.2 displays the values of the variable $x$, and the second column displays the values of $x^2$. Therefore,

$$\bar{x} = \frac{24}{6} = 4 \qquad s^2 = \frac{124 - 6(4)^2}{6} \qquad \sum_{i=1}^{n} x_i^2 = 124$$
$$n = 6$$
$$= \frac{28}{6} = 4.67$$

This is, of course, the same answer obtained in Example 2.

**Table 6.2**

| $x$ | $x^2$ |
|---|---|
| 5 | 25 |
| 7 | 49 |
| 1 | 1 |
| 2 | 4 |
| 3 | 9 |
| 6 | 36 |
| 24 | 124 |

---

When the data are presented in a frequency distribution, the formula for variance can be written as follows. As before, to find the standard deviation one computes the square root of the variance.

Variance of Frequency Distributions

Suppose that $x_1, x_2, \ldots, x_m$ have respective frequencies $f_1, f_2, \ldots, f_m$. Then

$$s^2 = \frac{x_1^2 f_1 + x_2^2 f_2 + \ldots + x_m^2 f_m - n\bar{x}^2}{n} = \frac{\sum\limits_{i=1}^{m} x_i^2 f_i - n\bar{x}^2}{n}$$

where $n = f_1 + f_2 + \ldots + f_m$.

---

## EXAMPLE 5

Find the variance and the standard deviation of the data tabulated in the first two columns of Table 6.3.

**Table 6.3**

| $x$ | $f$ | $xf$ | $x^2$ | $x^2f$ |
|-----|-----|------|-------|--------|
| 2 | 3 | 6 | 4 | 12 |
| 4 | 4 | 16 | 16 | 64 |
| 6 | 6 | 36 | 36 | 216 |
| 8 | 4 | 32 | 64 | 256 |
| 10 | 3 | 30 | 100 | 300 |
| | Total  20 | 120 | | 848 |

**Solution**

$$\bar{x} = \frac{120}{20} = 6$$

$$s^2 = \frac{848 - (20)6^2}{20} \qquad \sum_{i=1}^{n} x_i^2 f_i = 848$$

$$n = 20$$

$$= 6.4$$

$$s = \sqrt{6.4} = 2.53$$

An important property of standard deviation is the fact that in a set of data most of the data are located within 2 standard deviations of the mean of the data. At least 75% of data lie within 2 standard deviations of the mean for any data set, and we will learn in the next section that, for many collections of data, 95% of the data lie within 2 standard deviations of the mean.

## EXAMPLE 6

Babe Ruth won 12 American League home run championships in his career. The number of home runs he hit to win those championships were:

| 11 | 29 | 54 | 59 | 41 | 46 |
|----|----|----|----|----|----|
| 47 | 60 | 54 | 46 | 49 | 46 |

The mean of this data is 45.2 and the standard deviation is 13.0. What percentage of the data lie within 2 standard deviations of the mean?

**Solution**

$$\text{Mean} + 2 \text{ standard deviations} = 45.2 + 2(13.0) = 71.2$$

$$\text{Mean} - 2 \text{ standard deviations} = 45.2 - 2(13.0) = 19.2$$

Eleven of the 12 values lie within 2 standard deviations of the mean, or 92% of the data lie in this interval.

Undoubtedly you are wondering, "How am I going to use this new idea of standard deviation?" There are many applications, one of which follows. One of the most common misuses of statistics consists of making inappropriate comparisons. Knowing both the mean and the standard deviation helps to improve comparisons.

**EXAMPLE 7**

Juan made scores of 90, 82, 70, 61, and 94 on 5 tests. Of course, Juan did his best work relative to the rest of the class on the last test.

Or did he? What do we know about how the other students scored? Maybe everyone in the class made a higher score than 90 on the last test.

Example 7 illustrates a need for standard scores by which to make comparisons. Such scores are called $z$ scores.

**DEFINITION**

**$z$ Scores**

A score or measurement, denoted by $x$, from a distribution with mean $\bar{x}$ and standard deviation $s$ has a corresponding $z$ score given by

$$z = \frac{x - \bar{x}}{s}$$

representing the number of standard deviations from the mean.

**EXAMPLE 8**

The average weight of bags of potato chips is 10 oz., with a standard deviation of 1 oz. One bag of potato chips is weighed and has a weight of 10.2 oz. Convert this measurement into standard units.

**Solution**

We know that $x = 10.2$ oz., $\bar{x} = 10$ oz., and $s = 1$ oz. Therefore,

$$z = \frac{x - \bar{x}}{s} = \frac{10.2 - 10}{1} = 0.2$$

The $z$ score is a measurement expressed in standard units or without units. For example, if $x$ and $\bar{x}$ are in feet, then $s$ is in feet, and the division eliminates the units. Consequently, $z$ scores are useful for comparing two sets of data with different units.

**EXAMPLE 9**

Teresa scores a 76 on the entrance test at school X and an 82 at school Y. At which school did she have the better score?

**Solution**

To answer this question, we need to know that the mean score at school X was 70 with a standard deviation of 12 while the mean score at school Y was 76 with a standard deviation of 16. The $z$ scores are then as follows:

$$\text{School X:} \quad z = \frac{76 - 70}{12} = 0.5$$

$$\text{School Y:} \quad z = \frac{82 - 76}{16} = 0.375$$

Because 0.5 is greater than 0.375, Teresa's score at school X was better than her score at school Y in comparison to the scores of others who took the test.

## Interquartile Range

Remember that the first quartile $Q_1$ of a set of data is the score below which 25% of the data lies while the third quartile $Q_3$ is the score above which 25% of the data lies. Therefore the interval between the first quartile $Q_1$ and the third quartile $Q_3$ contains the middle 50% or half of the data. The length of this interval provides a useful measure of the variation of the central portion of data. The **interquartile range** (IQR) is computed by subtracting the first quartile from the third quartile.

**DEFINITION**

> ### Interquartile Range
>
> The *interquartile range (IQR)* equals
>
> $$Q_3 - Q_1$$

## EXAMPLE 10

The following are 16 grades received on Professor Wheeler's probability test, arranged in increasing order. Find the median, $Q_1$, $Q_3$, and the interquartile range.

**Solution**

Remember that we can easily approximate $Q_1$ by computing the median of the scores below the median. Similarly, we can approximate $Q_3$ by computing the median of the scores above the median.

$$64 \quad 66 \quad 68 \quad 71 \quad 73 \quad 75 \quad 76 \quad 78 \quad 82 \quad 83 \quad 84 \quad 90 \quad 92 \quad 95 \quad 96 \quad 99$$

$$Q_1 = 72 \qquad \text{median} = 80 \qquad Q_3 = 91$$

$$IQR = 91 - 72 = 19$$

Therefore, a span of 19 points includes the middle 50% of the grades.

## Box and Whisker Plots

Sometimes the median, $Q_1$, $Q_3$, and the range are displayed in a diagram called a **box and whisker plot.** Figure 6.1 shows such a plot for the data in Example 10. The interquartile range is represented as the box. The lines extending from $Q_1$ and $Q_3$ to the lowest and highest scores are the whiskers. To characterize data points widely separated from the rest of the data, we define an outlier.

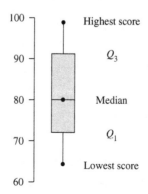

*FIGURE 6.1*

**DEFINITION**

> **Outlier**
>
> An *outlier* is any data point farther than 1.5 *IQRs* above $Q_3$ or farther than 1.5 *IQRs* below $Q_1$.

For the data in Example 10 we compute

$$Q_3 + 1.5(IQR) = 91 + 1.5(19) = 119.5$$

$$Q_1 - 1.5(IQR) = 72 - 1.5(19) = 43.5$$

Thus, for this set of data, there are no outliers.

Sometimes it is convenient to obtain the box and whisker plot from what we called in Section 6.1 a *stem and leaf plot*.

# EXAMPLE 11

The heights in feet of the 14 tallest buildings in Minneapolis are:

| 960 | 775 | 668 | 579 | 561 | 447 | 440 |
|-----|-----|-----|-----|-----|-----|-----|
| 416 | 403 | 366 | 356 | 355 | 340 | 337 |

Form a stem and leaf plot, and then make a box and whisker plot. Are there any outliers?

**Solution**

Hence,

$$IQR = 579 - 356 = 223$$

$$Q_3 + 1.5(IQR) = 913.5$$

$$Q_1 - 1.5(IQR) = 21.5$$

The value 960 is an outlier in this set of values.

**Table 6.4**

| Stems | Leaves |
|-------|--------|
| 9 | 60 |
| 8 |  |
| 7 | 75 |
| 6 | 68 |
| 5 | (79), 61 |
| 4 | 47, 40, 16, 03 |
| 3 | 66, (56), 55, 40, 37 |

$Q_3 = 579$

Median = 428

$Q_1 = 356$

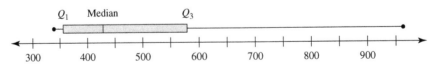

FIGURE 6.2

## Just for Fun

In a class of 6 students, each guesses the number of pennies in a jar. The 6 guesses are 52, 59, 62, 65, 49, and 42. One guess missed by −12, and the others by 5, 8, 11, −2, and −5. How can you use a statistic to help you find the number of pennies in the jar?

## EXERCISES

*For the given sets of observations in Exercises 1 through 6, find the range, the variance, and the standard deviation.*

1. 6, 8, 8, 14

2. 10, 12, 13, 14, 16

3. 16, 14, 12, 13, 15, 18, 24, 8, 10, 4

4. 18, 47, 64, 32, 41, 92, 84, 27, 14, 12

5. 9, 7, 16, 14, 12, 13, 14, 18, 24, 8, 10, 4

6. 10, 17, 18, 47, 64, 32, 41, 92, 84, 27, 14, 12

7. For Exercise 3, construct a box and whisker plot. List all outliers.

8. For Exercise 4, construct a box and whisker plot. List all outliers.

9. For Exercise 5, construct a stem and leaf plot. From this chart, find $Q_3 + 1.5(IQR)$ and $Q_1 - 1.5(IQR)$. Are there any outliers?

10. For Exercise 6, construct a stem and leaf plot. From this chart, find $Q_3 + 1.5(IQR)$ and $Q_1 - 1.5(IQR)$. Are there any outliers?

11. Compute $s^2$ for the following frequency distribution.

| $x$ | 10 | 14 | 18 | 22 |
|---|---|---|---|---|
| $f$ | 4 | 6 | 8 | 2 |

12. The mean of a population is 100, with a standard deviation of 10. Convert the following to $z$ scores.

(a) 110     (b) 80
(c) 71     (d) 120
(e) 140     (f) 40

13. Find the standard deviation of the following data.

| Class | Frequency |
|---|---|
| 0–2 | 10 |
| 3–5 | 20 |
| 6–8 | 30 |
| 9–11 | 40 |

14. Find the standard deviation of the given data.

| $x$ | Frequency |
|---|---|
| 1 | 10 |
| 4 | 20 |
| 7 | 30 |
| 10 | 40 |

15. Two instructors gave the same test to their classes. Both classes had a mean score of 72, but the scores of class A showed a standard deviation of 4.5, and those of class B showed a standard deviation of 9. Discuss the difference in the two classes' scores.

16. Joan decides to join the New Army to seek her fortune. She takes a battery of tests to determine placement into the appropriate corps. She makes 75 on the office work test and 80 on the outdoor activity test. The office work test has a mean of 60 and a standard deviation of 20, whereas the outdoor activity test has a mean 75 and a standard deviation of 10. Into which group should Joan be placed?

17. When Tran entered his profession in 1980, the average salary in the profession was $18,500, with a standard deviation of $1000. In 1990, the average salary in the profession was $28,000, with a standard deviation of $3000. Tran made $18,000 in 1980 and $27,000 in 1990. In which year did he do better in comparison with the rest of the profession?

18. The salaries of the 10 supervisors at the Aleo Automobile Works are $40,000, $43,000, $42,000, $41,000, $46,500, $41,500, $45,000, $47,000, $46,000, and $41,000. How many of the 10 supervisors have salaries within 1 standard deviation of the mean? How many have salaries within 2 standard deviations of the mean?

19. (Multiple choice) A student received a grade of 80 on a math final where the mean grade was 72 and the standard deviation was $s$. On the statistics final, he received a 90, where the mean grade was 80 and the standard deviation was 15. If the standardized scores (scores adjusted to a mean of 0 and a standard deviation of 1) were the same in each case, then $s =$

(a) 10    (b) 12    (c) 16    (d) 18    (e) 20

**20.** A teacher has just given an examination to his students. What does he know about his students' performance on the test if the distribution of scores has

   **(a)** a large range but a small standard deviation?
   **(b)** a mean higher than the median?
   **(c)** a mean lower than the median?
   **(d)** a small range but a large standard deviation?
   **(e)** a standard deviation of 0?

**21. (a)** What effect does adding the same amount to each observation have on the standard deviation? Test your conjecture by adding 5 to each member of the set 10, 12, 13, 14, 16.
   **(b)** What effect does dividing each entry by the same number have on the standard deviation? Check your conjecture by dividing 80, 85, 80, 70, and 80 by 5.
   **(c)** What effect does subtracting a number from each observation and then dividing each result by a number have on the standard deviation? Check your conjecture by subtracting 75 from each entry and then dividing by 5 for 80, 75, 80, 70, and 80.

**22.** A pollster tabulated the ages of 30 users of a vitamin pill designed to make the person who takes it feel young. The results are shown in the accompanying table. Find the standard deviation.

| Age | Frequency |
|-----|-----------|
| 20–29 | 1 |
| 30–39 | 2 |
| 40–49 | 4 |
| 50–59 | 5 |
| 60–69 | 9 |
| 70–79 | 6 |
| 80–89 | 3 |

**23.** In Exercise 14, Exercise Set 2 find $IQR$.

**24.** In Exercise 11(a), Exercise Set 2 find $IQR$.

**25.** It can be shown that, for any set of data, most of the values lie within 2 standard deviations on either side of the mean. Examine the data in the following table to verify this fact. First find the mean and standard deviation of each class.

| Height in Centimeters | | | | | |
|---|---|---|---|---|---|
| **Class I** | | | **Class II** | | |
| 156 | 158 | 182 | 168 | 180 | 183 |
| 178 | 159 | 176 | 180 | 187 | 190 |
| 160 | 176 | 174 | 176 | 176 | 178 |
| 166 | 160 | 172 | 188 | 186 | 174 |
| 189 | 187 | 154 | 179 | 192 | 188 |
| 153 | 180 | 198 | 176 | 179 | 181 |
| 159 | 162 | 176 | 173 | 174 | 180 |
| 180 | 166 | 192 | 178 | 176 | 175 |

   **(a)** How many of the values from Class I lie within 2 standard deviations on either side of the mean? What percent of Class I is this?
   **(b)** How many of the values from Class II lie within 2 standard deviations on either side of the mean? What percent of Class II is this?

**26.** The following table of data shows the miles per gallon reported by owners of 5 makes of 8-cylinder automobiles from different manufacturers.

| Manufacturer | | | | |
|---|---|---|---|---|
| **A** | **B** | **C** | **D** | **E** |
| 18 | 18 | 24 | 21 | 18 |
| 19 | 18 | 16 | 18 | 18 |
| 20 | 20 | 18 | 19 | 19 |
| 21 | 21 | 20 | 18 | 27 |
| 22 | 24 | 22 | 20 | 18 |
| 22 | 19 | 24 | 21 | 18 |

   **(a)** Which sample suggests the best gasoline mileage?
   **(b)** Which sample has the lowest standard deviation?

**27.** Starting with the original formula for variance on p. 408, show how this simplifies to the computational formula.

## Review Exercises

**28.** Consider the set of scores 1, 8, 16, 18, 20, 20, 21, 23, 24, 29. Find the following.
   **(a)** mean    **(b)** median    **(c)** mode    **(d)** range

**29.** The mean salary of all employees at the Brown Corporation is $30,000. Make up an example to show how this statistic may be misleading.

**30.** Find the mean for the following frequency distribution.

| $x$ | 10 | 20 | 30 | 40 | 50 |
|-----|----|----|----|----|----|
| $y$ | 4 | 6 | 8 | 4 | 3 |

# Test

*For questions 1 through 5, consider the following set of values.*

$$x: 3, 7, 11, 15, 19, 23$$

1. Find the mean.
2. Find the median.
3. Find the variance of $x$.
4. Find the interquartile range.
5. Draw a box and whisker plot.

*Questions 6 through 9 refer to the following frequency distribution:*

| Interval | Frequency |
|----------|-----------|
| 1–4 | 5 |
| 5–8 | 7 |
| 9–12 | 10 |
| 13–16 | 5 |
| 17–20 | 3 |

6. Draw a histogram.
7. Draw a frequency polygon.
8. Find the mean.
9. Find the standard deviation.

*Questions 10 through 18 refer to the following scores on a mathematics test.*

$$33, 37, 42, 48, 52, 53, 54, 55, 57, 59, 62, 62,$$
$$64, 64, 64, 68, 69, 71, 72, 72, 73, 73, 74, 74,$$
$$78, 79, 82, 83, 85, 87, 88, 89, 93, 96, 98$$

10. Approximate the interquartile range.
11. Find the mean.
12. Find the median.
13. Find the mode.
14. Make a stem and leaf chart.
15. Make a box and whisker plot.
16. Are there any outliers? Name them.
17. Make a grouped frequency distribution with class intervals 31–40, 41–50, . . . .
18. Find the standard deviation of your grouped frequency distribution in Exercise 17.

# ANSWERS TO SELECTED EXERCISES

## The Uses and Misuses of Statistics
## Section 6.1

1. **(a)** 14%
   **(b)** 21%
   **(c)** 19%
   **(d)** 75%

3. **(a)** Los Angeles
   **(b)** Chicago
   **(c)** Los Angeles

4. **(a)** 20
   **(c)** 16

5. **(a)**

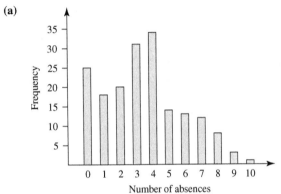

   **(b)**

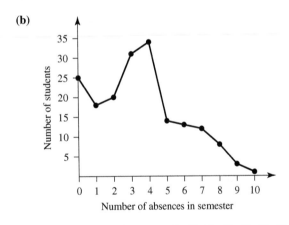

6. **(a)** 31, 37, 41, 42, 46, 51, 52, 53, 53, 57, 58, 59, 62, 68, 68, 69, 71, 73, 76, 82, 82, 96
   **(c)** 10

7. **(a)** 17, 22, 27, 32, 37

**(b)**

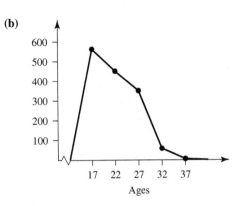

**(c)**

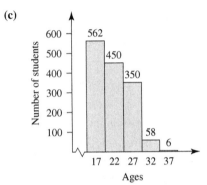

8. **(a)** 45% for Democratic candidates and 3% for Republican candidates; 16% and 30%.
   **(c)** 5.94 million dollars

9. Comments will vary
   **(a)** False. People with serious illnesses are more prone to die, and these people are usually in hospitals.
   **(b)** Misleading because there were many more drivers last year than soldiers in World War II.
   **(c)** The characteristics of the students and classes are not given.
   **(d)** Misleading because there are many more miles driven near home than on long trips.

10. **(a)**

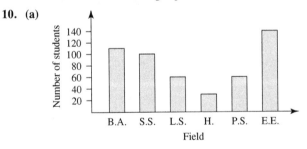

11. About 35

13. 1986–1990, approximately 110 cars

**15.**

```
0 | 7
1 | 2  4  5  6  6  6  6  7  7  8  8  9  9
2 | 1  4  6  8
3 | 2  5
4 | 1  3  8
5 | 1  9
6 | 1  1  2  2  5  5  8
7 | 2  2  3  3
```

The two groups who most heavily use the park are teenagers and people near retirement age.

**17.**

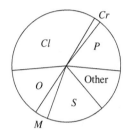

**19. (a)** 23

| **(b)** | | **(c)** |
| Class | Frequency | Class Marks |
|---|---|---|
| 40–43 | 5 | 41.5 |
| 44–47 | 12 | 45.5 |
| 48–51 | 20 | 49.5 |
| 52–55 | 6 | 53.5 |
| 56–59 | 5 | 57.5 |
| 60–63 | 2 | 61.5 |

**(d)**

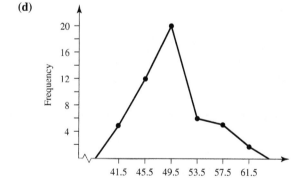

**21.**

| Class Marks | Class | Frequency |
|---|---|---|
| 167 | 160–174 | 3 |
| 182 | 175–189 | 6 |
| 197 | 190–204 | 9 |
| 212 | 205–219 | 6 |
| 227 | 220–234 | 2 |

**23.** Doubling the height and radius multiplies the volume by 8, not by 2. So, the visual presentation has 8 times the impact rather than 2, as it should be.

## Section 6.2

**1. (a)** Mean = 6  median = 5  mode = none
   **(b)** Mean = 7.5  median = 7  mode = 6
   **(c)** Mean = $3.\overline{7}$  median = 4  mode = 6
   **(d)** Mean = 3.625  median = 3.5  mode = 1
   **(e)** Mean = 17.125  median = 19.5  mode = 23
   **(f)** Mean = 14.67  median = 14  mode = 12

**3.** Mean = 27

**4. (a)** 5, 6, 7, 8
   **(c)** 3, 4, 4, 5
   **(e)** 3, 6, 6, 9

**5.** 520

**7. (a)** Median  **(b)** Mean  **(c)** Mode

**8. (a)** 273,23; 275; 302

**9. (a)** Mean = 82.725 kg
   **(b)** Median = 82 kg
   **(c)** Mode = 80 kg
   **(d)** The mean weight seems to be more useful. However, if you wish to downplay the extreme values you might use the median.

**11. (a)** 10, 13.5, 16
   **(b)** 18, 36.5, 64

**13.** 98% of those taking the test made lower scores than Aaron

**15.** $2.\overline{7}$

**17. (a)** The president is using the mean, which is misleading because the distribution is heavily skewed on one end.
   **(b)** The union is using the mode or median to show average.
   **(c)** The median or mode of $10,000 is more representative.

**19.** 99

**21.** 118.214

**23. (a)** No. He should have multiplied each salary by the frequency and divided by the sum of the frequencies (number of employees).
**(b)** $271.21

**25.** 8.75 million gal.

**26. (a)** $15,920

**27. (a)**

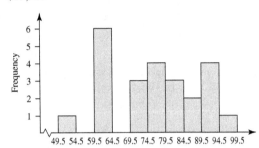

**(b)**

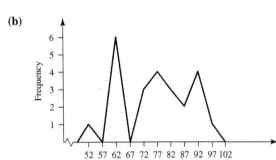

## Section 6.3

**1.** Range, 8; variance, 9; standard deviation, 3

**3.** Range, 20; variance, 27.44; standard deviation, 5.24

**5.** Range, 20; variance, 26.74; standard deviation 5.17

**7.**

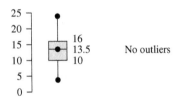

No outliers

**9.**

| Stems | Leaves |
|-------|--------|
| 0 | 4, 7, 8, 9 |
| 1 | 0, 2, 3, 4, 4, 6, 8 |
| 2 | 4 |

**11.** 13.44

**12. (a)** 1
**(c)** $-2.9$
**(e)** 4

**13.** 3

**15.** The scores in class B were much more scattered from the mean than those in class A.

**17.** 1990

**19. (b)** 12

**20. (a)** Most students scored near the average, but a small number had either extremely low or extremely high scores.

**21. (a)** No effect
**(b)** Reduces to $\frac{1}{5}$ the original
**(c)** Reduces to $\frac{1}{5}$ the original

**23.** $IQR = 26$

**25. (a)** Class I mean $+ 2s = 196.65$ mean $- 2s = 146.1$. All but one value is in this interval; 95.8%
**(b)** Class II mean $+ 2s = 191.64$ mean $- 2s = 168.11$. All but two values are in this interval; 91.7%

**26. (a)** $C$

**27.**
$$s^2 = \frac{\sum_{i=1}^{n}\left(x_i - \bar{x}\right)^2}{n} = \frac{\sum_{i=1}^{n}\left(x_i^2 - 2\bar{x}x_i + \bar{x}^2\right)}{n}$$

$$= \frac{\sum_{i=1}^{n}x_i^2 - 2\bar{x}\sum_{i=1}^{n}x_i + n\bar{x}^2}{n}$$

$$= \frac{\sum_{i=1}^{n}x_i^2 - 2\bar{x}n\dfrac{\sum_{i=1}^{n}x_i}{n} + n\bar{x}^2}{n}$$

$$= \frac{\sum_{i=1}^{n}x_i^2 - 2n\bar{x}^2 + n\bar{x}^2}{n} = \frac{\sum_{i=1}^{n}x_i^2 - n\bar{x}^2}{n}$$

**28. (a)** 18

**29.** Perhaps there are four employees with salaries of $20,000, $22,000, $24,000, and $19,000, and the president's salary is $65,000.

CPSIA information can be obtained
at www.ICGtesting.com
Printed in the USA
LVOW02s0800290616

494079LV00005B/18/P

9 781465 240